Henri Poincaré: Electrons to Special Relativity

Bruce D. Popp

Henri Poincaré: Electrons to Special Relativity

Translation of Selected Papers and Discussion

Bruce D. Popp
Norwood, MA, USA

ISBN 978-3-030-48041-7 ISBN 978-3-030-48039-4 (eBook)
https://doi.org/10.1007/978-3-030-48039-4

This work includes English-language translations of the following original French-language papers by Henri Poincaré, which belong to the public domain:

Poincaré, H. (1898). La mesure du temps. Revue de métaphysique et de morale, 6, 1–13.

Poincaré, H. (1900). La théorie de Lorentz et le principe de réaction. Archives néerlandaises des sciences exactes et naturelles, 5, 252–78.

Poincaré, H. (1905). Sur la dynamique de l'électron. Comptes rendus hebdomadaires de l'Académie des sciences de Paris, 140, 1504–1508.

Poincaré, H. (1906). Sur la dynamique de l'électron. Rendiconti del circolo matematico di Palermo, 21, 129–176.

Poincaré, H. (1908). La dynamique de l'électron. Revue générale des sciences pures et appliquées, 19, 386–402.

This Springer imprint is published by the registered company Springer Nature Switzerland AG
The registered company address is: Gewerbestrasse 11, 6330 Cham, Switzerland

Preface

The initial idea for this book goes back to a suggestion that, after completing translation of Henri Poincaré's monograph *Sur le problème des trois corps et les équations de la dynamique*, I translate what is often called the Palermo paper, *Sur la dynamique de l'électron* by Henri Poincaré. This seemed like a reasonable suggestion and my first thought was that preparing the translation would be an interesting, but minor, digression before starting a larger project that I had been anticipating for some time. The Palermo paper has frequently been compared to Einstein's 1905 paper on electrodynamics of moving bodies in discussions of the origin of special relativity.

My first impression was that the discussions of precedence for the theory of special relativity were sterile and uninteresting. ("Relativity postulate? Checkmark in the Poincaré column; checkmark in the Einstein column.") Further, as a physicist, it seemed clear to me that Poincaré and Einstein were looking to understand and solve rather different problems and while Poincaré's work could have led to a theory of special relativity by a different route, Einstein had gone directly to that target. Still, a new translation of Poincaré's article by a competent professional translator might have some value. Perhaps, I thought, I could post the translation on the internet and then move on to the project I had previously been thinking about.

After submitting the manuscript for my first book, I started translating the Palermo paper. I soon came to see that to understand the article I was translating (and thus produce an accurate and satisfying translation), I would need to look at a number of related articles by Poincaré from the same period and also an article from 1904 by Hendrik Lorentz. After further translating, reading, and researching, seven items in total were on my list of related articles by Poincaré; the list was later pruned to five. These are the articles translated in Part I of this book.

Of these five articles by Poincaré, three were published in 1905, 1906, and 1908 and were about the dynamics of the electron. The first of these appeared in the minutes of meetings of the Académie des Sciences de France and describes the work that Poincaré had in progress. The second article was the article referred to above as the Palermo paper because it was published in the proceedings of the Circolo matematico di Palermo. The third article is another look at these results published in the *Revue générale des sciences pures et appliquées* and has more of a popular science slant. The other two articles were earlier articles which discussed time and conservation of momentum by radiation, respectively.

This body of work reflects Poincaré's effort to understand and respond to two experimental results.

The first is the discovery of subatomic charged particles both in Cambridge with the measurement of the charge-to-mass ratio of electrons in cathode rays by J. J. Thomson and in Paris with the discovery of radiation from radioactive decay by Henri Becquerel, isolation of polonium and radium by the Curies, and characterization of the radiation by Becquerel, Ernest Rutherford, and Paul Villard. Cathode rays and beta rays were recognized as being the same particle but coming from different sources.

The second is the negative result of the Michelson-Morley experiment. Lorentz sought an explanation for why motion through the ether could not be detected and proposed the transformations named after him as an explanation. In contrast, Poincaré accepted the experimental result and thought that it indicated that a principle of relativity was a general law of nature, meaning that it is impossible to demonstrate absolute motion (for example motion relative to the ether). Poincaré understood the transformations not as a cause calling for an ad hoc assumption to explain an experimental result, but as a consequence of a principle of nature applying to electricity and magnetism and also to gravitation.

That is how I came to select the articles translated in Part I of this book. The articles respond to a need for a theory of electrons, the newly discovered particle, and present a classical theory of subatomic charged particles and a particle-based theory of electrodynamics. It later became clear that three letters from Poincaré to Lorentz were important for comparing their work and I added the translation of the letters.

While translating, certain subjects attracted my attention and interest. The first to come up was Poincaré's choice of notation (or it might be more accurate to say lack of notation) for vectors, vector products, and differential vector operators. Physics students from the last 50 years and longer taking an upper-level course in electrodynamics are familiar with the use of vector dot products, vector cross products, and nabla (also called dell, ∇); it even makes its way to the inside of the covers of Jackson, *Classical Electrodynamics*. Poincaré does not use these notational tools; he instead writes out the Cartesian components of three-dimensional vectors individually. I became curious about the notation used by his contemporaries, the timing of when the notation I was familiar with had become available for use and related questions. After discussion with a friend, I decided to retain in my translation the notation that Poincaré had used in his writing and to provide, in a separate place, examples of certain key equations rewritten in notation familiar to me. At first, this "separate place" was an appendix.

Over the months, as I proceeded with the translation, I came across several more subjects that caught my interest and led me to do further research and reading on subjects where it seemed useful to improve my understanding or satisfy my curiosity. This is a case of letting my thoughts wander and then assessing where they had gone.

As the number of subjects increased, I came to decide that these subjects needed to be organized and treated in chapters in a second part of the book. My approach to these subjects is that of a physicist looking at the work of an earlier generation of physicists to understand through the published literature the content of their work, their attitude, and their approach to what they were doing and how their agreement and differences advanced their understanding of nature and their science.

The chapters in Part II therefore represent my discussion of subjects that captured my interest and curiosity. I organized them in three main groups. They are the following.

Experiments on Electrons

In 1897 J. J. Thomson at Cavendish Laboratories, University of Cambridge, measured the charge-to-mass ratio of cathode rays; this showed that they were particles. His value for the ratio was about 1/1000 the value for the hydrogen ion in electrolysis experiments. This suggests that these particles were very light. He called them corpuscles. In 1898 he measured their charge using water drops formed in a cloud chamber by nucleation on ions produced by irradiation with X-rays; this charge was seen to be comparable to the charge carried by the hydrogen atom in electrolysis. In 1906 he was awarded the Nobel Prize in Physics for this work.

In 1896 A. Henri Becquerel in France discovered radiation from uranium. He shared the 1903 Nobel Prize in Physics with Pierre and Marie Curie; he was recognized for his discovery of spontaneous radioactivity. A component of this radiation is beta rays. Becquerel measured the charge-to-mass ratio of beta rays and identified them as the same particles as cathode rays.

Further work by Walter Kaufmann in 1901 and in the following years produced confirmation of the charge-to-mass ratio measurement and more accurate results.

Kaufmann was also able to measure the charge-to-mass ratio and the velocity of the electrons simultaneously. This exposed an important difference between the electrons in cathode rays and beta rays; the beta rays have much higher velocities. Some have velocities over 90% of the speed of light. Kaufmann was able to show for the high-speed electrons that the mass of the electrons observed in his apparatus was larger for higher speed electrons. Max Abraham in 1902 suggested an explanation for the dependence of mass on velocity.

This is the experimental start of subatomic physics and relativity.

Theory of Electrons

Before this discovery, the predominant understanding of electric charge and its role in electrodynamics (including Maxwell's equations) was based on analogy with the flow of a continuous fluid. There was also thought to be a continuous medium for electromagnetic radiation that was called the luminiferous ether. Just as water waves need water and sound waves need air as continuous media, it was thought that electromagnetic waves needed a continuous medium and this medium was called the ether. Repeated, increasingly sophisticated, attempts to detect effects of the ether resulting from the Earth's motion had produced negative results.

The discovery of the electron—a discrete, charge bearing, subatomic particle—required a new understanding of electrodynamics and Maxwell's equations in terms of discrete charged particles. Hendrik Lorentz had been working on this formulation of electrodynamics since about 1890 and published an important paper on it in 1904. The paper is reformatted and presented in Part III for the convenience of my readers. That paper notably explained why sensitive experiments had not been able to detect an effect due to the Earth moving through the ether; Lorentz proposed that the shape of moving particles changed in such a way that any effect due to their movement through the ether would be undetectable. This change of shape, a transformation, also provided an explanation for the dependence of electron mass on velocity. Lorentz showed his explanation fit Kaufmann's data better that Abraham's explanation. I extended Lorentz's reanalysis of Kaufmann's data and showed that Lorentz's explanation fit the data much better than he had thought. Further, I showed that Kaufmann's data gives a value for the rest-frame electron charge-to-mass ratio that is within 10% of the current accepted value.

In papers from 1905 and 1906, Henri Poincaré corrected and expanded on Lorentz's work. Whereas Lorentz argued that the transformations were necessary to explain why the ether could not be detected; Poincaré in contrast adopted a postulate stating that it is impossible to detect absolute motion. He showed that the transformations proposed by Lorentz were a consequence of this assumption, and named them "Lorentz transformations." Poincaré notably showed that the equations of electrodynamics are unchanged by the Lorentz transformations. He also showed that the Lorentz transformations are a group and that $x^2 + y^2 + z^2 - c^2t^2$ is an invariant of the group.

Poincaré took another step to look at the stability conditions for an electron. Lorentz and Abraham shared the electromagnetic world view that tried to explain mass and all of mechanics in terms of electromagnetic forces. Poincaré recognized that an electron with only electromagnetic forces could not be stable. The charge forming an electron would repel electrostatically. Poincaré calculated what force would be necessary to hold an electron together. The required force, whatever its origin, is now called the Poincaré stress. A wholly electromagnetic electron is not possible since a force of some other origin is needed to hold it together.

While Lorentz had conceived of the transformations as applying to electromagnetic forces and serving to explain the non-detection of the ether, Poincaré understood that the Lorentz transformations had to apply to all forces and bodies including gravitational forces and looked at the consequences for gravitation. He developed a theory of gravitation unchanged under Lorentz transformation.

And More

The subsequent chapters in Part II step back and take progressively broader views of Poincaré's work in different contexts. Poincaré was an exceptional mathematician. What can we learn about Poincaré as a physicist from the work translated in Part I? In that work, Poincaré had both good predictions and missed opportunities.

Next, I review the polemic about priority for special relativity and provide my perspective.

Poincaré's choice of notation for the equation of electrodynamics is conservative. For over 15 years of work in electricity and magnetism Poincaré continued to write out equations in Cartesian coordinates even though he certainly saw and understood the vector formalism developed and adopted by others.

The last chapter takes an even larger step back and looks at some subjects in translation, language, and culture suggested by other chapters.

Advice for Readers

The content of Part I is my translation of articles written by Poincaré over 100 years ago.

Poincaré often does not provide clear signposts about where he is going or why he is presenting a particular line of reasoning. Reading Poincaré therefore requires paying attention to the path that he is taking and stopping to appreciate the vistas when they come into view. For example, Poincaré derives the formula for relativistic composition of velocities in two different ways and does it with little fanfare. As they say, blink and you'll miss it.

In preparing the translation, I have tried to use, consistent with good judgment, terminology or a gloss familiar to a contemporary reader. When I have been successful, this is unobtrusive. There is one important exception: it became clear on reading these articles that when Poincaré writes "electron" he may mean a positively or negatively charged subatomic particle of unspecified charge-to-mass ratio (see, e.g., his discussion of channel rays in Part I, Chapter 6 on page 107), and he does not always mean the negatively charged lepton that we designate with this name.

Some of the hypotheses, reasoning, and explanations given by Poincaré have not stood the test of time. Since I have not provided any running commentary or summarization, readers of Part I will need to identify for themselves content which needs replacement or correction.

Care is warranted when looking at the secondary literature for information about the articles translated in Part I. As is noted in Part II, Chapter 11, there has been substantial polemic, spanning multiple decades, about priority for the theory of special relativity. This has left accumulations of toxic ink in different parts of the secondary literature about Poincaré's papers translated in Chapters 2 and 5. Consulting translations of Poincaré's original writing in Part I of this book instead of, or in combination with, the secondary literature may help you avoid toxic ink.

Let me point out two examples.

In Edmund Whitaker, *A History of the Theories of Aether and Electricity: the Modern Theories*, Thomas Nelson and Sons Limited, 1953, on page 51 with reference to the article by Poincaré translated in Chapter 2, Whittaker writes, "In 1900 Poincaré, referring to the fact that in free aether the electromagnetic momentum is $(1/c^2)$ times the Poynting flux of energy, suggested that electromagnetic energy might possess mass density equal to $1/c^2$ times the energy density—that is to say, $E = mc^2$ where E is energy and m is mass." Some elements of this statement are false—Poincaré was not discussing "free aether"[1]—and others are seriously misleading—"suggested... might possess." We can reasonably assume that this statement was meant to write that part of Einstein's work out of Whitaker's book. Reading the primary literature, translated here in Chapter 2, will help you understand what Poincaré's idea was and how far he did take it.

In Arthur I. Miller, *A Study of Henri Poincaré's "Sur la Dynamique de l'Électron"*, Arch. Hist. Exact Sci. **10**, 207–328 (1973), Miller purports to provide a study of the article translated in Part I Chapter 5 of this book. I suggest that you consider whether you agree with the statement on page 210, "Poincaré was not a highly innovative physicist." And on page 234 Miller wrote, "Abraham, Lorentz and Poincaré were committed proponents of the electromagnetic world-picture, they could neither discover nor accept the universal theory of relativity." This is only true of Abraham and Lorentz. Electron stability (and the need for Poincaré stresses) is one of the nails that Poincaré knowingly put in the coffin of the electromagnetic world view. (This is one of the points I discuss in Part II, Chapter 10.) What is the motive for the mischaracterization?

Other Translations

Other translations into English of Henri Poincaré *Sur la dynamique de l'électron* are available. I am aware of the following published translations.

Scott Walter, in J. Renn and M. Schemmel (eds.), *The Genesis of General Relativity Vol. 3: Theories of Gravitation in the Twilight of Classical Physics; Part I* (Boston Studies in the Philosophy of Science 250), Springer, (2007),

[1] Whitaker might not have given such prominence to Poincaré's writing if he had more accurately understood Poincaré's attitude to the ether. See Part II, Chapter 10.

p. 253–271. On December 26, 2016, this was also available at http://scottwalter.free.fr/papers/7pubsfr.html.

C. W. Kilmister, *Special Theory of Relativity*, Oxford: Pergamon, (1970), p. 145–185.

H. M. Schwartz, *American Journal of Physics, 39,* 1287–1294; *40*, 862–872, 1282–1287. In the first of the three parts, Schwartz writes, "However, direct translations are given only of selected portions of the text. Nor is the mathematical part reproduced verbatim whenever changes in arrangement of the argument facilitate its understanding. To the same end the notation is modernized throughout, including the employment of the vector formalism." It should be noted that in my present translation, the entire article is translated, the mathematical part is reproduced in full, and the mathematical notation used by Henri Poincaré is retained.

I am not aware of published translations into English of the other articles by Poincaré included in this book.

Norwood, MA, USA Bruce D. Popp

Acknowledgments

Bob Rosner suggested translating the Palermo paper, and Cathy Thygesen suggested creating an appendix with certain equations rewritten in familiar vector notation. I am very thankful for these suggestions and further discussions with each of them.

Help with German from my translation colleagues has been essential. I especially want to thank Carola Berger for abstracts of several articles by Walter Kaufmann and Max Abraham and help with a question about Walter Kaufmann's processing of his data. Nick Hartmann, Ilse Andrews, and Ken Kronenberg provided polished English translations of various sentences.

I am very grateful for Marian Comenetz's support, generosity with her time, and careful attention to editing.

Access to copies of correspondence between Lorentz and Poincaré has been provided by Noord-Hollands Archief, Haarlem, and the Archives Henri Poincaré. I appreciate the availability of this material online and the ease of accessing it. Hans van Felius at Noord-Hollands Archief was helpful in answering questions.

Other useful resources are acknowledged in Part II, Chapter 13, "Availability of Sources."

As always, support and encouragement from friends and family were a valuable addition to my patience and persistence. Among my friends and colleagues, I want to thank Marian Comenetz, Carola Berger, and Judy Lyons. I am profoundly grateful for my children and their partners, Miriam Popp, Andrew Popp and Cristina Aldrich, and Mireille Popp and Tyler Killingsworth.

Contents

Part I
Translation of Selected Papers

Chapter 1
The Measurement of Time

I. As long as we don't leave the domain of consciousness, the concept of time is relatively clear. Not only do we easily distinguish the present sensation from the memory of past sensations or the anticipation of future sensations, but we know perfectly well what we mean to say when we affirm that, of two conscious phenomena of which we have kept a memory, one came before the other or else that, of two expected conscious phenomena, one will come before the other.

When we state the two conscious facts are simultaneous, we mean to say that they enter so deeply into each other that analysis cannot separate them without damaging them.

The order in which we line up conscious phenomena has no arbitrariness. It is imposed on us and there is nothing about it that we can change.

I have only one observation to add. For a set of sensations to become a memory which could be ranked in time, it must have stopped being actual, that we might have lost the sense of its infinite complexity, without which it would remain real. It must have, to say it that way, crystallized around a center of associations of ideas which will be like a kind of label. It is only when they will have thus lost all life that we will be able to order our memories in time, like a botanist ordering dried flowers in their herbarium.

But, there can only be a finite number of these labels. On account of this, psychological time would be discontinuous. What is the origin of this idea that between two arbitrary moments there are other moments? How would we know that there were empty slots, if these slots were only revealed to us by their content?

II. But that isn't all; in this form we want not only to bring back the phenomena from our consciousness, but those for which other consciousnesses are the theater. Even more, we want to bring back to it physical facts, these little things with which we populate space and which no consciousness sees directly. It has to be done, because without that science could not exist. In a word, psychological time is given

---

Poincaré, H. (1898). La mesure du temps. *Revue de métaphysique et de morale, 6*, 1–13.

B. D. Popp, *Henri Poincaré: Electrons to Special Relativity*,
https://doi.org/10.1007/978-3-030-48039-4_1

to us and we want to create scientific and physical time. That is where the difficulty starts, or rather difficulties, because there are two of them.

Here are two consciousnesses which are like two mutually impenetrable worlds. By what right do we want to bring them into a single mold, to measure them with the same gauge? Isn't it as if one wanted to measure length with a gram or to weigh with a meter?

And furthermore, why do we speak of measurement? We may know that one fact is before another, but not *how much* earlier it is.

Therefore the two difficulties are:

1) Can we transform psychological time, which is qualitative into a quantitative time?
2) Can we reduce facts which happen in different worlds to a single measurement?

III. The first difficulty was commented on long ago; it was the subject of long discussions and we may say that the question is resolved.

We do not have the direct intuition of the equality of two time intervals. People who believe they have this intuition are tricked by an illusion.

When I say that from 12:00 to 1:00 the same time has passed as from 2:00 to 3:00, what does that statement mean?

The least thought shows that on its own it has none at all. It will only have what I want to give it via a definition which will have to comprise some degree of arbitrariness.

Psychologists could have gotten by with this definition; physicists and astronomers cannot; let's see how they got out of it.

To measure time, they make use of a pendulum and accept by definition that all beats of this pendulum are of equal length. But, this is only a first approximation; the running of the pendulum is changed by temperature, air resistance and barometric pressure. If we escape from these sources of error, a much better approximation will result. New sources, neglected until now, whether electric, magnetic or other, could manage to contribute small disturbances.

In fact, the best clocks must be corrected from time to time, and the corrections are done using astronomical observations; it is set up so that the sidereal clock marks the same time when the same star passes the meridian. In other words, it is the sidereal day; meaning the time for rotation of the Earth which is the constant unit of time. We accept, via a new definition substituted for the one which was drawn from the swinging of the pendulum, that two complete rotations of the Earth around its axis have the same length.

However, astronomers are still not content with this definition. Many think the tides act as a brake on our globe and that the rotation of the Earth becomes slower and slower. In this way the apparent acceleration of motion of the moon would be explained (it would seem to go faster than theory allows it to) because our clock, which is the Earth, would be slowing down.

IV. One could say that none of that matters much; most likely, our measuring instruments are imperfect, but it is sufficient that we could conceive of a perfect instrument. This ideal could never be reached, but it would be enough to have conceived it and have in that way given rigor to the definition of the unit of time.

Unfortunately that rigor has not yet been found. When we use a pendulum for measuring time, what is the postulate that we implicitly accept?

It is that *the length of two identical phenomena is the same*; or, if one prefers, that the same causes take the same time to produce the same effects.

And it is, at first glance, a good definition of the equality of two intervals.

However, let's be careful about it. Is it possible that one day experience will falsify our postulate?

Let me explain; I assume that at some point in the world the phenomenon α happens leading as a consequence at the end of some time to the effect α'. At another point in the world very far away from the first, the phenomenon β happens, which leads as a consequence to the effect β'. The phenomenon α and β are simultaneous as are the effects α' and β'.

At some later time, the phenomenon α occurs again under almost identical circumstances and simultaneously the phenomenon β occurs again also at a very far away point of the world and under nearly the same circumstances.

The effects α' and β' are also going to occur again. I assume that the effect α' takes place substantially before the effect β'.

If the experiment made us witnesses of such an event, our postulate would be falsified.

The experiment taught us that the first length $\alpha\alpha'$ is equal to the first length $\beta\beta'$ and that the second length $\alpha\alpha'$ is smaller than the second length $\beta\beta'$. In contrast, our postulate would require that the two lengths $\alpha\alpha'$ be equal to each other and also the two lengths $\beta\beta'$. The equality and inequality deduced from the experiment would be incompatible with the two equalities drawn from the postulate.

Now, can we affirm that the hypotheses that I just made are absurd? They hold nothing contrary to the principle of contradiction. Most likely they would not be able to occur without the principle of sufficient reason seeming to be violated. But, to justify such a fundamental definition, I would prefer another guarantee.

V. But that isn't all.

In physical reality, one cause does not produce one effect, but a multitude of distinct causes contribute to producing it, without us being able to discern the contribution of each of them.

Physicists seek to make this distinction; but they only do it approximately and whatever progress they may make, they will only ever do it approximately. It is approximately true that the motion of the pendulum is due solely to the attraction of the Earth; but in all rigor, even the attraction of Sirius could act on the pendulum.

Under these conditions, it is clear that the causes which one day produced a certain effect will only ever be approximately reproduced.

And then we will have to change our postulate and our definition. Instead of saying:

"The same causes take the same time to produce the same effects."We have to say:

"Nearly identical causes take about the same time to produce about the same effects."But our definition is therefore no more than close.

Further, as Calinon very rightly remarked in his recent monograph (*Études sur les diverses grandeurs*; Paris, Gauthier-Villars, 1897), "One of the circumstances of an arbitrary phenomenon is the rotational speed of the Earth; if this rotational speed varies, it constitutes, in the reproduction of these phenomena, a circumstance which no longer remains identical to itself. But assuming this rotational speed is constant is assuming that one knows how to measure time."

Our definition is therefore still not satisfactory; it is certainly not the definition that the astronomers, whom I talked about above, implicitly adopt when they state that the Earth's rotation is slowing down.

Coming from their mouth, what is the meaning of this statement? We can only understand it by analyzing the evidence that they give for their proposition.

They first sate that friction from the tides produces heat which reduces the energy. They therefore invoke the principle of conservation of energy.

They next state that the secular acceleration of the moon, calculated from Newton's law, would be smaller than the acceleration which is deduced from observations if the correction relating to the slowing of the Earth's rotation is not made.

They therefore invoke Newton's law.

In other words, they define duration in the following way: time must be defined such that Newton's law and the equations of motion are respected.

Newton's law is an experimental truth; as such, it is only approximate, which shows that we still only have a definition by approximation.

If we now suppose that another way to measure time is adopted, the experiments on which Newton's law is based would still retain their same meaning. Just the statement of the law would be different, because it would be translated into another language; it would obviously be much less simple.

In that way, the definition implicitly adopted by astronomers could be summarized as:

Time must be defined such that the equations of mechanics are as simple as possible.

In other words, there is no way to measure time which is truer than another; the one which is generally used is only more *convenient*.

We are not allowed to say about two clocks that one works well and the other does not; we can only say that it is better to refer to the indications of the first clock.

The difficulty with which we just concerned ourselves was, as already stated, frequently reported; among the most recent works where it was dealt with, I will cite, beyond the short work of Calinon, the treatise on mechanics from Andrade.

VI. The second difficulty has until now attracted much less attention; it is however entirely analogous to the preceding and so logically I should have talked about it earlier.

Two different psychological phenomena occur in two different consciousnesses; when I say that they are simultaneous, what do I mean?

When I say that a physical phenomenon, which occurs outside of any consciousness, is before or after a psychological phenomenon, what do I mean?

In 1572 Tycho Brahe noticed a new star in the heavens. An immense conflagration had occurred in some very distant star; but it had occurred much earlier—it took at least.

200 years before the light leaving that star had reached our Earth. This conflagration was therefore prior to the discovery of America.

So, when I say this, when I consider this gigantic phenomenon which perhaps had no witness, because the satellites of this star perhaps did not have inhabitants, when I state that this phenomenon is prior to the formation of the visual image of Hispaniola in the conscience of Christopher Columbus, what do I mean?

A little thought is sufficient for understanding that all these affirmations have no meaning on their own.

They can only be meaningful after an agreement.

VII. We must first ask ourselves how one can have the idea of bringing into a single framework all the mutually impenetrable worlds.

We wish to represent the external universe for ourselves and it is only at this price that we could believe we know it.

We will never have this representation, we know it: our infirmity is too great.

We want to know at least that it is possible to conceive of an infinite intelligence for which this representation would be possible, a sort of large consciousness which would see all and which would classify all *in its time*, as we classify, *in our time*, the little that we see.

This hypothesis is coarse and incomplete; because this supreme intelligence would only be a demigod; infinite in one sense, this intelligence would be limited in another because it would only have an imperfect memory of the past; and it couldn't have any other, because without that all memories would be equally present to it and there would be no time for it.

And however when we talk of time, for everything which happens outside of us, don't we unconsciously adopt this hypothesis; don't we place ourselves in the place of this imperfect god; and the atheists themselves, don't they put themselves in the place where God would be, if he existed?

What I just said perhaps shows us why we have sought to bring all physical phenomena back to a single framework. But that cannot substitute for a definition of simultaneity, because this hypothetical intelligence, if it even existed, would be incomprehensible for us.

We have to look for something else.

VIII. Ordinary definitions which are sufficient for psychological time, can no longer suffice for us. Two simultaneous psychological facts are linked so closely that analysis cannot separate them without damaging them. Is it the same for two physical facts? Isn't my present closer to my past from yesterday than the present of Sirius?

We also said that two facts must be regarded as simultaneous when the order of their succession can be freely inverted. It is obvious that this definition could not be suitable for two physical facts that occur very far apart, and that, as it involves them, we no longer even understand what this reversibility could be; further, it is first the succession itself that needs to be defined.

IX. Let us therefore seek to consider what is understood by simultaneity or anteriority and to do that let us analyze some examples.

I write a letter; it is next read by the friend to whom I sent it. Here are two facts which played out in two different consciousnesses. While writing this letter I had the visual image of it, and my friend in their turn had this same image while reading the letter.

Although these two facts happen in impenetrable worlds, I do not hesitate to regard the first as earlier than the second, because I believe that one is the cause of the other.

I hear thunder and I conclude that there had been an electric discharge; I do not hesitate to consider the physical phenomenon as earlier than the audible impression experienced by my consciousness, because I believe that one is the cause of the other.

Here is the rule that we follow, and the only one that we could follow; when one phenomenon appears to us as the cause of another, we regard it as earlier.

We therefore define time by the cause; but most often, when two facts appear connected to us by a constant relation, how do we recognize which is the cause and which is the effect? We accept that the earlier fact, the antecedent, is the cause of the other, the consequent. It is then by time that we define the cause. How do we extricate ourselves from this circular reasoning?

Sometimes we say *post hoc, ergo propter hoc*; sometimes *propter hoc, ergo post hoc*; will we ever get out of this vicious circle?

X. Let us therefore look, not at how one manages to get out of it, because one can't completely do that, but how one tries to get out of it.

I perform a voluntary act A and I next experience a sensation D, which I regard as a consequence of act A; additionally, for an arbitrary reason, I infer that this consequence is not immediate, but that outside my consciousness two facts B and C, of which I was not a witness, occur and in such a way that B is the effect of A, that C is that of B and D is that of C.

But why that? If I think that I'm right to regard the four facts A, B, C and D as related to each other by a causality chain, why arrange them in causal order A, B, C and D and also in chronological order A, B, C and D instead of in some other order?

I see clearly that in the action A I have the feeling of having been active, whereas in experiencing the sensation D, I have that of having been passive. That is why I regard A as the initial cause and D as the ultimate effect; it is why I order A at the beginning of the chain and D at the end; but why put B before C instead of C before B?

If one is asked this question, one ordinarily responds: it is well known that it is B which is the cause of C because one *always* sees B occur before C. These two phenomena, when they are witnessed, occur in a certain order; when the analogous phenomena occur without a witness, there is no reason for this order to be inverted.

That seems likely, but be careful about it; we never directly know the physical phenomena B and C; what we know are sensations B′ and C′ produced respectively by B and C. Our consciousness immediately teaches us that B′ proceeds C′ and we allow that B and C are in succession in the same order.

This rule in fact seems wholly natural and just the same one is often led to exceptions from it. We hear the sound of thunder only a few seconds after the electric discharge from the cloud. With two lightning strikes, one far away, the other nearby, couldn't the first be earlier than the second, even though the noise from the second reaches us before that from the first?

XI. Another difficulty: do we have the right to speak of the cause of a phenomenon? If all the parts of the universe are connected to some extent, one arbitrary phenomenon will not be the effect arising from a single cause, but the result of infinitely many causes; it is, we often say, the consequence of the state of the universe at an earlier moment.

How can the rules applicable to such complex circumstances be stated? And however it is only at this price that the rules can be generalized and made rigorous.

In order to not lose ourselves in this infinite complexity, let us make a simpler hypothesis; consider three bodies, for example the Sun, Jupiter and Saturn; but for more simplicity look at them as reduced to material points and isolated from the rest of the world.

The positions and speeds of the three bodies at a given moment suffice for determining their positions and their speeds at the following moment and consequently at an arbitrary moment. Their positions at the moment t determine their positions at the moment $t + h$, as well as their positions at the moment $t - h$.

There's even more; the position of Jupiter at a moment t, combined with Saturn at a moment $t + a$, determines the position of Jupiter at an arbitrary moment and that of Saturn at an arbitrary moment.

The set of positions the Jupiter occupies at the moment $t + \varepsilon$ and Saturn occupies at the moment $t + a + \varepsilon$ is related to the set of positions which Jupiter occupies at the moment t and Saturn occupies that the moment $t + a$ by laws fully as precise as those of Newton, however much more complicated.

Henceforth why not regard one of these sets as the cause of the other, which would lead to considering as simultaneous the moment t for Jupiter and the moment $t + a$ for Saturn?

For that, there can only be reasons of convenience and simplicity; which are very powerful, it is true.

XII. Let us move to less artificial examples; so we can understand the definition implicitly accepted by scholars, let us look at their work and try to find according to what rules they seek simultaneity.

I will take two simple examples; the measurement of the speed of light and the determination of longitudes.

When an astronomer tells me that a stellar phenomenon, which his telescope revealed to him at that moment, had however occurred 50 years ago, I try to understand what he means and for that, I will first ask him how he knows it, meaning how he had measured the speed of light.

He started by *allowing* that light has a constant speed, and in particular that its speed is the same in all directions. It is a postulate without which no measurement of this speed could be tried. This postulate can never however be directly verified by experiment; it could be contradicted by it, if the results of various measurements were not consistent. We will have to think that we are fortunate that this contradiction has not happened and that the small discrepancies which can happen can be easily explained.

The postulate, in any case satisfying the principle of sufficient reason, was accepted by everyone; what I want to hold on to is that it provides us a new rule in the search for simultaneity, completely different from what we had stated earlier.

Having accepted this postulate, let's see how the speed of light is measured. It is known that Roemer made use of the eclipses of the satellites of Jupiter and sought how much the event was delayed from prediction.

But, how is this predicted? It is predicted by using astronomical laws, for example Newton's law of gravitation.

Couldn't the observed facts be explained just as well if a value slightly different from the adopted value were given to the speed of light, and if it were accepted that Newton's law is an approximation? Only, one would be led to replace Newton's law by another more complicated.

Thus, a value for the speed of light is adopted such that astronomical laws compatible with this value are as simple as possible.

When sailors or geographers determine a longitude, they have to solve exactly the same problem that concerns us: without being in Paris, they must calculate the time in Paris.

How do they go about it?

Either, they carry a chronometer with them set to Paris. The qualitative problem of simultaneity is referred back to the quantitative problem of the measurement of time. I'm not going back to the relative difficulties of this latter problem because I described them at length above.

Or else, they observe an astronomical phenomenon such as a lunar eclipse and accept that this phenomenon is observed simultaneously at all places on the Earth.

This is not entirely true, because the propagation of light is not instantaneous; if absolute accuracy is desired, it would be necessary to make a correction according to a complicated rule.

Or else, finally, they can make use of the telegraph. To start with, it is clear that receiving the signal at Berlin, for example, comes after the transmission of this same signal from Paris. It is the rule of cause-and-effect analyzed above.

But later, how much later? In general, one neglects the length of the transmission and considers both events as simultaneous. But, to be rigorous, one would have again to make a small correction by a complicated calculation; it isn't done in practice, because it would be much smaller than the observational errors; its theoretical need remains undiminished from our point of view, which is that of a rigorous definition.

I want to take away two things from this discussion:

1) The rules applied vary widely.
2) It is difficult to separate the qualitative problem of simultaneity from the quantitative problem of the measurement of time; either one makes use of a chronometer, or one has to include the transmission speed, like that of light, because one could not measure such a speed without *measuring* a time.

XIII. It is appropriate to conclude.

We do not have direct intuition of simultaneity any more than that of the equality of two durations.

If we believe that we have this intuition, it is an illusion.

We compensate for this by using some rules that we apply nearly always without our being aware of it.

But what is the nature of these rules?

No general rule, no rigorous rule, but a multitude of small rules applicable to each specific case.

These rules are not imposed on us and one could take pleasure in inventing others; however, one would not be able to set them aside without greatly complicating the statement of the laws of physics, mechanics and astronomy.

We therefore choose these rules, not because they are true, but because they are the most convenient and we could summarize them by stating:

"The simultaneity of two events, or the order of their succession, the equality of two durations, must be defined such that the statement of the natural laws is as simple as possible. In other words, all these rules, all these definitions are solely the fruit of an unconscious opportunism."

H. Poincaré

Chapter 2
Lorentz's Theory and the Conservation of Momentum

Most likely it will seem strange that at a moment raised to the glory of Lorentz I return to the considerations that I previously presented as an objection to his theory. I could state that the following pages instead are of a nature to reduce this objection and not deepen it.

But I'm not going to make use of that excuse because I have one that is 100 times better. *Good theories are flexible*. Those which have a rigid form and which cannot be adapted without collapsing truly have too little vitality. But if the theory shows us some true relations, it can be dressed in a thousand various forms and it will resist all the assaults and what makes up its essence will not change. This is what I explained in the seminar that I gave recently at the *Congrès international de physique* in Paris.

Good theories overcome all objections; objections that are only specious do not get a hold on them and they triumph even over serious objections but they triumph over them by changing.

The objections help them, which is far from harming them, because the objections allow them to develop all the hidden virtues which were in them. Just so, Lorentz's theory is of that kind, and that is the only excuse that I wish to make.

I'm not asking the reader to forgive me for that, but to forgive me for having presented at such length ideas with so little novelty.

Part 1

Let us first quickly review the calculation by which it was established that in Lorentz's theory, the principal that for every action there is an equal and opposite reaction is no longer true, at least when one wants to apply it only to matter.

Poincaré, H. (1900). La théorie de Lorentz et le principe de réaction. *Archives néerlandaises des sciences exactes et naturelles, 5*, 252–278.

B. D. Popp, *Henri Poincaré: Electrons to Special Relativity*,
https://doi.org/10.1007/978-3-030-48039-4_2

Let us seek the resultant of all the ponderomotive forces applied to all the electrons located inside a certain volume. This resultant, or instead its projection on the x-axis, is represented by the integral:

$$X = \int \left[\eta\gamma - \xi\beta + \frac{4\pi f}{K_0}\right]\rho \mathrm{d}\tau$$

where the integration extends over all elements $\mathrm{d}\tau$ of the volume considered, and where the ξ, η and ζ represent the components of the electron velocity.

Because of the equations:

$$\rho\eta = -\frac{\mathrm{d}g}{\mathrm{d}t} + \frac{1}{4\pi}\left(\frac{\mathrm{d}\alpha}{\mathrm{d}z} - \frac{\mathrm{d}\gamma}{\mathrm{d}x}\right); \ \rho\zeta = -\frac{\mathrm{d}h}{\mathrm{d}t} + \frac{1}{4\pi}\left(\frac{\mathrm{d}\beta}{\mathrm{d}x} - \frac{\mathrm{d}\alpha}{\mathrm{d}y}\right); \ \rho = \sum\frac{\mathrm{d}f}{\mathrm{d}x}$$

and by adding and subtracting the term:

$$\frac{\alpha}{4\pi}\frac{\mathrm{d}\alpha}{\mathrm{d}x},$$

I can write:

$$X = X_1 + X_2 + X_3 + X_4,$$

where:

$$X_1 = \int \left(\beta\frac{\mathrm{d}h}{\mathrm{d}t} - \gamma\frac{\mathrm{d}g}{\mathrm{d}t}\right)\mathrm{d}\tau$$

$$X_2 = \frac{1}{4\pi}\int \left(\alpha\frac{\mathrm{d}\alpha}{\mathrm{d}x} + \beta\frac{\mathrm{d}\alpha}{\mathrm{d}y} + \gamma\frac{\mathrm{d}\alpha}{\mathrm{d}z}\right)\mathrm{d}\tau$$

$$X_3 = \frac{-1}{4\pi}\int \left(\alpha\frac{\mathrm{d}\alpha}{\mathrm{d}x} + \beta\frac{\mathrm{d}\beta}{\mathrm{d}x} + \gamma\frac{\mathrm{d}\gamma}{\mathrm{d}x}\right)\mathrm{d}\tau$$

$$X_4 = \frac{4\pi}{K_0}\int f\sum\frac{\mathrm{d}f}{\mathrm{d}x}\mathrm{d}\tau$$

Integration by parts gives

$$X_2 = \frac{1}{4\pi}\int \alpha(l\alpha + m\beta + u\gamma)\mathrm{d}\omega - \frac{1}{4\pi}\int \alpha\left(\frac{\mathrm{d}\alpha}{\mathrm{d}x} + \frac{\mathrm{d}\beta}{\mathrm{d}y} + \frac{\mathrm{d}\gamma}{\mathrm{d}z}\right)\mathrm{d}\tau$$

$$X_3 = \frac{-1}{8\pi}\int l(\alpha^2 + \beta^2 + \gamma^2)\mathrm{d}\omega$$

where the double integrals extend over all the elements $d\omega$ of the surface which limits the volume under consideration, and where l, m and n designate the directional cosines of the normal to this element.

Observing that

$$\frac{d\alpha}{dx}+\frac{d\beta}{dy}+\frac{d\gamma}{dz}=0,$$

it can be seen that one can write:

$$X_2+X_3=\frac{1}{8\pi}\int\left[l\left(\alpha^2-\beta-\gamma^2\right)+2m\alpha\beta+2n\alpha\gamma\right]d\omega. \quad (1)$$

Let us now transform X_4.

Integration by parts gives:

$$X_4=\frac{4\pi}{K_0}\int\left(lf^2+mfg+nfh\right)d\omega-\frac{4\pi}{K_0}\int\left(f\frac{df}{dx}+g\frac{df}{dy}+h\frac{df}{dz}\right)d\tau.$$

I call X_4' and X_4'' the two integrals of the right-hand side, such that

$$X_4=X_4'-X_4''.$$

If one uses the equations:

$$\frac{df}{dy}=\frac{dg}{dx}+\frac{K_0}{4\pi}\frac{d\gamma}{dt}$$
$$\frac{df}{dz}=\frac{dh}{dx}-\frac{K_0}{4\pi}\frac{d\beta}{dt}$$

we can write:

$$X_4''=\mathrm{Y}+\mathrm{Z}$$

where

$$Y=\frac{4\pi}{K_0}\int\left(f\frac{df}{dx}+g\frac{dg}{dx}+h\frac{dh}{dx}\right)d\tau$$
$$Z=\int\left(g\frac{d\gamma}{dt}-h\frac{d\beta}{dt}\right)d\tau.$$

Next one finds:

$$Y = \frac{2\pi}{K_0} \int l(f^2 + g^2 + h^2)\mathrm{d}\omega$$

$$X_1 - Z = \frac{\mathrm{d}}{\mathrm{d}t} \int (\beta h - \gamma g)\mathrm{d}\tau.$$

Finally one has:

$$X = \frac{\mathrm{d}}{\mathrm{d}t} \int (\beta h - \gamma)\mathrm{d}\tau + (X_2 + X_3) + \left(X_4' - Y\right), \tag{2}$$

where $X_2 + X_3$ is given by formula (1), whereas:

$$X_4' - Y = \frac{2\pi}{K_0} \int \left[l(f^2 - g^2 - h^2) + 2mfg + 2nfh\right]\mathrm{d}\omega.$$

This term, $(X_2 + X_3)$, represents the projection on the x-axis of a pressure acting on the differential elements $\mathrm{d}\omega$ of the surface delimiting the volume under consideration. It is immediately recognized that this pressure is nothing other than *Maxwell's magnetic pressure*, introduced by this scientist in a well-known theory.

Similarly the term, $\left(X_4' - Y\right)$, represents the effect of *Maxwell's electrostatic pressure*.

In the absence of the first term:

$$\frac{\mathrm{d}}{\mathrm{d}t} \int (\beta h - \gamma)\mathrm{d}\tau$$

the ponderomotive force would therefore be nothing other than the result of the Maxwell pressures.

If our integrals are extended to all space, the double integrals disappear and there only remains:

$$X = \frac{\mathrm{d}}{\mathrm{d}t} \int (\beta h - \gamma)\mathrm{d}\tau$$

If therefore we call M one of the material masses considered and call V_x, V_y and V_z the components of its velocity, it should hold, if conservation of momentum were applicable, that:

$$\sum MV_x = \text{const.}; \quad \sum MV_y = \text{const.}; \quad \sum MV_z = \text{const.}$$

In contrast, we have:

$$\sum MV_x + \int (\gamma g - \beta h)\mathrm{d}\tau = \text{const.}$$
$$\sum MV_y + \int (\alpha h - \gamma f)\mathrm{d}\tau = \text{const.}$$
$$\sum MV_z + \int (\beta f - \alpha g)\mathrm{d}\tau = \text{const.}$$

Observe that

$$\gamma g - \beta h, \quad \alpha h - \gamma f, \quad \beta f - \alpha g$$

are the three components of the *Poynting vector*.

If one sets:

$$J = \frac{1}{8\pi}\sum \alpha^2 + \frac{2\pi}{K_0}\sum f^2,$$

Poynting's equation in fact gives us:

$$\int \frac{\mathrm{d}J}{\mathrm{d}t}\mathrm{d}\tau = \frac{1}{K_0}\int \begin{vmatrix} l & m & n \\ \alpha & \beta & \gamma \\ f & g & h \end{vmatrix} \mathrm{d}\omega + \frac{4\pi}{K_0}\int \rho \sum f\xi \mathrm{d}\tau. \qquad (3)$$

The first integral of the left-hand side represents, as is known, the quantity of electromagnetic energy which enters the volume in consideration through its surface as radiation and the second term represents the quantity of electromagnetic energy which is created inside the volume by transformation of energy of other kinds. Electromagnetic energy can be regarded as a fictitious fluid whose density is $K_0 J$ and which moves in space according to Poynting's laws. Except, it has to be allowed that this fluid is not indestructible and that in the element of volume $\mathrm{d}\tau$ during a unit time a quantity of it equal to $\frac{4\pi}{K_0}\rho \mathrm{d}\tau \sum f\xi$ is destroyed (or that an equal quantity of opposite sign is created if this expression is negative); this is what prevents us, in our reasoning, from completely comparing our fictitious fluid to a real fluid.

The quantity of this fluid which passes during unit time through a unit surface oriented perpendicularly to the x, y or z axis is equal to:

$$K_0 J U_x, \quad K_0 J U_y, \quad K_0 J U_z,$$

Here U_x, U_y and U_z are the components of the velocity of the fluid. By comparison with Poynting's formula, it is found that:

$$K_0 J U_x = \gamma g - \beta h$$
$$K_0 J U_y = \alpha h - \gamma f$$
$$K_0 J U_z = \beta f - \alpha g$$

such that our formulas become:

$$\begin{aligned} \sum MV_x + \int K_0 J U_x \mathrm{d}\tau &= \text{const.} \\ \sum MV_y + \int K_0 J U_y \mathrm{d}\tau &= \text{const.} \\ \sum MV_z + \int K_0 J U_z \mathrm{d}\tau &= \text{const.} \end{aligned} \tag{4}$$

They state that the momentum of the matter itself plus that of our fictitious fluid is given by a constant factor.

In Ordinary Mechanics, if the momentum is constant, it is concluded that the motion of the center of gravity is straight and uniform.

But here we do not have the possibility of concluding that the center of gravity of the system formed by the matter and our fictitious fluid has a straight and uniform motion; this is because this fluid is not indestructible.

The position of the center of gravity of the fictitious fluid depends on the integral over all space

$$\int xJ \mathrm{d}\tau.$$

The derivative of this integral is:

$$\int x \frac{\mathrm{d}J}{\mathrm{d}t} \mathrm{d}\tau = -\int x \left(\frac{\mathrm{d}JU_x}{\mathrm{d}x} + \frac{\mathrm{d}JU_y}{\mathrm{d}y} + \frac{\mathrm{d}JU_z}{\mathrm{d}z} \right) \mathrm{d}\tau - \frac{4\pi}{K_0} \int \rho x \sum f\xi \mathrm{d}\tau$$

Now, the first integral on the right-hand side becomes, by integration by parts:

$$\int JU_x \mathrm{d}\tau$$

$$\text{or } \frac{1}{K_0}\left(C - \sum MV_x\right)$$

where C designates the constant from the right-hand side of the first equation (4).

Let us then represent the total mass of the matter by M_0, the coordinates of its center of gravity by X_0, Y_0 and Z_0, the total mass of the fictitious fluid by M_1, its center of gravity by X_1, Y_1 and Z_1, the total mass of the system (matter plus fictitious fluid) by M_2, and its center of gravity by X_2, Y_2 and Z_2, such that one has:

$$M_2 = M_0 + M_1, \quad M_2 X_2 = M_0 X_0 + M_1 X_1,$$

$$\frac{\mathrm{d}}{\mathrm{d}t}(M_0 X_0) = \sum M V_x, \quad K_0 \int x J \mathrm{d}\tau = M_1 X_1.$$

It then follows:

$$\frac{\mathrm{d}}{\mathrm{d}t}(M_2 X_2) = C - 4x \int \rho x \sum f \xi \mathrm{d}\tau \tag{3}$$

Here is how equation (3) could be stated in ordinary language.

If there is in no way any creation or destruction of electromagnetic energy, the last term disappears; hence the center of gravity of the system formed by matter and electromagnetic energy (regarded as a fictitious fluid) has a straight and uniform motion.

Let us now assume that at certain points there has been destruction of electromagnetic energy which was transformed into nonelectric energy. It would be necessary to consider the system formed not only by matter and electromagnetic energy, but by the nonelectric energy coming from the transformation of the electromagnetic energy.

But it must be agreed that this nonelectric energy remains at the point where the transformation occurred and that it is not subsequently carried along by the matter where it is ordinarily located. There is nothing in this convention which should surprise us because it is only a matter of a mathematical fiction. If this convention is adopted, the motion of the center of gravity of the system is still straight and uniform.

To extend the statement to the case where there is not only destruction, but also creation of energy, it is sufficient to assume that at each point there is some amount of nonelectric energy, at the expense of which electromagnetic energy is formed. We will then retain the preceding agreement, meaning that instead of localizing the nonelectric energy as is ordinarily done, we regard it as immobile. Under this condition, the center of gravity will again move in a straight line.

Let us now return to equation (2) by assuming the integrals extend over an infinitesimal volume. It will then mean that the resultant of the Maxwell pressures which are exerted on the surface of the volume is in equilibrium with:

1) The forces of nonelectric origin applied to the matter which is located in this volume;
2) The inertial forces of this matter;
3) The inertial forces of the fictitious fluid enclosed in this volume.

To define this inertia of the fictitious fluid, it is appropriate that the fluid that was created at an arbitrary point by transformation of the energy, first arises without velocity and that it takes on its velocity from the already existing fluid; if therefore, the quantity of fluid increases, but its velocity remains constant, there will nonetheless be some inertia to overcome because the new fluid takes on the velocity of the former fluid; the velocity of the assembly would decrease if an arbitrary cause didn't become involved to keep it constant. Similarly when there is destruction of electromagnetic energy, the fluid must lose its velocity before being destroyed by ceding it to the remaining fluid.

Because the equilibrium holds for an infinitesimal volume, it will hold for a finite volume. If in fact we decompose it into infinitesimal volumes, the equilibrium will hold for each of them. To move to a finite volume, the collection of forces applied to the various infinitesimal volumes has to be considered; solely among the Maxwell pressures only the forces exerted on the surface of the total finite volume will be retained, and those exerted on the surface elements which separate two infinitesimal contiguous volumes will be eliminated. This elimination will in no way change the equilibrium, because the pressures eliminated in that way are pairwise equal and oppositely directed.

The equilibrium will therefore again occur for the finite volume.

It will therefore occur for all space. But in this case, it is not necessary to consider either the Maxwell pressures which are zero at infinity, or the forces of nonelectric origin which are in balance because of the Newton's third law applicable to the force is considered in Ordinary Mechanics.

The two kinds of inertial forces are therefore in equilibrium; hence there is a double consequence:

1) The principle of the conservation of the projections of the momentum applies to the system of matter and fictitious fluid; we also find equation (4) again.
2) The principle of the conservation of the moments of the momentum or in other words, the *conservation of angular momentum* applies to the system of matter and fictitious fluid. This is a new consequence which supplements the data provided by equation (4).

Since, from the perspective which interests us, electromagnetic energy therefore behaves like a fluid that has inertia, it has to be concluded that if an arbitrary device after having produced electromagnetic energy transmits it by radiation in some direction, then this device will have to *recoil* like an artillery piece which fires a projectile.

Of course, this recoil will not occur if the producing device transmits energy equally in all directions; in contrast it will occur if this symmetry does not exist and if the electromagnetic energy produced is sent in a single direction, as happens for example if the device is a Hertz exciter placed at the focus of a parabolic mirror.

It is easy to evaluate numerically the magnitude of this recoil. If the device has a mass of 1 kg and if it sends 3 million J in a single direction with the speed of light, the velocity due to the recoil is 1 cm/s. In other words if the energy produced by a 3000 W machine is sent in a single direction a force of 1 dyne would be needed to keep the machine in place despite the recoil.

It is obvious that such a weak force could not be detected by experiment. But one could imagine that if, however impossible, sufficiently sensitive measurement devices to show it were available, it would in that way be possible to prove that the conservation of momentum is not applicable to matter alone, and that it would confirm Lorentz's theory and condemn other theories.

Things aren't like that; Hertz's theory and in general all the other theories lead to the same calculation as that for Lorentz.

Just now, I used the example of a Hertz exciter whose radiation would be made parallel by a parabolic mirror, I could have taken a simpler example borrowed from optics; a beam of parallel light rays strikes a mirror perpendicularly and after reflection returns in the opposite direction. Energy first propagates from left to right, for example, and is then returned from right to left by the mirror.

The mirror thus must *recoil,* and the recoil is easily calculated by the preceding considerations.

It is easy to recognize the problem which was already handled by Maxwell in sections 792 and 793 of his work. It also calls for the recoil of the mirror just like what we have deduced from Lorentz's theory.

If we go deeper into the study of the mechanism for this recoil, this is what we find. Let us consider an arbitrary volume and apply equation (2); this equation teaches us that the force of electromagnetic origin which acts on electrons—meaning on the matter contained in the volume—is equal to the resultant of the Maxwell pressures increased by a correction factor which is the derivative of the integral:

$$\int (\beta h - \gamma g) \mathrm{d}\tau.$$

If the regime is established, this integral as constant and the correction term is zero.

The recoil called for by Lorentz's theory is that which is due to the Maxwell pressure. Now, all theories call for Maxwell pressure; all theories therefore call for the same recoil.

Part 2

But then a new question comes up. We called for the recoil in Lorentz's theory because this theory is contrary to the conservation of momentum. Among the other theories, there are some, like Hertz's theory, which do conserve momentum. How is it that they lead to the same recoil?

Let me give the explanation for this paradox right away and leave the justification of this explanation until later. In Lorentz's theory and in Hertz's theory the device which produces the energy and sends it in one direction recoils, but this energy, thus radiated, propagates by passing through some medium, such as air, for example.

In Lorentz's theory, when the air receives the energy thus radiated, it does not undergo any mechanical action; nor does it receive any either when this energy leaves it after having passed through it. In contrast, in Hertz's theory, when the air receives the energy, it is pushed forward and then it recoils when this energy leaves it. The motion of the air that the energy passes through thus balances, from the perspective of conservation of momentum, the motion of the devices which produce this energy. In Lorentz's theory, this compensation does not happen.

In fact, let's go back to Lorentz's theory and our equation (2) and apply it to a homogeneous dielectric. It is known how Lorentz represents a dielectric medium; this medium would contain electrons capable of small motions, and these motions would produce the dielectric polarization to which the effect would be added, from some perspectives, to that of the electric displacement itself.

Let X, Y and Z be the components of this polarization. It then holds:

$$\frac{\mathrm{d}X}{\mathrm{d}t}\mathrm{d}\tau = \sum \rho\xi, \quad \frac{\mathrm{d}Y}{\mathrm{d}t}\mathrm{d}\tau = \sum \rho\eta, \quad \frac{\mathrm{d}Z}{\mathrm{d}t}\mathrm{d}\tau = \sum \rho\zeta. \tag{5}$$

The summations on the right-hand sides are extended to all electrons contained inside the element $\mathrm{d}\tau$ and these equations can be regarded as the definition of the dielectric polarization itself.

For the expression for the resultant of the ponderomotive forces (which I no longer designate with X in order to avoid any confusion with polarization), we found the integral:

$$\int \rho\left[\eta\gamma - \zeta\beta + \frac{4\pi f}{K_0}\right]\mathrm{d}\tau$$

or

$$\int \rho\eta\gamma\mathrm{d}\tau - \int \rho\zeta\beta\mathrm{d}\tau + \frac{4\pi}{K_0}\int \rho f\mathrm{d}\tau.$$

The first two integrals can be replaced by

$$\int \gamma\frac{\mathrm{d}Y}{\mathrm{d}t}\mathrm{d}\tau, \quad \int \beta\frac{\mathrm{d}Z}{\mathrm{d}t}\mathrm{d}\tau$$

because of equation (5). As for the third integral, it is zero because the total charge of an element of dielectric containing some number of electrons is zero. Our ponderomotive force therefore reduces to:

$$\int\left(\gamma\frac{\mathrm{d}Y}{\mathrm{d}t} - \beta\frac{\mathrm{d}Z}{\mathrm{d}t}\right)\mathrm{d}\tau.$$

If I designate the force due to the various Maxwell pressures by Π, such that

$$\Pi = (X_2 + X_3) + (X_4' - Y)$$

then our equation (2) becomes:

$$\Pi = \int \left(\gamma \frac{\mathrm{d}Y}{\mathrm{d}t} - \beta \frac{\mathrm{d}Z}{\mathrm{d}t}\right) \mathrm{d}\tau + \frac{\mathrm{d}}{\mathrm{d}t} \int (\gamma g - \beta h) \mathrm{d}\tau. \tag{2'}$$

Additionally there is a relation like this

$$a \frac{\mathrm{d}^2 X}{\mathrm{d}t^2} + bX = f \tag{A}$$

where a and b are two constants characteristic of the medium; from this it is easily deduced that:

$$X = (n^2 - 1)f \tag{B}$$

and even

$$Y = (n^2 - 1)g, \quad Z = (n^2 - 1)h$$

where n is the index of refraction for the color considered.

One could be led to replace the relation (A) by others more complicated, for example if one needed to assume more complex ions. It doesn't matter, because one would always be led to equation (B).

To go farther, we are going to assume a plane wave propagating in the direction of the x axis towards positive x, for example. If the wave is polarized in the x-z plane, one will have

$$X = f = \alpha = Z = h = \beta = 0$$

and

$$\gamma = ng \frac{4\pi}{\sqrt{K_0}}.$$

Incorporating all of these relations, (2′) first becomes

$$\Pi = \int \gamma \frac{\mathrm{d}Y}{\mathrm{d}t} \mathrm{d}\tau + \int \gamma \frac{\mathrm{d}g}{\mathrm{d}t} \mathrm{d}\tau + \int g \frac{\mathrm{d}\gamma}{\mathrm{d}t} \mathrm{d}\tau,$$

where the first integral represents the ponderomotive force. But if the proportions,

$$\frac{g}{1} = \frac{Y}{n^2 - 1} = \frac{\gamma}{n\left(\frac{4\pi}{\sqrt{K_0}}\right)}$$

are considered, our equation becomes

$$\frac{\sqrt{K_0}}{4\pi}\Pi = n\left(n^2 - 1\right)\int g\frac{\mathrm{d}g}{\mathrm{d}t}\mathrm{d}\tau + n\int g\frac{\mathrm{d}g}{\mathrm{d}t}\mathrm{d}\tau + n\int g\frac{\mathrm{d}g}{\mathrm{d}t}\mathrm{d}\tau. \tag{6}$$

But to draw something from this formula, it has to be seen how the energy is distributed and propagates in the dielectric medium. The energy is divided into three parts: 1) electric energy, 2) magnetic energy, 3) mechanical energy due to the motion of the ions. The expressions for these three parts are respectively:

$$\frac{2\pi}{K_0}\sum f^2, \quad \frac{1}{8\pi}\sum \alpha^2, \quad \frac{2\pi}{K_0}\sum fX$$

and in the case of a plane waves, they are proportional to each other as:

$$1, \quad n^2, \quad n^2 - 1\,.$$

In the preceding analysis, we had what we called the momentum of the electromagnetic energy play a role. It is clear that the density of our fictitious fluid will be proportional to the sum of the two parts (electric and magnetic) of the total energy and that the third part, which is purely mechanical, will have to be set aside. But what velocity is it appropriate to give to this fluid? At first, one might think that it is the wave propagation velocity, meaning $1/(n\sqrt{K_0})$. But it is not so simple. At each point the electromagnetic energy and the mechanical energy are proportional; if therefore at one point the electromagnetic energy comes to decrease, the mechanical energy will also decrease, meaning that it will partially transform into electromagnetic energy; there will therefore be creation of the fictitious fluid.

For a moment, designate the density of the fictitious fluid by ρ and its velocity, which I assume to be parallel to the x-axis, by ξ; I assume that all our functions depend only on x and t, since the plane of the wave is perpendicular to the x-axis. The continuity equation is then written

$$\frac{\mathrm{d}\rho}{\mathrm{d}t} + \frac{\mathrm{d}\rho\xi}{\mathrm{d}x} = \frac{\delta\rho}{\mathrm{d}t}$$

where $\delta\rho$ is the quantity of fictitious fluid created during time $\mathrm{d}t$. Now, this quantity is equal to the quantity of mechanical energy destroyed, which is to the quantity of electromagnetic energy destroyed, meaning $-\mathrm{d}\rho$, as $n^2 - 1$ is to $n^2 + 1$; hence

$$\frac{\delta\rho}{n^2 - 1} = -\frac{\mathrm{d}\rho}{n^2 + 1};$$

such that our equation becomes

$$\frac{\mathrm{d}\rho}{\mathrm{d}t}\frac{2n^2}{n^2 + 1} + \frac{\mathrm{d}\rho\xi}{\mathrm{d}x} = 0.$$

If ξ is a constant, this equation shows us that the propagation velocity is equal to

$$\xi\frac{n^2 + 1}{2n^2}.$$

If the propagation velocity is $1/(n\sqrt{K_0})$, it will then hold that

$$\xi = \frac{2n^2}{(n^2 + 1)\sqrt{K_0}}$$

If the total energy is J', the electromagnetic energy will be $J = \frac{n^2+1}{2n^2}J'$ and the momentum of the fictitious fluid will be:

$$K_0 J\xi = K_0\frac{n^2 + 1}{2n^2}J'\xi = \frac{J'\sqrt{K_0}}{n} \tag{7}$$

because the density of the fictitious fluid is equal to the energy multiplied by K_0.

Hence in equation (6) the first term of the right-hand side represents the ponderomotive force, meaning the derivative of the momentum of the matter of the dielectric, while the last two terms represent the derivative of the momentum of the fictitious fluid. These two momentums are therefore related to each other as $n^2 - 1$ and 2.

So let Δ be the density of the dielectric material, and W_x, W_y and W_z be the components of its velocity. Let's go back to equation (4). The first term $\sum MV_x$ represents the momentum of all the real matter; we will break it down in two parts. The first part, which we will continue to designate by $\sum MV_x$, will represent the momentum of the energy-producing devices; the second part will represent the momentum of the dielectrics; it will be equal to

$$\int \Delta . W_x \mathrm{d}\tau$$

such that equation (4) will become

$$\sum MV_x + \int (\Delta.W_x + K_0 JU_x)\mathrm{d}\tau = \text{const.} \tag{4'}$$

According to what we just saw, it will follow that:

$$\frac{\Delta.W_x}{n^2 - 1} = \frac{K_0 JU_x}{2}.$$

Further, let us designate, as above, the total energy by J'; let us also distinguish the real velocity of the fictitious fluid, meaning that which results from Poynting's law and which we have designated by U_x, U_y and U_z, and the apparent velocity of the energy, meaning what would be deduced from the propagation speed of the waves and that we will designate by U'_x, U'_y and U'_z. It results from equation (7) that:

$$JU_x = J'U'_x$$

Equation (4′) can be written in the form:

$$\sum MV_x + \int (\Delta.W_x + K_0 J'U'_x)\mathrm{d}\tau = \text{const.}$$

Equation (4′) shows the following: if the device radiates energy in a single direction *in the vacuum*, it experiences a recoil which is composed solely, from the perspective of the conservation of momentum, by the motion of the fictitious fluid.

But, if the radiating, instead of occurring in the vacuum, is done in a dielectric, this recoil will be compensated in part by the motion of the fictitious fluid and in part by the motion of the dielectric matter, and the fraction of the recoil of the producing device which will thus be compensated by the motion of the dielectric, meaning by the motion of real matter, will be, I state, this fraction ${}^{n^2-1}/_{n^2-1}$.

That is what follows from Lorentz's theory; how do we now switch to Hertz's theory?

The content of Mossotti's ideas on the makeup of dielectrics is known.

Dielectrics, other than the vacuum, are formed of small conducting spheres (or more generally small conducting bodies) separated from each other by an un-polarizable insulating medium analogous to the vacuum. How does one go from that to Maxwell's ideas? One imagines that the vacuum itself had the same makeup: it was not un-polarizable, but formed of conducting cells, separated by partitions formed of an ideal, insulating and un-polarizable matter. The specific inductive power of the vacuum was therefore greater than that of un-polarizable ideal matter (likewise in the primitive understanding of Mossotti, the inductive power of the dielectrics was greater than that of the vacuum, and for the same reason). And the ratio of the first of these powers to the second was even larger as the space occupied by the conducting cells was larger compared to the space occupied by the insulating partitions.

Finally, move to the limit; by regarding the inductive power of the insulating matter as infinitesimal, and at the same time the insulating partitions as infinitesimally thin, such that the space occupied by these partitions is infinitesimal, the inductive power of the vacuum remains finite. *This transition to the limit leads us to Maxwell's theory.*

All this is well known and I will limit myself to quickly reviewing it. So, *between Lorentz's theory and Hertz's theory there is the same relation as between Mossotti's theory and Maxwell's theory.*

In fact let us assume that we could attribute the same makeup to the vacuum as Lorentz attributes to ordinary dielectrics; meaning that we consider it as an un-polarizable medium in which electrons can undergo small motions.

Lorentz's formulas will still be applicable, *only K_0 will no longer represent the inductive power of the vacuum, but that of ideal un-polarizable medium.* Now move to the limit by assuming K_0 is infinitesimal; it will of course be necessary to compensate for this hypothesis by multiplying the number of electrons such that the inductive powers of the vacuum and the other dielectrics remain finite.

The theory to which this transition to the limit leads is none other than Hertz's theory.

Let V be the speed of light in the vacuum. In Lorentz's basic theory, it is equal to $1/\sqrt{K_0}$; but that is no longer so in the modified theory, where it is equal to

$$\frac{1}{n_0\sqrt{K_0}},$$

where n_0 is the index of refraction of the vacuum relative to the un-polarizable ideal medium. If n designates the index of refraction of a dielectric relative to the ordinary vacuum, its index relative to this ideal medium will be nn_0 and the speed of light in this dielectric will be

$$\frac{V}{n} = \frac{1}{nn_0\sqrt{K_0}}.$$

In Lorentz's formulas, n must therefore be replaced by nn_0.

For example, the dragging of waves in Lorentz's theory is represented by the Fresnel formula,

$$v\left(1 - \frac{1}{n^2}\right)$$

In the modified theory it would be

$$v\left(1-\frac{1}{n^2{n_0}^2}\right)$$

If we move to the limit, then we must make $K_0=0$, hence $n_0=\infty$; therefore in Hertz's theory, the drag will be v, meaning that it will be total. This consequence, contrary to Fizeau's experiment, is sufficient to condemn Hertz's theory, such that the only interest of these considerations is as a curiosity.

Let us however go back to our equation (4′). It teaches us that the fraction of the recoil which is compensated by the motion of the dielectric matter is equal to

$$\frac{n^2-1}{n^2+1}.$$

In the modification of Lorentz's theory, this fraction will be:

$$\frac{n^2{n_0}^2-1}{n^2{n_0}^2+1}.$$

If we move to the limit by making $n_0=\infty$, this fraction is equal to 1, such that the recoil is entirely compensated by the motion of the dielectric matter. In other words, in Hertz's theory the principle of reaction is not violated and applies only to matter.

This is what would be seen again using equation (4′); if in the limit K_0 is zero, the term $\int K_0 J' U'_x \mathrm{d}\tau$, which represents the momentum of the fictitious fluid, also becomes zero, such that considering the momentum of the real matter is sufficient.

This consequence follows: *to experimentally demonstrate that conservation of momentum is violated in reality as it is in Lorentz's theory, it would not be sufficient to show that the energy-producing devices experience a recoil,* which would be difficult enough, *it would additionally be necessary to show that the recoil is not compensated by the motion of the dielectrics and in particular by the air through which the electromagnetic waves pass.* This would obviously be much more difficult still.

A final remark on this subject. Let us assume that the medium through which the waves pass is magnetic. A portion of the wave energy will be found in mechanical form. If μ is the magnetic permeability of the medium, the *total* magnetic energy will be:

$$\frac{\mu}{8\pi}\int\sum\alpha^2\mathrm{d}\tau$$

but only a fraction, specifically:

$$\frac{1}{8\pi}\int\sum\alpha^2 \mathrm{d}\tau$$

will strictly speaking be magnetic energy; the other part:

$$\frac{\mu-1}{8\pi}\int\sum\alpha^2 \mathrm{d}\tau$$

will be *mechanical* energy used to bring the particle currents to a shared orientation perpendicular to the field, against the elastic force which tends to bring these currents into the equilibrium orientation that they take in the absence of a magnetic field.

An analysis can be applied to these media just like the preceding analysis and where the mechanical energy would play the same role as the mechanical energy played in the case of dielectrics. It would in that way be recognized that if non-dielectric (I mean whose dielectric power would be the same as the vacuum) magnetic media existed, the matter of these media would undergo a mechanical action subsequent to the passage of the waves such that the recoil of the producing devices would in part be compensated by the motions of these media, as it is by the media of the dielectrics.

To get out of this case that does not occur in nature, we assume a medium that is both dielectric and magnetic where the fraction of the recoil compensated by the motion of the medium would be stronger than for a nonmagnetic medium of the same dielectric power.

Part 3

Why is the conservation of momentum obvious to our thinking? It is important to consider this in order to see whether the preceding paradoxes can actually be considered as an objection to Lorentz's theory.

If this principle, in most cases, is obvious to us, it is because its negation would lead to perpetual motion; is that the case here?

Let A and B be two arbitrary bodies, acting on each other, but take away any external action; if the action of one were not equal to the reaction of the other, they could be attached to each other by a rod of invariable length such that they behave like *a single* solid body. Since the forces applied to this solid do not produce equilibrium, the system would start in motion and this motion would go on endlessly by accelerating, *on one condition however,* that the mutual action of the two bodies depend only on their *relative* position and *relative* velocity, but is independent of their *absolute* position and *absolute* velocity.

More generally, for an arbitrary conservative system, let U be its potential energy, m be the mass of one of the points of the system, x', y' and z' be the components of its velocity; the equation for the total energy will be:

$$\sum \frac{m}{2}\left(x'^2 + y'^2 + z'^2\right) + U = \text{const.}$$

Now refer the system to moving axes driven with translational velocity v parallel to the x-axis: let x', y$'$ and z' be the components of the relative velocity relative to these axes, it will follow:

$$x' = x_1' + r, \quad y' = y_1', \quad z' = z_1'.$$

and consequently:

$$\sum \frac{m}{2}\left((x_1' + r)^2 + y_1'^2 + z_1'^2\right) + U = \text{const.}$$

Because the *principle of relative motion*, U only depends on the *relative* position of the points of the system, the laws of relative motion do not differ from those of absolute motion and the equation for the total energy and the relative motion is written

$$\sum \frac{m}{2}\left(x_1'^2 + y_1'^2 + z_1'^2\right) + U = \text{const.}$$

By subtracting the two equations one from the other, it is found that

$$r \sum m x_1' + \frac{r^2}{2} \sum m = \text{const.} \tag{8}$$

or

$$\sum m x_1' = \text{const.} \tag{9}$$

which is the analytical expression of conservation of momentum.

The conservation of momentum therefore appears to us to be a consequence of the conservation of energy and the principle of relative motion. This last principle is necessarily obvious to our thinking when applied to an isolated system.

But in the case that concerns us, it does not involve an isolated system, because we only consider the matter itself, aside from which there is still the ether. If all material objects are driven in a shared translation, as for example in the translation of the Earth, the phenomena can differ from what they would be if this translation did not exist because the ether might not be dragged in this translation. The principle of relative motion thus understood and applied to matter alone is so lacking in obviousness to our thinking that experiments were done to show the translation of the Earth. These experiments, it is true, gave negative results but one is somewhat surprised by it.

However one question still comes up. These experiments, as I said, have given a negative result and Lorentz's theory explains this negative result. It seems that the principle of relative motion, which was not required *a priori*, is verified *a posteriori* and that the conservation of momentum should follow from it; and however that is not how it is, how did that happen?

It is because in reality, what we have called the principle of relative motion was only imperfectly confirmed as Lorentz's theory shows. It is due to a composition of effects, but:

1) This compensation only occurred by neglecting v^2 unless some additional hypothesis is made that I will not discuss for the moment.

It is however not significant for our purpose, because if v^2 is neglected, equation (8) will directly yield equation (9), meaning conservation of momentum.

2) For this compensation to happen, the phenomena have to be referred not to real time t, but to some *local time* t' defined in the following manner.

I assume that observers placed at different points, set their watches using a light signal; that they seek to correct these signals for the transmission time but neglecting the translational motion driving them and consequently believing that the signals are transmitted equally quickly in both directions, they limit themselves to crossing the observations by sending a signal from A to B and then another signal from B to A. The local time t' is the time marked by the watches set in that way.

If then $V = 1/\sqrt{K_0}$ is the speed of light, and v the translation of the Earth that I assume parallel to the positive x-axis, then:

$$t' = t - \frac{vx}{V^2}$$

3) The apparent energy propagates in relative motion according to the same laws as real energy in absolute motion, but the apparent energy is not exactly equal to the corresponding real energy.
4) In relative motion, the bodies producing electromagnetic energy are subject to an additional apparent force which does not exist in absolute motion.

We are going to look at how these various circumstances resolve the contradiction that I just reported.

Let us imagine an electric energy producing device arranged such that the energy produced is sent in a single direction. This will, for example, be a Hertz exciter provided with a parabolic mirror.

First at rest, the exciter sends energy in the direction of the x-axis and this energy is precisely equal to what is expended in the exciter. As we have seen, the device *recoils* and acquires some velocity.

If we refer everything to mobile axes linked to the exciter, the apparent phenomenon will have to be, except for the exceptions made above, the same as if the exciter were at rest; it is therefore going to radiate an *apparent* quantity of energy which will be equal to the energy expended in the exciter.

Next it will again experience an impulse due to the recoil, and as it is no longer at rest, but already has some velocity, this impulse will produce some work and the total energy of the exciter will increase.

If therefore the real radiated electromagnetic energy were equal to the apparent electromagnetic energy—meaning, as I just stated, the energy expended in the exciter—then the total energy increase of the device would have been obtained without any expenditure. This is contrary to the principle of conservation of energy. If therefore a recoil is produced, it is because the apparent energy is not equal to the real energy and the phenomena in relative motion are not exactly the same as in absolute motion.

Let us now look at things a little more closely. Let v' be the velocity of the exciter, v that of the mobile axes, which I no longer assume to be linked to the exciter, and V that of the radiation; all these velocities are parallel to the positive x-axis. For simplification, we will assume that the radiation has the form of a polarized plane waves, which gives us the equations:

$$f = h = \alpha = \beta = 0,$$

$$4\pi \frac{dg}{dt} = -\frac{d\gamma}{dx}, \quad -\frac{1}{4\pi V^2}\frac{d\gamma}{dt} = \frac{dg}{dx}, \quad V\frac{d\gamma}{dx} + \frac{d\gamma}{dt} = 0$$

hence:

$$\gamma = 4\pi V g.$$

The real energy contained in the unit volume will be:

$$\frac{\gamma^2}{8\pi} + 2\pi V^2 g^2 = 4\pi V^2 g^2.$$

Now let us look at what happens with the apparent motion relative to the mobile axes. The apparent electric and magnetic fields are:

$$g' = g - \frac{v}{4\pi V^2}\gamma, \quad \gamma' = \gamma - 4\pi v g\,.$$

We therefore have for the apparent energy in a unit volume (neglecting v^2 but not vv'):

$$\frac{\gamma'^2}{8\pi} + 2\pi V^2 g'^2 = \left(\frac{\gamma^2}{8\pi} - vg\gamma\right) + 2\pi V^2\left(g^2 - \frac{vg\gamma}{2\pi V^2}\right)$$

or else

$$4\pi V^2 g^2 - 2vg\gamma = 4\pi V^2 g^2\left(1 - \frac{2v}{V}\right).$$

The equations of apparent motion are additionally written

$$4\pi\frac{\mathrm{d}g'}{\mathrm{d}t'} = -\frac{\mathrm{d}\gamma'}{\mathrm{d}x'}, \quad -\frac{1}{4\pi V^2}\frac{\mathrm{d}\gamma'}{\mathrm{d}t'} = \frac{\mathrm{d}g'}{\mathrm{d}x'}$$

which shows that the apparent speed of propagation is still V.

Let T be the length of the emission; what will be the length actually occupied by the disturbance in space?

The leading edge of the disturbance left at time 0 from point 0 and at time t it is located at point Vt; the trailing edge left at time T, not at point 0, but from point $v'T$, because the exciter from which it emanated has moved during the time T with a velocity v'. This trailing edge is therefore at the moment t at point $v'T + V(t - T)$. The actual length of the disturbance is therefore

$$L = Vt - [v'T + V(t - T)] = (V - v')T.$$

What is now the apparent length? The leading edge left at local time 0 from point 0; at local time t' its abscissa relative to the mobile axes will be Vt'. The trailing edge left at time T from point $v'T$ whose abscissa relative to the mobile axes is $(v' - v)T$; the corresponding local time is

$$T\left(1 - \frac{vv'}{V^2}\right).$$

At local time t', it is at point x, where x is given by the equations:

$$t' = t - \frac{vx}{V^2}, \quad x = v'T + V(t - T)$$

hence, by neglecting v^2:

$$x = [v'T + V(t' - T)]\left(1 + \frac{v}{V}\right).$$

The abscissa of this point relative to the mobile axes will be

$$x - vt' = (v'T - VT)\left(1 + \frac{v}{V}\right) + Vt'.$$

The apparent length of the perturbation will therefore be

$$L' = Vt' - (x - vt') = (V - v')T\left(1 + \frac{v}{V}\right) = L\left(1 + \frac{v}{V}\right).$$

The total real energy (per unit section) is therefore

$$\left(\frac{\gamma^2}{8\pi} + 2\pi V^2 g^2\right)L = 4\pi V^2 g^2 L,$$

and the apparent energy is

$$\left(\frac{\gamma'^2}{8\pi} + 2\pi V^2 g'^2\right)L' = 4\pi V^2 g^2 L\left(1 - \frac{2v}{V}\right)\left(1 + \frac{v}{V}\right) = \\ = 4\pi V^2 g^2 L\left(1 - \frac{v}{V}\right).$$

If therefore $J\mathrm{d}t$ represents the real energy radiated during the time $\mathrm{d}t$, then $J\mathrm{d}t\left(1 - \frac{v}{V}\right)$ will represent the apparent energy.

Let $D\mathrm{d}t$ be the energy expended in the exciter, it is the same in real motion and in apparent motion.

It is still necessary to account for the recoil. The force of the recoil multiplied by $\mathrm{d}t$ is equal to the increase of the momentum of the fictional fluid, meaning equal to

$$\mathrm{d}tK_0JV = \frac{J}{V}\mathrm{d}t$$

because the quantity of fluid created is $\mathrm{d}tK_0J$ and its velocity is V. The work done by the recoil is therefore:

$$-\frac{v'J\mathrm{d}t}{V}.$$

In the apparent motion, v' needs to be replaced by $v' - v$ and J by $J\left(1 - \frac{v}{V}\right)$. The apparent work due to the recoil is therefore:

$$-\frac{(v' - v)J\mathrm{d}t}{V}\left(1 - \frac{v}{V}\right) = J\mathrm{d}t\left(-\frac{v'}{V} + \frac{v}{V} + \frac{vv'}{V^2}\right).$$

Finally in the apparent motion, the apparent additional force that I talked about above (4) must be accounted for. This additional force is equal to

$$-\frac{vJ}{V^2}$$

and its work, by neglecting v^2 is $-\frac{vv'}{V^2}Jd\tau$.

Having laid that out, the equation for the total energy in the real motion is written:

$$J - D - \frac{v'J}{V} = 0. \tag{10}$$

The first term represents the radiated energy, the second the energy expenditure and the third the work from the recoil.

The equation for the total energy in apparent motion will be written:

$$J\left(1 - \frac{v}{V}\right) - D + J\left(-\frac{v'}{V} + \frac{v}{V} + \frac{vv'}{V^2}\right) - \frac{vv'}{V^2}J = 0 \tag{11}$$

The first term represents the apparent radiated energy, the second the energy expenditure, the third the apparent work from the recoil and the fourth the work from the apparent additional force.

The agreement of equations (10) and (11) removes the appearance of contradiction reported above.

If therefore, in Lorentz's theory, the recoil can take place without violating the principle of conservation of energy, it is because the apparent energy for an observer carried along with the mobile axes is not equal to the actual energy. Let us therefore assume that our exciter undergoes a motion of recoil and that the observer is carried along in this motion ($v' = v < 0$), the exciter would appear immobile to this observer and it would seem to the observer that the exciter radiates as much energy as at rest. But in reality it will radiate less of it and this is what compensates the work of the recoil.

I could have assumed that the mobile axes are invariably linked to the exciter, meaning $v = v'$, but my analysis would not then have been able to show the role of the apparent additional force. To do that, I had to assume v' much larger than v such that I could neglect v^2 without neglecting vv'.

I could have also shown the need for the apparent additional forces in the following way:

The actual recoil is J/V; in the apparent motion, J has to be replaced by $J(1 - {}^{v}\!/\!_{V})$ such that the apparent recoil is

$$\frac{J}{V} - \frac{Jv}{V^2}$$

To supplement the real recoil, an apparent additional force

$$-\frac{Jv}{V^2}$$

has to be added to the apparent recoil (I put the – because the recoil, as its name indicates, occurs in the negative direction).

The existence of the apparent additional force is therefore a necessary consequence of the phenomenon of recoil.

Thus, according to Lorentz's theory, the principle of conservation of momentum must not apply to matter alone; the principle of relative motion must not apply to matter alone either. What needs to be noted is that there is an intimate and necessary connection between these two facts.

It would therefore be sufficient to experimentally establish one of these two for the other to be established ipso facto. It would undoubtedly be less difficult to prove the second; but it is already nearly impossible because, for example, Mr. Liénard calculated that with a machine generating 100 kW, the apparent additional force would only be 1/600 dyne.

An important consequence follows from this correlation between these two facts; it is that Fizeau's experiment is itself already contrary to conservation of momentum. If in fact, as this experiment indicates the dragging of waves is only partial, it is because the *relative* propagation of the waves in a moving medium does not follow the same law as the propagation in a medium at rest; meaning that the principle of relative motion does not apply to matter alone and it must have to undergo at least one correction specifically that which I spoke of above (observation 2) and which consists of referring everything to our "local time". If this correction is not compensated by others, one would have to conclude the conservation of momentum is not true either for matter alone.

In that way, all theories which do not respect this principle would be condemned as a group, *unless we agree to profoundly modify all our ideas on electrodynamics*. That is an idea that I have developed at greater length in a previous article (*Éclairage Électrique*, volume V, number 40).

Chapter 3
Three Letters to H. A. Lorentz

First Letter

My dear colleague,

I greatly regret the circumstances that prevented me first from hearing your presentation and then from talking with you during your stay in Paris.

For some time, I have studied your paper electromagnetic phenomena in a system, moving with any velocity smaller than that of light,[1] in greater detail; the importance of this paper is very great and I had already indicated the main results in my presentation in St. Louis.

I agree with you on all the essential points; however, there are some differences in detail.

Thus, on page 813,[2] instead of setting:

$$\frac{1}{kl^3}\rho = \rho'; \;\; k^2 u_x = u'_x, \;\; k^2 u_y = u'_y,$$

It seems to me that it needs to be:

$$\frac{1}{kl^3}\rho(1+\varepsilon v_x) = \rho' \quad \frac{1}{kl^3}\rho(v_x+\varepsilon) = \rho' u'_x$$

where $\varepsilon = -w/c$, or $\varepsilon = -w$ if we choose units such that $c = 1$.

These handwritten letters from Henri Poincaré to Hendrik Lorentz, in French, are transcribed in Kox, A. J. (2008). *The Scientific Correspondence of H. A. Lorentz* (Vol. 1). New York: Springer Science+Business Media, pp. 176–9, letters 126–8 and are held by the Noord-Hollands Archief (http://noord-hollandsarchief.nl/bronnen/archieven?mivast=236&mizig=210&miadt=236&miaet=1&micode=364&minr=721571&miview=inv2). The letters are undated, but from the content they must have been written between late April and early June 1905.

B. D. Popp, *Henri Poincaré: Electrons to Special Relativity*,
https://doi.org/10.1007/978-3-030-48039-4_3

It seems to me that this modification has to be made if the apparent charge on the electron is to be conserved.

The formulas (10) on page 813 are then modified and for the last term instead of

$$l^2 \cdot \frac{w}{c^2}\left(u_y' D_y' + u_z' D_z'\right), \quad -\frac{l^2}{k} \cdot \frac{w}{c^2} u_x' D_y', \quad -\frac{l^2}{k} \cdot \frac{w}{c^2} u_x' D_z'$$

I find

$$l^2 \cdot \frac{w}{c^2}\left(u_x' D_x' + u_y' D_y' + u_z' D_z'\right), \quad 0, \quad 0$$

This is the Liénard force, that you also find but with difficulties. And then the question comes up of knowing whether this force is compensated or uncompensated.

This shows that there are the following relations between the actual forces X, Y, Z and the apparent forces X', Y' and Z':

$$X' = A\left(X + \varepsilon \sum X v_x\right), \quad Y' = BY, \quad Z' = BZ$$

where A and B are coefficients and $A\varepsilon \sum Xv_x$ represents the Liénard force.

If all the forces are of electrical origin, the equilibrium conditions (or from the modified d'Alembert principle) give:

$$X = Y = Z = 0$$

hence

$$X' = Y' = Z' = 0$$

If not all the forces are of electrical origin, there will again have to be compensating forces provided that they behave just as if they were of electrical origin.

But there is another thing.

You assume $l = 1$.

Langevin assumes $kl^3 = 1$.

I had assumed $kl = 1$ to retain the unit of time, but that led me to unallowable consequences.

On the other hand, I come up with contradictions (between the formulas for action and energy) with all hypotheses other than those of Langevin.

The reasoning by which you establish that $l = 1$ does not seem conclusive to me, or rather it doesn't any longer and leaves l undetermined when I do the calculation by changing the formulas on page 813 as I told you.

What do you think of all this, do you want me to provide you more details or is what I've given you sufficient. In any case, please excuse me for taking up your time.

Your devoted colleague,

Poincaré.

Second Letter

My dear colleague,

Thank you for your friendly letter. Since I wrote to you my ideas have changed on a few points. I find like you that $l = 1$, by a different route.

Let $-\varepsilon$ be the speed of translation with that of light taken as unity.

$$k = \left(1 - \varepsilon^2\right)^{-1/2}$$

one then has the transformation

$$x' = kl(x + \varepsilon t), \quad t' = kl(t + \varepsilon x), \quad y' = ly, \quad z'; = lz.$$

This transformation forms a group. Let two composed transformations correspond to

$$k, l, \varepsilon$$

and

$$k', l', \varepsilon'$$

their resultant will correspond to

$$k'', l'', \varepsilon''$$

hence:

$$k'' = \left(1 - \varepsilon''^2\right)^{-1/2}, \quad l'' = ll', \quad \varepsilon'' = \frac{\varepsilon + \varepsilon'}{1 + \varepsilon\varepsilon'}$$

If we now take

$$l = \left(1 - \varepsilon^2\right)^m, \quad l' = \left(1 - \varepsilon'^2\right)^m$$

we will only have:

$$l'' = \left(1 - \varepsilon''^2\right)^m$$

when $m = 0$.

On the other hand, I can find agreement between the calculation of the masses by means of the electromagnetic inertia and by means of least action and by means of the energy only under Langevin's hypothesis.

I hope to clarify this contradiction very soon; I will keep you up-to-date with my efforts.

Your devoted colleague,
Poincaré.

Third Letter

My dear colleague,

I have continued the research that I told you about. My results fully confirmed yours in the sense that perfect compensation (which prevents the experimental determination of absolute movement only happens completely in the hypothesis where $l = 1$. However, for this hypothesis to be allowable, it has to be accepted that each electron is subject to additional forces for which the work is proportional to the changes in its volume.

Hence if you prefer, each electron behaves as if it was a hollow capacitor subject to a constant internal pressure (additionally negative) and independent of the volume. Under these conditions, the compensation is complete.

I am happy to find myself in full agreement with you and to thus have arrived at the perfect intelligence of your beautiful work.

You're very devoted colleague,
Poincaré.

Translator's Notes

1. Lorentz, H. A. (1904). Electromagnetic phenomena in a system moving with any velocity smaller than that of light. *Proceedings KNAW [Royal Netherlands Academy of Arts and Sciences], 6 (1903–4)*, 809–831. This paper is also available in the Part III of this book.
2. In Part III, these are equations 7 and 8 on page 263.

Chapter 4
Electricity – On the Dynamics of the Electron

Note from **Henri Poincaré**

On first consideration it seemed that the aberration of light and the optical phenomena associated with it were going to provide us a means for determining the absolute motion of the Earth or more accurately its motion, not with respect to the other stars, but with respect to the ether. It isn't anything like that; experiments taking into account only the first power of the operation first failed and the explanation was easily found; but Michelson, having conceived an experiment where terms in the square of the aberration could be shown, was no more successful. It appears that this impossibility of showing absolute motion is a general law of nature.

An explanation was proposed by Lorentz, who introduced the hypothesis of a contraction of all bodies in the direction of Earth's motion; this contraction would account for Michelson's experiment and all those done until now, but it would leave room for other experiments yet more delicate, and easier to conceive than to execute, which could be able to show absolute motion of the Earth. But, if the impossibility of such an observation were considered to be highly probable, one could then anticipate that these experiments, if one could manage to conduct them, would again give a negative result. Lorentz sought to supplement and amend his hypothesis so as to bring it into agreement with the postulate that it is *completely* impossible to determine absolute motion. He did manage to do this in his article entitled *Electromagnetic Phenomena in a System Moving with Any Velocity Smaller than that of Light* (Proceedings of the Amsterdam Academy, May 27, 1904).

The importance of the question led me to take it up again; the results that I obtained are in agreement with those of Lorentz on all important points; I was only led to amend and supplement them in some points of detail.

Poincaré, H. (1905). Sur la dynamique de l'électron. *Comptes rendus de l'Académie des Sciences, 140*, 1504–1508.

B. D. Popp, *Henri Poincaré: Electrons to Special Relativity*,
https://doi.org/10.1007/978-3-030-48039-4_4

The essential point, established by Lorentz, is that the electromagnetic field equations are not altered by a specific transformation (that I will give the name *Lorentz*) and which has the following form:

$$x' = kl(x + \varepsilon t), \ \ y' = ly, \ \ z' = lz, \ \ t' = kl(t + \varepsilon x), \tag{1}$$

where x, y and z are the coordinates and t the time before the transformation, and after the transformation they are x', y' z', and t'. Additionally, ε is a constant which defines the transformation

$$k = \frac{1}{\sqrt{1-\varepsilon^2}}$$

and l is an arbitrary function of ε. It can be seen that in this transformation the x axis plays a specific role, but a transformation can obviously be constructed where this role would be played by an arbitrary straight line passing through the origin. The set of all these transformations, joined with the set of all spatial rotations, must form a group; but, for that to be true, it must be that $l = 1$. One is therefore led to assume that $l = 1$ and that is a consequence that Lorentz had reached by another route.

Let ρ be the charge density of the electron and ξ, η and ζ the speed of the electron before the transformation; for the same quantities ρ', ξ', η' and ζ' after the transformation, it will hold that

$$\rho' = \frac{k}{l^3}\rho(1 + \varepsilon\xi), \quad \rho'\xi' = \frac{k}{l^3}\rho(\xi + \varepsilon), \quad \rho'\eta' = \frac{\rho\eta}{l^3}, \quad \rho'\zeta' = \frac{\rho\zeta}{l^3}. \tag{2}$$

These formulas differ a little from those which Lorentz had found.

Now let X, Y, Z, and X', Y', Z' be the three components of the force before and after the transformation; *the force is referred to a unit volume;* I find:

$$X' = \frac{k}{l^5}\left(X + \varepsilon \sum X\xi\right), \quad Y' = \frac{Y}{l^5}, \quad Z' = \frac{Z}{l^5}. \tag{3}$$

These formulas also differ a little from those of Lorentz; the additional term in $\sum X\xi$ recalls a result previously obtained by Liénard.

If we now designate the components of the force referred no longer to the unit volume, but to the unit mass of the electron, by X_1, Y_1, Z_1, and X'_1, Y'_1, Z'_1, we will have:

$$X'_1 = \frac{k}{l^5}\left(X_1 + \varepsilon \sum X_1\xi\right), \quad Y'_1 = \frac{\rho}{\rho'}\frac{Y_1}{l^5}, \quad Z'_1 = \frac{\rho}{\rho'}\frac{Z_1}{l^5}. \tag{4}$$

Lorentz was also led to assume that the moving electron takes the shape of a flattened ellipsoid; it is also the hypothesis made by Langevin, only, whereas Lorentz assumed the two of the axes of the ellipsoid remain constant, which agrees with his hypothesis that $l = 1$. Langevin assumed that it is the volume which remains constant. Both authors have shown that these two hypotheses agree with the experiments of Kaufmann[1] and also the fundamental Abraham's hypothesis[2] (spherical electron). Langevin's hypothesis would have the advantage of being sufficient in itself, because it is sufficient to regard the electron as deformable and incompressible in order to explain why it takes an ellipsoidal shape when in motion. But I show, in agreement on that matter with Lorentz, that it is incapable of agreeing with the impossibility of an experiment showing absolute motion. That requires, as I stated, that $l = 1$ is the only hypothesis for which the set of Lorentz transformations forms a group.

But with Lorentz's hypothesis, the agreement of the formulas is not the only thing; at the same time a possible explanation of the contraction of the electron is obtained by assuming that *the deformable and compressible electron is subject to a kind of constant external pressure whose work is proportional to the variations of volume*.

I show, by an application of the principle of least action, that under these conditions, the compensation is complete, if it is assumed that inertia is an exclusively electromagnetic phenomenon, as is generally accepted since Kaufmann's experiment, and that apart from the constant pressure which I just mentioned and which acts on the electron, all forces are of electromagnetic origin. In that way there is an explanation for the impossibility of showing absolute motion and for the contraction of all bodies in the direction of the Earth's motion.

But that is not all: Lorentz, in the work cited, thought it necessary to supplement his hypothesis by assuming that all forces, whatever their origin might be, were affected by a translation in the same way as the electromagnetic forces and consequently the effect produced on their components by the Lorentz transformation is still defined by equations (4).

A closer examination of this hypothesis and in particular a look at what modifications it means we would have to make to the laws of gravitation is important. That is what I'm looking to determine; I was first led to assume that the propagation of gravity is not instantaneous but occurs at the speed of light. This seems to contradict a result obtained by Laplace who stated that this propagation is, if not instantaneous, at least much faster than that of light. But, in reality, the question posed by Laplace is considerably different from the one that we are looking at here. For Laplace, the introduction of a finite propagation speed was the *only* modification that he made to Newton's law. Here, in contrast, this modification is accompanied by several others; it is therefore possible, and in fact it happens, that a partial compensation arises among them.

When we will therefore talk about the position or speed of the attracting body, it will involve that position or that speed at the moment when the *gravitational wave* leaves this body; when we will talk about the position or speed of the attracted body, it will involve that position or that speed at the moment when the gravitational wave coming from the other body reaches this attracted body; it is clear that the first moment is prior to the second.

If therefore x, y and z are projections onto the three axes of the vector which joins the two positions and if the speed of the attracted body is ξ, η and ζ, and that of the attracting body ξ_1, η_1 and ζ_1, then the three components of the attraction (which I will again be able to call X_1, Y_1 and Z_1) will be functions of x, y, z, ξ, η, ζ, ξ_1, η_1 and ζ_1. I now ask whether it is possible to determine these functions in a way that they are affected by the Lorentz transformation according to equations (4) and that they agree with the ordinary law of gravitation whenever the speeds ξ, η, ζ, ξ_1, η_1 and ζ_1 are small enough that their squares can be neglected compared to the square the speed of light.

The response must be affirmative. It is found that the corrected attraction is made up of two forces, one parallel to the vector x, y and z, and the other parallel to the speed ξ_1, η_1 and ζ_1.

The divergence from the ordinary law of gravitation is, as I just stated, of order ξ^2; if, as Laplace did, it is only assumed that the speed of propagation is that of light, this divergence would be of order ξ, meaning 10,000 times larger. It is not therefore absurd to assume, at first glance, that astronomical observations are not sufficiently precise for detecting such a small divergence as we were imagining. But only an in-depth discussion will be able to decide that.

Translator's Notes

1. This appears to be a reference to Kaufmann, W. (1901). Die magnetische und electrische Ablenkbarkeit der Becquerelstrahlen und die scheinbare Masse der Elektronen. *Nachrichten von der Königl. Gesellschaft der Wissenschaften zu Göttingen, 2*, 143–155.
2. This could be a reference to Abraham, M. (1902). Dynamik des Electrons. *Nachrichten von der Gesellschaft der Wissenschaften zu Göttingen*, 20–41 that Poincaré (1906) cites in §6 above eq. 5 (page 77 in this book) or to Abraham, M. (1903). Prinzipien der Dynamik des Eleckrons. *Annalen der Physik, Ser. 4 vol. 10 supplement*, 105–179.

Chapter 5
On the Dynamics of the Electron

Introduction

On first consideration it seemed that the aberration of light and the optical and electrical phenomena associated with it were going to provide us a means for determining the absolute motion of the Earth or more accurately its motion, not with respect to the other stars, but with respect to the ether. Fresnel[1] had already tried it, but he soon recognized that the motion of the Earth did not change the laws of refraction and reflection. Analogous experiments, like that of the water-filled telescope and all those where only first-order terms in the aberration were considered were to give only negative results; the explanation for this was soon found. But, Michelson, who had imagined an experiment sensitive to the terms depending on the square of the aberration, failed in turn.

It seems that this impossibility of showing the absolute motion of the Earth experimentally could be a general law of Nature; we are naturally led to accept this law, that we will call the *Relativity Postulate* and to allow it without restriction. Should this postulate, until now in agreement with experiment, later be confirmed or rejected by more precise experiments, it is in any case interesting to look at what its consequences might be.

An explanation was proposed by Lorentz and Fitz Gerald, who introduced the hypothesis of a contraction experienced by all bodies in the direction of motion of the Earth and proportional to the square of the aberration; this contraction, which we will call the *Lorentz contraction*, took into account the Michelson experiment and all those which had been done until now. The hypothesis would become insufficient, however, if the relativity postulate were to be accepted in its full generality.

H. Poincaré (Paris)
Session of July 23, 1905
Poincaré, H. (1906). Sur la dynamique de l'électron. *Rendiconti del circolo matematico di Palermo, 21*, 129–176.

B. D. Popp, *Henri Poincaré: Electrons to Special Relativity*,
https://doi.org/10.1007/978-3-030-48039-4_5

Lorentz sought to supplement it and amend it so as to bring it into full agreement with this postulate. This is what he succeeded in doing in his article entitled *Electromagnetic Phenomena in a System Moving with Any Velocity Smaller than that of Light* (Proceedings of the Amsterdam Academy, May 27, 1904)[2].

The importance of the question led me to take it up again; the results that I obtained are in agreement with those of Lorentz on all important points; I was only led to amend and supplement them in some points of detail. The differences, which are of secondary importance, will be seen later.

Lorentz's idea can be summarized as follows: if one can, without any visible phenomenon being modified, give any system a shared translation, it is because the equations of the electromagnetic environment are not altered by certain transformations, which we will call *Lorentz transformations*; two systems, the one stationary and the other in translation, thus become the exact image of each other.

Langevin[1] had sought to modify Lorentz's idea; for both authors, the moving electron takes the form of a flattened ellipsoid, but for Lorentz two of the axes of the ellipsoid remain constant and in contrast for Langevin it is the volume of the ellipsoid which remains constant. Both authors additionally showed that these two hypotheses agree with Kaufmann's experiments and also with Abraham's primitive hypothesis (undeformable spherical electron)[3].

The advantage of Langevin's theory is that it does not call on electromagnetic forces and binding forces; but it is incompatible with the relativity postulate. That is what Lorentz had shown; it is what I found in turn by another route by calling on the principles of group theory.

That means it's necessary to go back to Lorentz's theory; but to keep it and avoid intolerable contradictions, a special force has to be assumed which explains both the contraction and the two constant axes. I sought to determine this force, and I found that *it could be compared to a constant external pressure acting on the deformable and compressible electron and its work is proportional to the variations in the volume of this electron.*

If the inertia of matter were then exclusively of electromagnetic origin, as is generally accepted since Kaufmann's experiment, and if all the forces are of electromagnetic origin other than this constant pressure that I just spoke of, then the relativity postulate can be established with full rigor. That is what I show by a very simple calculation based on the principle of least action.

But that isn't all. Lorentz, in the work cited, thought it necessary to supplement his hypothesis such that the postulate is still true when there are forces other than electromagnetic forces. According to him, all forces, whatever their origin, are affected by the Lorentz transformation (and consequently by a translation) in the same way as the electromagnetic forces.

[1]Langevin had been anticipated by Bucherer from Bonn, who came out with the same idea before him. (*See*: Bucherer, *Mathematische Einführung in die Elektronentheorie*; August 1904. Teubner, Leipzig).

It is important to examine this hypothesis more closely and in particular to seek what modifications it would force us to make to the laws of gravitation.

First it is found that it would force us to assume that the propagation of gravitation is not instantaneous but occurs at the speed of light. One could think that this is a sufficient reason to reject the hypothesis, since Laplace had proven that it could not be so. But in reality, the effect of this propagation is in large part compensated by a different cause, such that there is no contradiction between the proposed law and astronomical observations.

Would it be possible to find a law, which satisfies the condition imposed by Lorentz and at the same time reduced to Newton's law any time that the speeds of the stars are small enough that their squares can be neglected (as well as the product of the accelerations by the distances) compared to the square of the speed of light?

The answer to this question must be affirmative as will be seen later.

Is the law thus modified compatible with astronomical observations?

At first glance, it seems so, but the question will only be settled by an in-depth discussion.

But even if we accept that this discussion is settled in favor of the new hypothesis, what will we have to conclude from it? If the attraction propagates with the speed of light, that cannot be because of a fortuitous occurrence, that must be because it is a function of the ether; and then it will be necessary to look into the nature of this function and associate other functions of the fluid with it.

We cannot be satisfied with formulas that are simply juxtaposed and which only happen to agree by lucky chance; said another way, it has to happen because these formulas are mutually involved. The mind would only be satisfied when it believes that it sees the reason for this agreement to such an extent that it has the illusion that it could have anticipated it.

But the question can also be presented from another point of view so that a comparison will be better understood. Let us imagine an astronomer before Copernicus who was thinking about the Ptolemaic system; he would notice that for all the planets one of the two circles, epicycle or deferent, is traversed in the same time. That cannot be by chance; there is therefore some unknown mysterious link between all the planets.

Copernicus, by simply changing the coordinate system regarded as fixed, made this appearance disappear; each planet now describes only one circle and the periods of revolution become independent (until Kepler reestablished the link between them that was thought to have been destroyed).

It is possible that there is something analogous here; if we were to accept the relativity postulate, we would find in the law of gravitation and in the electromagnetic laws a common number which would be the speed of light. We would find it again in other forces of arbitrary origin which can only be explained in two ways:

Either there is nothing in the world that is not of electromagnetic origin.

Or else, this part which would be, to state it that way, shared by all physical phenomena would only be an appearance, something which would arise from our measurement methods. How do we make our measurements? We would start to say, by transporting one or another of the objects regarded as invariable solids; but that is no longer true in the current theory, if the Lorentz contraction is accepted. In this theory, two equal lengths are, by definition, two lengths that light takes the same time to traverse.

Perhaps it would suffice to renounce this definition so that Lorentz's theory was as completely overthrown as was the Ptolemaic system by the intervention of Copernicus. If that were to happen one day, that would not prove that the effort made by Lorentz was pointless, because Ptolemy, whatever we might think of it, was not useless to Copernicus.

I too have not hesitated to publish these few partial results even though at this moment the whole theory itself might seem to be in danger from the discovery of magnetocathode rays.

§1 – Lorentz Transformation

Lorentz adopted a specific system of units so as to make the factors of 4π disappear from the formulas. I will do the same and additionally I will choose the units of length and time such that the speed of light is equal to one. Under these conditions, by calling: f, g, h the electric displacement; α, β, γ the magnetic force; F, G, H the vector potential; ψ the scalar potential; ρ the electric charge density; ξ, η, ζ the electron velocity; and u, v, w the current, the fundamental formulas become:[4]

$$\begin{gathered} u=\frac{\mathrm{d}f}{\mathrm{d}t}+\rho\xi=\frac{\mathrm{d}\gamma}{\mathrm{d}y}-\frac{\mathrm{d}\beta}{\mathrm{d}z}, \quad \alpha=\frac{\mathrm{d}H}{\mathrm{d}y}-\frac{\mathrm{d}G}{\mathrm{d}z}, \quad f=-\frac{\mathrm{d}F}{\mathrm{d}t}-\frac{\mathrm{d}\psi}{\mathrm{d}x}, \\ \frac{\mathrm{d}\alpha}{\mathrm{d}t}=\frac{\mathrm{d}g}{\mathrm{d}z}-\frac{\mathrm{d}h}{\mathrm{d}y}, \quad \frac{\mathrm{d}\rho}{\mathrm{d}t}+\sum\frac{\mathrm{d}\rho\xi}{\mathrm{d}x}=0, \sum\frac{\mathrm{d}f}{\mathrm{d}x}=\rho, \quad \frac{\mathrm{d}\psi}{\mathrm{d}t}+\sum\frac{\mathrm{d}F}{\mathrm{d}x}=0, \\ \square=\Delta-\frac{\mathrm{d}^2}{\mathrm{d}t^2}=\sum\frac{\mathrm{d}^2}{\mathrm{d}x^2}-\frac{\mathrm{d}^2}{\mathrm{d}t^2}, \quad \square\psi=-\rho, \quad \square F=-\rho\xi. \end{gathered} \tag{1}$$

An element of matter of volume $\mathrm{d}x\mathrm{d}y\mathrm{d}z$ experiences a mechanical force whose components $X\mathrm{d}x\mathrm{d}y\mathrm{d}z$, $Z\mathrm{d}x\mathrm{d}y\mathrm{d}z$, $Y\mathrm{d}x\mathrm{d}y\mathrm{d}z$ are determined from the formula:

$$X=\rho f+\rho(\eta\gamma-\zeta\beta). \tag{2}$$

These equations are subject to a remarkable transformation discovered by Lorentz and which is of interest because it explains why no experiment is able to let us know the absolute motion of the universe. Let us set:

$$x'=kl(x+\varepsilon t), t'=kl(t+\varepsilon x), y'=ly, z'=lz, \tag{3}$$

where l and ε are arbitrary constants, and where

$$k = \frac{1}{\sqrt{1 - \varepsilon^2}}.$$

If we then set:

$$\square' = \sum \frac{d^2}{dx'^2} - \frac{d^2}{dt'^2},$$

it will follow:

$$\square' = \square l^{-2}$$

Now consider a sphere driven in a motion of uniform translation with the electron and let:

$$(x - \xi t)^2 + (y - \eta t)^2 + (z - \zeta t)^2 = r^2$$

be the equation of this mobile sphere whose volume will be $\frac{4}{3}\pi r^3$.

The transformation will change it into an ellipsoid whose equation is easy to find. It is in fact easily deduced from equations (3):

$$x = \frac{k}{l}(x' - \varepsilon t'), \quad t = \frac{k}{l}(t' - \varepsilon x'), \quad y = \frac{y'}{l}, \quad z = \frac{z'}{l}. \tag{3'}$$

The equation for the ellipsoid then becomes:

$$k^2(x' - \varepsilon t' + \varepsilon \xi x')^2 + (y' - \eta k t' + \eta k \varepsilon x')^2 + (z' - \zeta k t' + \zeta k \varepsilon x')^2 = l^2 r^2.$$

This ellipsoid moves with a uniform motion; for $t' = 0$, it reduces to

$$k^2 x'^2 (1 + \varepsilon \xi)^2 + (y' + \eta k \varepsilon x')^2 + (z' + \zeta k \varepsilon x')^2 = l^2 r^2$$

and its volume is:

$$\frac{4}{3}\pi r^3 \frac{l^3}{k(1 + \xi \varepsilon)}.$$

If we want the charge of an electron to be unchanged by the transformation and if we call ρ' the new electric charge density, it will follow:

$$\rho' = \frac{k}{l^3}(\rho + \varepsilon \rho \xi). \tag{4}$$

What will the new speeds ξ', η', ζ' be? It will have to be:

$$\xi' = \frac{dx'}{dt'} = \frac{d(x+\varepsilon t)}{d(t+\varepsilon x)} = \frac{\xi+\varepsilon}{1+\varepsilon\xi},$$
$$\eta' = \frac{dy'}{dt'} = \frac{dy}{kd(t+\varepsilon x)} = \frac{\eta}{k(1+\varepsilon\xi)}, \quad \zeta' = \frac{\zeta}{k(1+\varepsilon\xi)}$$

hence

$$\rho'\xi' = \frac{k}{l^3}(\rho\xi+\varepsilon\rho), \quad \rho'\eta' = \frac{1}{l^3}\rho\eta, \quad \rho'\zeta' = \frac{1}{l^3}\rho\zeta. \tag{4'}$$

Here is where I need to indicate for the first time a divergence from Lorentz. Lorentz set (up to differences in notation; *loc. cit.*, page 813, formulas 7 and 8[5]):

$$\rho' = \frac{1}{kl^3}\rho, \quad \xi' = k^2(\xi+\varepsilon), \quad \eta' = k\eta, \quad \zeta' = k\zeta.$$

That way the formulas:

$$\rho'\xi' = \frac{k}{l^3}(\rho\xi+\varepsilon\rho), \quad \rho'\eta' = \frac{1}{l^3}\rho\eta, \quad \rho'\zeta' = \frac{1}{l^3}\rho\zeta;$$

are found, but the value of ρ' is different.

It needs to be noted that the formulas (4) and (4′) satisfy the continuity condition

$$\frac{d\rho'}{dt'} + \sum \frac{d\rho'\xi'}{dx'} = 0.$$

In fact, let λ be an undetermined quantity and D the functional determinant of

$$t+\lambda\rho, \quad x+\lambda\rho\xi, \quad y+\lambda\rho\eta, \quad z+\lambda\rho\zeta \tag{5}$$

with respect to t, x, y and z. It will follow:

$$D = D_0 + D_1\lambda + D_2\lambda^2 + D_3\lambda^3 + D_4\lambda^4.$$

with $D_0 = 1$ and $D_1 = d\rho/dt + \sum d\rho\xi/dx = 0$.

Let $\lambda' = l^2\lambda$, we see that the four functions

$$t' + \lambda'\rho', \quad x' + \lambda'\rho'\xi', \quad y' + \lambda'\rho'\eta', \quad z' + \lambda'\rho'\zeta' \tag{5'}$$

are related to the functions (5) by the same linear relations as the former variables to the new variables. If the functional determinant of the functions (5′) with respect to the new variables is therefore designated D', it will follow:

$$D' = D, \ \ D' = D'_0 + D'_1\lambda' + D'_2\lambda'^2 + D'_3\lambda'^3 + D'_4\lambda'^4,$$

hence:

$$D'_0 = D_0 = 1, D'_1 = l^{-2}D_1 = 0 = \frac{d\rho'}{dt'} + \sum \frac{d\rho'\xi'}{dx'}.$$

which was to be proven.

With the hypothesis from Lorentz, this condition would not be fulfilled, because ρ' does not have the same value.

We will now define new vector and scalar potentials so as to satisfy the conditions:

$$\Box'\psi' = -\rho', \quad \Box' F' = -\rho'\xi' . \tag{6}$$

From that we next draw:

$$\psi' = \frac{k}{l}(\psi + \varepsilon F), \quad F' = \frac{k}{l}(F + \varepsilon\psi), \quad G' = \frac{1}{l}G, \quad H' = H. \tag{7}$$

These formulas do differ from those of Lorentz, but in the final analysis the divergence only bears on the definitions.

We will choose new electric and magnetic fields so as to satisfy the equations:

$$f' = -\frac{dF'}{dt'} - \frac{d\psi'}{dx'}, \quad \alpha' = \frac{dH'}{dy'} - \frac{dG'}{dz'}. \tag{8}$$

It is easy to see that:

$$\frac{d}{dt'} = \frac{k}{l}\left(\frac{d}{dt} - \varepsilon\frac{d}{dx}\right), \quad \frac{d}{dx'} = \frac{k}{l}\left(\frac{d}{dx} - \varepsilon\frac{d}{dt}\right), \quad \frac{d}{dy'} = \frac{1}{l}\frac{d}{dy}, \quad \frac{d}{dz'} = \frac{1}{l}\frac{d}{dz}$$

and from that to conclude:

$$\begin{aligned} f' &= \frac{1}{l^2}f, \quad g' = \frac{k}{l^2}(g + \varepsilon\gamma), \quad h' = \frac{k}{l^2}(h - \varepsilon\beta), \\ \alpha' &= \frac{1}{l^2}\alpha, \quad \beta' = \frac{k}{l^2}(\beta - \varepsilon h), \quad \gamma' = \frac{k}{l^2}(\gamma + \varepsilon g). \end{aligned} \tag{9}$$

These formulas are identical to Lorentz's.

Our transformation does not alter equations (1). In fact, the continuity condition, and also equations (6) and (8), already provided us some of equations (1) (except for accenting of the letters).

Equations (6) connected with the continuity condition give us:

$$\frac{d\psi'}{dt'} + \sum \frac{dF'}{dx'} = 0. \tag{10}$$

It remains to establish that:

$$\frac{df'}{dt'} + \rho'\xi' = \frac{d\gamma'}{dy'} - \frac{d\beta'}{dz'}, \quad \frac{d\alpha'}{dt'} = \frac{dg'}{dz'} - \frac{dh'}{dy'}, \quad \sum \frac{df'}{dx'} = \rho'$$

and it can be easily seen that these are necessary consequences of equations (6), (8) and (10).

We must now compare the forces before and after the transformation.

Let X, Y, Z be the force before and X', Y', Z' be the force after the transformation; with all of them referred to a unit volume. In order for X' to satisfy the same equations as before the transformation, it must hold that:

$$\begin{aligned} X' &= \rho' f' + \rho'(\eta'\gamma' - \zeta'\beta'), \\ Y' &= \rho' g' + \rho'(\zeta'\alpha' - \xi'\gamma'), \\ Z' &= \rho' h' + \rho'(\xi'\beta' - \eta'\alpha'), \end{aligned}$$

or, by replacing the quantities by their values (4), (4′) and (9) while making use of equations (2):

$$\begin{aligned} X' &= \frac{k}{l^5}\left(X + \varepsilon \sum X\xi\right), \\ Y' &= \frac{1}{l^5} Y, \\ Z' &= \frac{1}{l^5} Z. \end{aligned} \tag{11}$$

If we represent the force referred, no longer to the unit volume, but now to the unit electrical charge of the electron, by X_1, Y_1, Z_1 and the same qualities after the transformation by X'_1, Y'_1, Z'_1 we will have:

$$X_1 = f + \eta\gamma - \zeta\beta, \quad X'_1 = f' + \eta'\gamma' - \zeta'\beta', \quad X = \rho X_1, \quad X' = \rho' X'_1$$

and we will have the equations:

$$
\begin{aligned}
X_1' &= \frac{k}{l^5}\frac{\rho}{\rho'}\left(X_1 + \varepsilon \sum X_1 \xi\right), \\
Y_1' &= \frac{1}{l^5}\frac{\rho}{\rho'} Y_1, \\
Z_1' &= \frac{1}{l^5}\frac{\rho}{\rho'} Z_1.
\end{aligned} \tag{11'}
$$

Lorentz had found (within the difference of notation, page 813, formula (10)):

$$
\begin{aligned}
X_1 &= l^2 X_1' - l^2 \varepsilon(\eta' g' + \zeta' h'), \\
Y_1 &= \frac{l^2}{k} Y_1' + \frac{l^2 \varepsilon}{k} \xi' g', \\
Z_1 &= \frac{l^2}{k} Z_1' + \frac{l^2 \varepsilon}{k} \xi' h',
\end{aligned} \tag{11''}
$$

Before going farther, the cause of this significant divergence must be found. It obviously means that the formulas for ξ', η', ζ' are not the same, even though the formulas for the electric and magnetic fields are the same.

If the inertia of the electrons is exclusively of electromagnetic origin and if additionally they are only subject to forces of electromagnetic origin, then the equilibrium condition requires that inside the electrons it hold:

$$X = Y = Z = 0.$$

Hence, in light of equations (11), these relations are equivalent to

$$X' = Y' = Z' = 0.$$

The equilibrium conditions of the electrons are therefore unchanged by the transformation.

Unfortunately, such a simple assumption is not allowable. If, in fact, one supposes that $\xi = \eta = \zeta = 0$, the conditions $X = Y = Z = 0$ would lead to $f = g = h = 0$, and consequently $\sum \frac{df}{dx} = 0$, meaning $\rho = 0$. One would arrive at analogous results in the most general case. One therefore has to accept that in addition to electromagnetic forces, there are either other forces or binding. One must then look at what conditions these forces or binding must satisfy for the equilibrium of the electrons to be undisturbed by the transformation. This will be taken up in a subsequent section.

§2 – Principle of Least Action

The way Lorentz deduced his equations from the principle of least action is known[6]. Although I have nothing essential to add to it, I will however go back over the question because I prefer to present it in a slightly different form which will be useful for my purpose. I will set:

$$J = \int \left[\frac{\sum f^2}{2} + \frac{\sum \alpha^2}{2} - \sum Fu \right] \mathrm{d}t \mathrm{d}\tau, \tag{1}$$

by assuming that f, α, F, u, etc. are subject to the following conditions and to those which could be deduced from them by symmetry:

$$\sum \frac{\mathrm{d}f}{\mathrm{d}x} = \rho, \quad \alpha = \frac{\mathrm{d}H}{\mathrm{d}y} - \frac{\mathrm{d}G}{\mathrm{d}z}, \quad u = \frac{\mathrm{d}f}{\mathrm{d}t} + \rho\xi\,. \tag{2}$$

As for the integral J, it must be extended to:

1) the entire space with respect to the element of volume, $\mathrm{d}\tau = \mathrm{d}x\mathrm{d}y\mathrm{d}z$;
2) the limits included between $t = t_0$, $t = t_1$ with respect to time, t.

According to the principle of least action, the integral J must be a minimum if the various quantities which appear in it are subject to:

1) conditions (2);
2) the condition that the state of the system is fixed at the two limit epochs $t = t_0$, $t = t_1$.

This last condition allows us to transform our integrals by integration by parts over time. If we in fact have an integral of the form

$$\int A \frac{\mathrm{d}B\delta C}{\mathrm{d}t} \mathrm{d}t \mathrm{d}\tau,$$

where C is one of the quantities which define the state of the system and δC is its variation, it will be equal (by integrating by parts with respect to time) to:

$$\int |AB\delta C|_{t=t_0}^{t=t_1} \mathrm{d}\tau - \int \frac{\mathrm{d}A}{\mathrm{d}t} \mathrm{d}B\delta C.$$

Since the state of the system is determined at the two limit epochs, $\delta C = 0$ for $t = t_0$, $t = t_1$; therefore the first integral which relates to these two epochs is zero; and only the second remains.

We can similarly integrate by parts relative to x, y or z; we have in fact

$$\int A\frac{dB}{dx}dxdydzdt = \int ABdydzdt - \int B\frac{dA}{dx}dxdydzdt.$$

Since our integrals extend to infinity, in the first integral on the right-hand side x must be made equal to $\pm\infty$; therefore, since we always assume that all our functions become zero at infinity, this integral must be zero and it will follow

$$\int A\frac{dB}{dx}d\tau dt = -\int B\frac{dA}{dx}d\tau dt.$$

If the system were assumed subject to binding, it would be necessary to add a binding condition to the conditions imposed on the various quantities appearing in the integral J.

First give F, G, H increments δF, δG, δH; hence:

$$\delta\alpha = \frac{d\delta H}{dy} - \frac{d\delta G}{dz}.$$

One should have:

$$\delta J = \int\left[\sum\alpha\left(\frac{d\delta H}{dy} - \frac{d\delta G}{dz}\right) - \sum u\delta F\right]dtd\tau = 0,$$

or, by integrating by parts,

$$\delta J = \int\left[\sum\left(\delta G\frac{d\alpha}{dz} - \delta H\frac{d\alpha}{dy}\right) - \sum u\delta F\right]dtd\tau$$
$$= -\int\sum\delta F\left(u - \frac{d\gamma}{dy} + \frac{d\beta}{dz}\right)dtd\tau = 0,$$

hence, by equating the coefficient of the arbitrary δF to zero,

$$u = \frac{d\gamma}{dy} - \frac{d\beta}{dz}. \tag{3}$$

This relation gives us (with an integration by parts):

$$\int\sum Fud\tau = \int\sum F\left(\frac{d\gamma}{dy} - \frac{d\beta}{dz}\right)d\tau = \int\sum\left(\beta\frac{dF}{dz} - \gamma\frac{dF}{dy}\right)d\tau$$
$$= \int\sum\alpha\left(\frac{dH}{dy} - \frac{dG}{dz}\right)d\tau,$$

or

$$\int \sum F u \mathrm{d}\tau = \int \sum \alpha^2 \mathrm{d}\tau$$

hence finally:

$$J = \int \left(\frac{\sum f^2}{2} - \frac{\sum \alpha^2}{2} \right) \mathrm{d}t \mathrm{d}\tau. \tag{4}$$

Now, and because of the relation (3), δJ is independent of δF and consequently of $\delta\alpha$; let us now vary the other variables.

It follows, by returning to the expression (1) for J,

$$\delta J = \int \left(\sum f \delta f - \sum F \delta u \right) \mathrm{d}t \mathrm{d}\tau.$$

But f, g, h are subject to the first of the conditions (2), such that

$$\sum \frac{\mathrm{d}\delta f}{\mathrm{d}x} = \delta\rho, \tag{5}$$

and which it is appropriate to write:

$$\delta J = \int \left[\sum f \delta f - \sum F \delta u - \psi \left(\sum \frac{\mathrm{d}\delta f}{\mathrm{d}x} - \delta\rho \right) \right] \mathrm{d}t \mathrm{d}\tau. \tag{6}$$

From the principles of calculus of variations, we learn that the calculation must be done as if, ψ being an arbitrary function, δJ were represented by the expression (6) and as if the variations were no longer subject to the condition (5).

We will have additionally

$$\delta u = \frac{\mathrm{d}\delta f}{\mathrm{d}t} + \delta\rho\xi,$$

hence, after integration by parts,

$$\delta J = \int \sum \delta f \left(f + \frac{\mathrm{d}F}{\mathrm{d}t} + \frac{\mathrm{d}\psi}{\mathrm{d}x} \right) \mathrm{d}t \mathrm{d}\tau + \int \left(\psi \delta\rho - \sum F \delta\rho\xi \right) \mathrm{d}t \mathrm{d}\tau. \tag{7}$$

If we first assume that the electrons experience no variation, $\delta\rho = \delta\rho\xi = 0$ and the second integral is zero. Since δJ must become zero, it must follow that:

$$f + \frac{\mathrm{d}F}{\mathrm{d}t} + \frac{\mathrm{d}\psi}{\mathrm{d}x} = 0. \tag{8}$$

In the general case, there rests, therefore:

$$\delta J = \int \left(\psi \delta\rho - \sum F\delta\rho\xi\right) \mathrm{d}t\mathrm{d}\tau. \tag{9}$$

The forces which act on the electrons remain to be determined. To do that we will have to assume that a complementary force $-X\mathrm{d}\tau$, $-Y\mathrm{d}\tau$, $-Z\mathrm{d}\tau$ is applied to each element of the electron and write that this force is in equilibrium with the forces of electromagnetic origin. Let U, V, W be the components of the displacements of the element $\mathrm{d}\tau$ of the electron; this displacement is considered from an arbitrary initial position. Let δU, δV, δW be the variations of this displacement; the virtual work corresponding to the complementary force will be:

$$-\int \sum X\delta U \mathrm{d}\tau,$$

such that the equilibrium condition that we just talked about will be written:

$$\delta J = -\int \sum X\delta U \mathrm{d}\tau\mathrm{d}t. \tag{10}$$

This is a matter of transforming δJ. To do that, we start by looking for the continuity equation expressing that the charge of an electron is conserved by the variation.

Let x_0, y_0, z_0 be the initial position of an electron. Its current position will be

$$x = x_0 + U, \quad y = y_0 + V, \quad z = z_0 + W.$$

We will additionally introduce an auxiliary variable ε, which will produce the variations of our various functions, such that for an arbitrary function A, we will have:

$$\delta A = \delta\varepsilon \frac{\mathrm{d}A}{\mathrm{d}\varepsilon}.$$

It will in fact be useful to be able to switch from the notation of calculus of variations to that of ordinary differential calculus, or vice versa.

It will be possible to regard our functions: first as depending on five variables x, y, z, t, ε, such that the position does not change when only t and ε change—we will designate their derivatives by the ordinary d; second as depending on five variables x_0, y_0, z_0, t, ε, such that a single electron is always followed when only t and ε vary—we will then designate their derivatives by round ∂. We will then have:

$$\xi = \frac{\partial U}{\partial t} = \frac{\mathrm{d}U}{\mathrm{d}t} + \xi\frac{\mathrm{d}U}{\mathrm{d}x} + \eta\frac{\mathrm{d}U}{\mathrm{d}y} + \zeta\frac{\mathrm{d}U}{\mathrm{d}z} = \frac{\partial x}{\partial t}. \tag{11}$$

We now designate by Δ the functional determinant of x, y, z, relative to x_0, y_0, z_0:

$$\Delta = \frac{\partial(x,y,z)}{\partial(x_0,y_0,z_0)}.$$

If, with ε, x_0, y_0, z_0 remaining constant, we give an increase ∂t to t, there will result for x, y, z increases ∂x, ∂y, ∂z and for Δ an increase of $\partial\Delta$ and it will hold that:

$$\partial x = \xi\partial t, \quad \partial y = \eta\partial t, \quad \partial z = \zeta\partial t,$$
$$\Delta + \partial\Delta = \frac{\partial(x+\partial x, y+\partial y, z+\partial z)}{\partial(x_0,y_0,z_0)};$$

hence

$$1 + \frac{\partial\Delta}{\Delta} = \frac{\partial(x+\partial x, y+\partial y, z+\partial z)}{\partial(x,y,z)} = \frac{\partial(x+\xi\partial t, y+\eta\partial t, z+\zeta\partial t)}{\partial(x,y,z)}.$$

From which one can deduce:

$$\frac{1}{\Delta}\frac{\partial\Delta}{\partial \mathrm{t}} = \frac{\mathrm{d}\xi}{\mathrm{d}x} + \frac{\mathrm{d}\eta}{\mathrm{d}y} + \frac{\mathrm{d}\zeta}{\mathrm{d}z}. \tag{12}$$

Since the mass[7] of each electron is invariant, we will have:

$$\frac{\partial\rho\Delta}{\partial t} = 0, \tag{13}$$

hence:

$$\frac{\partial\rho}{\partial t} + \sum\rho\frac{\mathrm{d}\xi}{\mathrm{d}x} = 0, \quad \frac{\partial\rho}{\partial t} = \frac{\mathrm{d}\rho}{\mathrm{d}t} + \sum\xi\frac{\mathrm{d}\rho}{\mathrm{d}x}, \quad \frac{\mathrm{d}\rho}{\mathrm{d}t} + \sum\frac{\mathrm{d}\rho\xi}{\mathrm{d}x} = 0.$$

Such are the various forms of the equation of continuity as it relates to the variable t. We find the analogous forms as it relates to the variable ε. Let:

$$\delta U = \frac{\partial U}{\partial\varepsilon}\delta\varepsilon, \quad \delta V = \frac{\partial V}{\partial\varepsilon}\delta\varepsilon, \quad \delta W = \frac{\partial W}{\partial\varepsilon}\delta\varepsilon;$$

it will follow:

$$\delta U = \frac{\mathrm{d}U}{\mathrm{d}\varepsilon}\delta\varepsilon + \delta U\frac{\mathrm{d}U}{\mathrm{d}x} + \delta V\frac{\mathrm{d}U}{\mathrm{d}y} + \delta W\frac{\mathrm{d}U}{\mathrm{d}z}, \tag{11'}$$

$$\frac{1}{\Delta}\frac{\partial\Delta}{\partial\varepsilon} = \sum\frac{\partial U}{\partial\varepsilon}, \quad \frac{\partial\rho\Delta}{\partial\varepsilon} = 0, \tag{12'}$$

$$\delta\varepsilon\frac{\partial\rho}{\partial\varepsilon}+\sum\rho\frac{\mathrm{d}\delta U}{\mathrm{d}x}=0,\quad \frac{\partial\rho}{\partial\varepsilon}=\frac{\mathrm{d}\rho}{\mathrm{d}\varepsilon}+\sum\frac{\delta U}{\delta\varepsilon}\frac{\mathrm{d}\rho}{\mathrm{d}x},\quad \delta\rho+\frac{\mathrm{d}\rho\delta U}{\mathrm{d}x}=0. \tag{13'}$$

The difference between the definition of $\delta U=\frac{\partial U}{\partial\varepsilon}\delta\varepsilon$ and that of $\delta\rho=\frac{\mathrm{d}\rho}{\mathrm{d}\varepsilon}\delta\varepsilon$ will be noted; it will be noted that it is in fact this definition of δU which is appropriate for the formula (10).

That last equation is going to allow us to transform the first term of (9); in fact we find:

$$\int\psi\delta\rho\mathrm{d}t\mathrm{d}\tau=-\int\psi\sum\frac{\mathrm{d}\rho\delta U}{\mathrm{d}x}\mathrm{d}t\mathrm{d}\tau$$

or, by integrating by parts,

$$\int\psi\delta\rho\mathrm{d}t\mathrm{d}\tau=\int\sum\rho\frac{\mathrm{d}\psi}{\mathrm{d}x}\delta U\mathrm{d}t\mathrm{d}\tau. \tag{14}$$

We now propose to determine:

$$\delta(\rho\xi)=\frac{\mathrm{d}(\rho\xi)}{\mathrm{d}\varepsilon}\delta\varepsilon.$$

We observe the $\rho\Delta$ can only depend on x_0, y_0, z_0; in fact, if an element of electron is considered whose initial position is a rectangular parallelepiped whose edges are $\mathrm{d}x_0$, $\mathrm{d}y_0$, $\mathrm{d}z_0$, then the charge of this element is

$$\rho\Delta\mathrm{d}x_0\mathrm{d}y_0\mathrm{d}z_0$$

and, since this charge needs to remain constant, it follows that:

$$\frac{\partial\rho\Delta}{\partial t}=\frac{\partial\rho\Delta}{\partial\varepsilon}=0. \tag{15}$$

From that it is deduced:

$$\frac{\partial^2\rho\Delta}{\partial t\partial\varepsilon}=\frac{\partial}{\partial\varepsilon}\left(\rho\Delta\frac{\partial U}{\partial t}\right)=\frac{\partial}{\partial t}\left(\rho\Delta\frac{\partial U}{\partial\varepsilon}\right). \tag{16}$$

For an arbitrary function A it is known from the equation of continuity that,

$$\frac{1}{\Delta}\frac{\partial A\Delta}{\partial t}=\frac{\mathrm{d}A}{\mathrm{d}t}+\sum\frac{\mathrm{d}A\xi}{\mathrm{d}x}$$

and similarly

$$\frac{1}{\Delta}\frac{\partial A\Delta}{\partial\varepsilon}=\frac{\mathrm{d}A}{\mathrm{d}\varepsilon}+\sum\frac{\mathrm{d}A\frac{\partial U}{\partial\varepsilon}}{\mathrm{d}x}$$

Therefore it follows:

$$\frac{1}{\Delta}\frac{\partial}{\partial\varepsilon}\left(\rho\Delta\frac{\partial U}{\partial t}\right)=\frac{\mathrm{d}\rho\frac{\partial U}{\partial t}}{\mathrm{d}\varepsilon}+\frac{\mathrm{d}\left(\rho\frac{\partial U}{\partial t}\frac{\partial U}{\partial\varepsilon}\right)}{\mathrm{d}x}+\frac{\mathrm{d}\left(\rho\frac{\partial U}{\partial t}\frac{\partial V}{\partial\varepsilon}\right)}{\mathrm{d}y}+\frac{\mathrm{d}\left(\rho\frac{\partial U}{\partial t}\frac{\partial W}{\partial\varepsilon}\right)}{\mathrm{d}z}\tag{17}$$

$$\frac{1}{\Delta}\frac{\partial}{\partial t}\left(\rho\Delta\frac{\partial U}{\partial\varepsilon}\right)=\frac{\mathrm{d}\rho\frac{\partial U}{\partial\varepsilon}}{\mathrm{d}t}+\frac{\mathrm{d}\left(\rho\frac{\partial U}{\partial t}\frac{\partial U}{\partial\varepsilon}\right)}{\mathrm{d}x}+\frac{\mathrm{d}\left(\rho\frac{\partial V}{\partial t}\frac{\partial U}{\partial\varepsilon}\right)}{\mathrm{d}y}+\frac{\mathrm{d}\left(\rho\frac{\partial W}{\partial t}\frac{\partial U}{\partial\varepsilon}\right)}{\mathrm{d}z}\tag{17'}$$

The right-hand sides of (17) and (17′) must be equal and, recalling that

$$\frac{\partial U}{\partial t}=\xi,\quad \frac{\partial U}{\partial\varepsilon}\delta\varepsilon=\delta U,\quad \frac{\mathrm{d}\rho\xi}{\mathrm{d}\varepsilon}\delta\varepsilon=\delta\rho\xi,$$

it follows that:

$$\begin{aligned}\delta\rho\xi&+\frac{\mathrm{d}(\rho\xi\delta U)}{\mathrm{d}x}+\frac{\mathrm{d}(\rho\xi\delta V)}{\mathrm{d}y}+\frac{\mathrm{d}(\rho\xi\delta W)}{\mathrm{d}z}\\&=\frac{\mathrm{d}(\rho\delta U)}{\mathrm{d}t}+\frac{\mathrm{d}(\rho\xi\delta U)}{\mathrm{d}x}+\frac{\mathrm{d}(\rho\eta\delta U)}{\mathrm{d}y}+\frac{\mathrm{d}(\rho\zeta\delta U)}{\mathrm{d}z}\end{aligned}\tag{18}$$

We now transform the second term from (9) and get:

$$\begin{aligned}\int\sum F\delta\rho\xi\mathrm{d}t\mathrm{d}\tau=\int\Bigg[&\sum F\frac{\mathrm{d}(\rho\delta U)}{\mathrm{d}t}+\sum F\frac{\mathrm{d}(\rho\eta\delta U)}{\mathrm{d}y}+\sum F\frac{\mathrm{d}(\rho\zeta\delta U)}{\mathrm{d}z}\\&-\sum F\frac{\mathrm{d}(\rho\xi\delta V)}{\mathrm{d}y}-\sum F\frac{\mathrm{d}(\rho\xi\delta W)}{\mathrm{d}z}\Bigg]\mathrm{d}t\mathrm{d}\tau.\end{aligned}$$

By integrating by parts, the right-hand side becomes:

$$\int\left[-\sum\rho\delta U\frac{\mathrm{d}F}{\mathrm{d}t}-\sum\rho\eta\delta U\frac{\mathrm{d}F}{\mathrm{d}y}-\sum\rho\zeta\delta U\frac{\mathrm{d}F}{\mathrm{d}z}+\sum\rho\xi\delta V\frac{\mathrm{d}F}{\mathrm{d}y}+\sum\rho\xi\delta W\frac{\mathrm{d}F}{\mathrm{d}z}\right]\mathrm{d}t\mathrm{d}\tau.$$

Now remark that:

$$\sum\rho\xi\delta V\frac{\mathrm{d}F}{\mathrm{d}y}=\sum\rho\zeta\delta U\frac{\mathrm{d}H}{\mathrm{d}x},\quad \sum\rho\xi\delta W\frac{\mathrm{d}F}{\mathrm{d}z}=\sum\rho\eta\delta U\frac{\mathrm{d}G}{\mathrm{d}x}.$$

If, in fact, in both sides of these relations, the sums are expanded, they become identities; and let us recall that

$$\frac{\mathrm{d}H}{\mathrm{d}x} - \frac{\mathrm{d}F}{\mathrm{d}z} = -\beta, \quad \frac{\mathrm{d}G}{\mathrm{d}x} - \frac{\mathrm{d}F}{\mathrm{d}y} = \gamma,$$

the right-hand side in question will become:

$$\int \left[-\sum \rho \delta U \frac{\mathrm{d}F}{\mathrm{d}t} + \sum \rho \gamma \eta \delta U - \sum \rho \beta \zeta \delta U \right] \mathrm{d}t \mathrm{d}\tau,$$

such that finally:

$$\delta J = \int \sum \rho \delta U \left(\frac{\mathrm{d}\psi}{\mathrm{d}x} + \frac{\mathrm{d}F}{\mathrm{d}t} + \beta\zeta - \gamma\eta \right) \mathrm{d}t \mathrm{d}\tau = \int \sum \rho \delta U (-f + \beta\zeta - \gamma\eta) \mathrm{d}t \mathrm{d}\tau.$$

By equating the coefficients of δU in both sides of (10), it follows:

$$X = f - \beta\zeta + \gamma\eta$$

This is equation (2) from the previous section.

§3 – Lorentz Transformation and the Principle of Least Action

We are going to see if the principle of least action gives us the reason for the success of the Lorentz transformation. First it needs to be seen what this transformation does to the integral:

$$J = \int \left(\frac{\sum f^2}{2} - \frac{\sum \alpha^2}{2} \right) \mathrm{d}t \mathrm{d}\tau$$

(formula 4 from §2).

We first find

$$\mathrm{d}t' \mathrm{d}\tau' = l^4 \mathrm{d}t \mathrm{d}\tau$$

because x', y', z', t' are related to x, y, z, t by linear relations whose determinant is equal to l^4; it next follows:

$$\begin{aligned} l^4 \sum f'^2 &= f^2 + k^2(g^2 + h^2) + k^2\varepsilon^2(\beta^2 + \gamma^2) + 2k^2\varepsilon(g\gamma - h\beta) \\ l^4 \sum \alpha'^2 &= \alpha^2 + k^2(\beta^2 + \gamma^2) + k^2\varepsilon^2(g^2 + h^2) + 2k^2\varepsilon(g\gamma - h\beta) \end{aligned} \tag{1}$$

(formulas 9 from §1), hence:

$$l^4\left(\sum f'^2 - \sum \alpha'^2\right) = \sum f^2 - \sum \alpha^2;$$

such that if one sets:

$$J' = \int \left(\frac{\sum f'^2}{2} - \frac{\sum \alpha'^2}{2}\right) \mathrm{d}t' \mathrm{d}\tau'$$

it follows:

$$J' = J.$$

For this equality to be justified, it is however necessary that the limits of integration be the same; until now we have allowed t to vary from t_0 to t_1; and x, y, z to vary from $-\infty$ to $+\infty$. As such, the integration limits would be changed by the Lorentz transformation; but nothing prevents us from assuming $t_0 = -\infty$, $t = +\infty$; with these conditions, the limits of the same for J and for J'.

We now need to compare the following two equations analogous to equation (10) from §2:

$$\delta J = -\int \sum X \delta U \mathrm{d}\tau \mathrm{d}t$$
$$\delta J' = -\int \sum X' \delta U' \mathrm{d}\tau' \mathrm{d}t'. \tag{2}$$

To do that, we first need to compare $\delta U'$ to δU.

Consider an electron whose initial coordinates are x_0, y_0, z_0; at the moment t, these coordinates will be:

$$x = x_0 + U, \quad y = y_0 + V, \quad z = z_0 + W.$$

If the corresponding electron is considered after the Lorentz transformation, its coordinates will be

$$x' = kl(x + \varepsilon t), \quad y' = ly, \quad z' = lz,$$

where

$$x' = x_0 + U', \quad y' = y_0 + V', \quad z' = z_0 + W';$$

but it will only reach these coordinates at the moment

$$t' = kl(t + \varepsilon x)$$

If we were to make our variables undergo variations δU, δV, δW and at the same time we were to give t an increase δt, the coordinates x, y, z will undergo a total increase

$$\delta x = \delta U + \xi \delta t, \quad \delta y = \delta V + \eta \delta t, \quad \delta z = \delta W + \zeta \delta t\,.$$

We will also have:

$$\delta x' = \delta U' + \xi' \delta t', \quad \delta y' = \delta V' + \eta' \delta t', \quad \delta z' = \delta W' + \zeta' \delta t'\,,$$

and because of the Lorentz transformation:

$$\delta x' = kl(\delta x + \varepsilon \delta t), \quad \delta y' = l\delta y, \quad \delta z' = l\delta z, \quad \delta t' = kt(\delta t + \varepsilon \delta x),$$

hence, by assuming $\delta t = 0$, the relations:

$$\begin{aligned}
&\delta x' = \delta U' + \xi' \delta t' = kl\delta U, \\
&\delta y' = \delta V' + \eta' \delta t' = l\delta V, \\
&\delta t' = kl\varepsilon \delta U.
\end{aligned}$$

We observe that

$$\xi' = \frac{\xi + \varepsilon}{1 + \xi\varepsilon}, \quad \eta' = \frac{\eta}{k(1 + \xi\varepsilon)};$$

it will follow, by replacing $\delta t'$ with its value,

$$\begin{aligned}
&kl(1 + \xi\varepsilon)\delta U = \delta U'(1 + \xi\varepsilon) + (\xi + \varepsilon)kl\varepsilon\delta U, \\
&l(1 + \xi\varepsilon)\delta V = \delta V'(1 + \xi\varepsilon) + \eta l\varepsilon\delta U.
\end{aligned}$$

If we recall the definition of k, we can draw from it that:

$$\begin{aligned}
\delta U &= \frac{k}{l}\delta U' + \frac{k\varepsilon}{l}\xi\delta U', \\
\delta V &= \frac{1}{l}\delta V' + \frac{k\varepsilon}{l}\eta\delta U',
\end{aligned}$$

and similarly that

$$\delta W = \frac{1}{l}\delta W' + \frac{k\varepsilon}{l}\zeta\delta U';$$

hence

$$\sum X\delta U = \frac{1}{l}(kX\delta U' + Y\delta V' + Z\delta W') + \frac{k\varepsilon}{l}\delta U' \sum X\xi \tag{3}$$

Hence, because of equations (2) it must be that:

$$\int \sum X'\delta U' \mathrm{d}t' \mathrm{d}\tau' = \int \sum X\delta U \mathrm{d}t \mathrm{d}\tau = \frac{1}{l^4} \int \sum X\delta U \mathrm{d}t' \mathrm{d}\tau'$$

By replacing $\sum X\delta U$by its value (3) and identifying, it follows:

$$X' = \frac{k}{l^5}X + \frac{k\varepsilon}{l^5}\sum X\xi \quad Y' = \frac{1}{l^5}Y, \quad Z' = \frac{1}{l^5}Z.$$

These are equations (11) from §1. The principle of least action therefore leads us to the same result as the analysis from §1.

If we refer back to formulas (1), we see that $\sum f^2 - \sum \alpha^2$ is unchanged by the Lorentz transformation, up to a constant factor; it is not the same for the expression $\sum f^2 + \sum \alpha^2$ which appears in the energy. If we limit ourselves to the case where ε is sufficiently small that its square can be neglected such that $k = 1$ and if we also assume $l = 1$, we find:

$$\sum f'^2 = \sum f^2 + 2\varepsilon(g\gamma - h\beta),$$
$$\sum \alpha'^2 = \sum \alpha^2 + 2\varepsilon(g\gamma - h\beta),$$

or, by addition,

$$\sum f'^2 + \sum \alpha'^2 = \sum f^2 + \sum \alpha^2 + 4\varepsilon(g\gamma - h\beta).$$

§4 – The Lorentz Group

It is important to note that the Lorentz transformation do form a group.

In fact, if one sets:

$$x' = kl(x + \varepsilon t), \quad y' = ly, \quad z' = lz, \quad t' = kl(t + \varepsilon x),$$

and additionally

$$x'' = k'l'(x' + \varepsilon' t'), \quad y'' = l'y' \quad z'' = l'z', \quad t'' = k'l'(t' + \varepsilon' x'),$$

with

$$k^{-2} = 1 - \varepsilon^2, \quad k'^{-2} = 1 - \varepsilon'^2$$

it will follow:

$$x'' = k''l''(x + \varepsilon'' t), \quad y'' = l''y \quad z'' = l''z, \quad t'' = k''l''(t + \varepsilon'' x),$$

with

$$\varepsilon'' = \frac{\varepsilon + \varepsilon'}{1 + \varepsilon\varepsilon'}, \quad l'' = ll', \quad k'' = kk'(1 + \varepsilon\varepsilon') = \frac{1}{\sqrt{1 - \varepsilon''^2}}.$$

If we give l the value 1 and we assume that ε is infinitesimal,

$$x' = x + \delta x, \quad y' = y + \delta y, \quad z' = z + \delta z, \quad t' = t + \delta t,$$

it will follow:

$$\delta x = \varepsilon t, \quad \delta y = \delta z = 0, \quad \delta t = \varepsilon x.$$

That is the infinitesimal generating transformation of the group, which I will call the T_1 transformation and which can be written using the Lie notation:

$$t\frac{\mathrm{d}\varphi}{\mathrm{d}x} + x\frac{\mathrm{d}\varphi}{\mathrm{d}t} = T_1.$$

If we assume $\varepsilon = 0$ and $l = 1 + \delta l$, we would in contrast find

$$\delta x = x\delta l, \quad \delta y = y\delta l, \quad \delta z = z\delta l, \quad \delta t = t\delta l$$

and we will have another infinitesimal transformation T_0 of the group (supposing that l and ε are regarded as independent variables) and with the Lie notation it would be:

$$T_0 = x\frac{\mathrm{d}\varphi}{\mathrm{d}x} + y\frac{\mathrm{d}\varphi}{\mathrm{d}y} + z\frac{\mathrm{d}\varphi}{\mathrm{d}z} + t\frac{\mathrm{d}\varphi}{\mathrm{d}t}.$$

But we could give the particular role that we had given to the x-axis to the y-axis or the z-axis; in that way one would have two other infinitesimal transformations:

$$T_2 = t\frac{d\varphi}{dy} + y\frac{d\varphi}{dt}$$
$$T_3 = t\frac{d\varphi}{dz} + z\frac{d\varphi}{dt}$$

which would not alter the Lorentz equations either.

One can form the combinations imagined by Lie, such as

$$[T_1, T_2] = x\frac{d\varphi}{dy} - y\frac{d\varphi}{dx};$$

but it is easy to see that this transformation is equivalent to a change of coordinate axes, the axes turning a very small angle around the z-axis. We shouldn't therefore be surprised if a similar change leaves the form of the Lorentz equations unchanged, since the equations are obviously independent of the choice of axes.

We are therefore led to consider a continuous group that we will call the *Lorentz group* in which will allow as infinitesimal transformations:

1) the transformation T_0 which will be permutable with all the others;
2) the three transformations T_1, T_2, T_3; and
3) the three rotations $[T_1, T_2]$, $[T_2, T_3]$, $[T_3, T_1]$.

An arbitrary transformation of this group could always be broken down into a transformation of the form:

$$x' = lx, \quad y' = ly, \quad z' = lz, \quad t' = lt$$

and a linear transformation which does not change the quadratic form:

$$x^2 + y^2 + z^2 - t^2.$$

We can also generate our group in another way. Any transformation of the group could be regarded as a transformation of the form:

$$x' = kl(x + \varepsilon t), \quad y' = ly, \quad z' = lz, \quad t' = kl(t + \varepsilon x) \tag{1}$$

preceded and followed by a suitable rotation.

But for our purposes, we should only consider a part of the transformations from this group; we should assume that l is a function of ε, and it will be a matter of choosing this function such that this part of the group, which I will call P, again forms a group.

Turning the system 180° around the y-axis, we should find a transformation which will have to again belong to P. Now this amounts to changing the sign of x, x', z and z'; in that way it is found that:

$$x' = kl(x - \varepsilon t), \quad y' = ly, \quad z' = lz, \quad t' = kl(t - \varepsilon x) \tag{2}$$

Thus l is not changed when ε is changed to $-\varepsilon$.

On the other hand, if P is a group, the inverse substitution of (1), which is written:

$$x' = \frac{k}{l}(x - \varepsilon t), \quad y' = \frac{y}{l}, \quad z' = \frac{z}{l}, \quad t' = \frac{k}{l}(t - \varepsilon x), \tag{3}$$

should also belong to P; it will therefore have to be identical to (2), meaning that

$$l = \frac{1}{l}.$$

It will therefore have to be that $l = 1$.

§5 – Langevin Waves

Langevin[8] put the formulas which define the electromagnetic field produced by the motion of a single electron in a particularly elegant form.

Return to the equations

$$\Box\psi = -\rho, \qquad \Box F = -\rho\xi. \tag{1}$$

It is known that they can be integrated by delayed potentials and that one finds:

$$\psi = \frac{1}{4\pi}\int\frac{\rho_1 \mathrm{d}\tau_1}{r}, \quad F = \frac{1}{4\pi}\int\frac{\rho_1\xi_1 \mathrm{d}\tau_1}{r}. \tag{2}$$

In these formulas one has:

$$\mathrm{d}\tau_1 = \mathrm{d}x_1\mathrm{d}y_1\mathrm{d}z_1, \quad r^2 = (x - x_1)^2 + (y - y_1)^2 + (z - z_1)^2$$

while ρ_1 and ξ_1 are values of ρ and ξ at the point x_1, y_1, z_1 and at the moment

$$t_1 = t - r.$$

Let x_0, y_0, z_0 be the coordinates of a differential element of an electron at the moment t; and

$$x_1 = x_0 + U, \quad y_1 = y_0 + V, \quad z_1 = z_0 + W$$

be its coordinates at the moment t_1.

U, V, W are functions of x_0, y_0, z_0 such that we will be able to write:

$$dx_1 = dx_0 + \frac{dU}{dx_0}dx_0 + \frac{dU}{dy_0}dy_0 + \frac{dU}{dz_0}dz_0 + \xi_1 dt_1;$$

and if one assumes t to be constant, and also x, y and z:

$$dt_1 = +\sum \frac{x - x_1}{r} dx_1.$$

We can then write:

$$dx_1\left(1 + \xi_1 \frac{x_1 - x}{r}\right) + dy_1 \xi_1 \frac{y_1 - y}{r} + dz_1 \xi_1 \frac{z_1 - z}{r}$$
$$= dx_0\left(1 + \frac{dU}{dx_0}\right) + dy_0 \frac{dU}{dy_0} + dz_0 \frac{dU}{dz_0}$$

with the two other equations that can be deduced by circular permutation.

We therefore have:

$$d\tau_1 \left| 1 + \xi_1 \frac{x_1 - x}{r}, \quad \xi_1 \frac{y_1 - y}{r}, \quad \xi_1 \frac{z_1 - z}{r} \right| = d\tau_0 \left| 1 + \frac{dU}{dx_0}, \quad \frac{dU}{dy_0}, \quad \frac{dU}{dz_0} \right| \qquad (3)$$

by setting

$$d\tau_0 = dx_0 dy_0 dz_0.$$

We will study the determinants which appear on both sides of (3) and start with the left-hand side; on trying to expand it, one sees that the terms of second and third degree in ξ_1, η_1, ζ_1 disappear and that the determinant is equal to

$$1 + \xi_1 \frac{x_1 - x}{r} + \eta_1 \frac{y_1 - y}{r} + \zeta_1 \frac{z_1 - z}{r} = 1 + \omega,$$

where ω designates the radial component of the speed ξ_1, η_1, ζ_1, meaning the component directed along the radius vector going from the point x, y, z to the point x_1, y_1, z_1.

In order to get the second determinant, I consider the coordinates of various molecules of the electron at a moment t_1' which is the same for all the differential elements, but in such a way that for the differential element that I consider one can have $t_1 = t_1'$. The coordinates of a differential element will then be:

$$x_1' = x_0 + U', \quad y_1' = y_0 + V', \quad z_1' = z_0 + W',$$

where U', V', W' are what U, V, W become when t_1 is replaced in them by t_1'; as t_1' is the same for all the differential elements, it will hold:

$$dx_1' = dx_0\left(1 + \frac{dU'}{dx_0}\right) + dy_0\frac{dU'}{dy_0} + dz_0\frac{dU'}{dz_0}$$

and consequently

$$d\tau_1' = d\tau_0\left|1 + \frac{dU'}{dx_0}, \quad \frac{dU'}{dy_0}, \quad \frac{dU'}{dz_0}\right|,$$

by setting

$$d\tau_1' = dx_1' dy_1' dz_1'$$

But the element of electric charge is

$$d\mu_1 = \rho_1 d\tau_1'$$

and additionally *for the differential element considered*, one has $t_1 = t_1'$ and consequently $\frac{dU'}{dx_0} = \frac{dU}{dx_0}$, etc.; we can therefore write:

$$d\mu_1 = \rho_1 d\tau_0\left|1 + \frac{dU'}{dx_0}, \quad \frac{dU'}{dy_0}, \quad \frac{dU'}{dz_0}\right|,$$

such that equation (3) will become:

$$\rho_1 d\tau_1(1+\omega) = d\mu_1$$

and equations (2):

$$\psi = \frac{1}{4\pi}\int\frac{d\mu_1}{r(1+\omega)}, \quad F = \int\frac{\xi_1 d\mu_1}{r(1+\omega)}.$$

If we are dealing with a single electron, our integrals will reduce to a single element, provided that only points x, y, z are considered that ae sufficiently far away so that r and ω have substantially the same value for all points of the electron. The potentials ψ, F, G, H will depend on the position of this electron and also its speed, because not only do the ξ_1, η_1, ζ_1 appear in the numerator in F, G, H, but the radial component ω appears in the denominator. It is of course its position and its velocity at the moment t_1 that are involved.

The partial derivatives of ψ, F, G, H with respect to t, x, y, z (and consequently the electric and magnetic fields) will furthermore depend on its acceleration. Additionally, they will depend on it linearly, because in these derivatives this acceleration comes in following a single differentiation.

In that way, Langevin was led to distinguish the terms in the electric and magnetic fields that do not depend on the acceleration (which he calls the speed wave) and those that are proportional to the acceleration (which he calls the acceleration wave).

The Lorentz transformation makes the calculation of these two waves easier. We can in fact apply this transformation to the system such that the speed of the single electron under consideration become zero. We will take the direction of this velocity for the x-axis before the transformation, such that, at the moment t,

$$\eta_1 = \zeta_1 = 0,$$

and we will take $\varepsilon = -\xi_1$, such that

$$\xi_1' = \eta_1' = \zeta_1' = 0.$$

We can therefore reduce the calculation of the two waves to the case where the electron velocity is zero. We start with the velocity wave; we can first remark that this wave is the same as if the motion of the electron were uniform.

If the velocity of electron is zero, it follows:

$$\omega = 0, \quad F = G = H = 0, \quad \psi = \frac{\mu_1}{4\pi r},$$

where μ_1 is the electric charge of the electron. The velocity having been brought to zero by the Lorentz transformation, we therefore have:

$$F' = G' = H' = 0, \quad \psi = \frac{\mu_1}{4\pi r'},$$

where r' is the distance from the point x', y', z' to the point x_1', y_1', z_1', and consequently:

$$\alpha' = \beta' = \gamma' = 0,$$
$$f' = \frac{\mu_1 (x' - x_1')}{4\pi r'^3}, \quad g' = \frac{\mu_1 (y' - y_1')}{4\pi r'^3}, \quad h' = \frac{\mu_1 (z' - z_1')}{4\pi r'^3}.$$

We now do the inverse Lorentz transformation to find the actual field corresponding to a velocity $-\varepsilon$, 0, 0. By referring to equations (9) and (3) from §1:

$$\alpha = 0, \quad \beta = \varepsilon h, \quad \gamma = -\varepsilon g,$$
$$f = \frac{\mu_1 k l^3}{4\pi r'^3}(x + \varepsilon t - x_1 - \varepsilon t_1), \quad g = \frac{\mu_1 k l^3}{4\pi r'^3}(y - y_1), \quad h = \frac{\mu_1 k l^3}{4\pi r'^3}(z - z_1), \tag{4}$$

It can be seen that the magnetic field is perpendicular to the x-axis (direction of the velocity) and to the electric field and that the electric field is directed towards the point:

$$x_1 + \varepsilon(t_1 - t), \quad y_1, \quad z_1, \tag{5}$$

If the electron were to continue to move with a straight and uniform motion with the speed that it had at the moment t_1, meaning with the velocity $-\varepsilon$, 0, 0, this point (5) would be the one that it would occupy at the moment t.

Now switch to the acceleration wave; by using the Lorentz transformation, we can refer its determination to the case where the velocity is zero. This is the case which occurs if an electron is imagined to execute very small amplitude oscillations, but very fast, such that the displacements and the velocities are infinitesimal but the accelerations are finite. This brings us back to the field which was studied in the celebrated paper by Hertz, *Die Kräfte elektrischer Schwingungen nach der Maxwell'schen Theorie*, that considered a very distant point. Under these conditions:

1) The electric and magnetic fields are equal to each other.
2) They are perpendicular to each other.
3) They are perpendicular to the normal to the spherical wavefront, meaning to the sphere whose center is at the point x_1, y_1, z_1.

I state that these three properties will still be present when the velocity is not zero, and for that, it is sufficient for me to prove that they are unchanged by the Lorentz transformation.

In fact, let A be the shared strength of the two fields; let:

$$(x - x_1) = r\lambda, \quad (y - y_1) = r\mu, \quad (z - z_1) = r\nu, \quad \lambda^2 + \mu^2 + \nu^2 = 1.$$

These properties will be expressed by the equalities:

$$A^2 = \sum f^2 = \sum \alpha^2, \quad \sum f\alpha = 0, \quad \sum f(x - x_1) = 0, \quad \sum \alpha(x - x_1) = 0,$$
$$\sum f\lambda = 0, \quad \sum \alpha\lambda = 0;$$

which means again that:

$$\begin{array}{ccc} \frac{b}{A}, & \frac{g}{A}, & \frac{h}{A} \\ \frac{\alpha}{A}, & \frac{\beta}{A}, & \frac{\gamma}{A} \\ \lambda, & \mu, & \nu \end{array}$$

are the directional cosines of the three rectangular directions and from that the relations are deduced:

$$f = \beta\nu - \gamma\mu, \quad \alpha = h\mu - g\nu,$$

or

$$fr = \beta(z - z_1) - \gamma(y - y_1), \quad \alpha r = h(y - y_1) - g(z - z_1), \tag{6}$$

along with the equations that can be deduced from them by symmetry.

If we take up equations (3) from §1, we find:

$$\begin{aligned} x' - x'_1 &= kl[(x - x_1) + \varepsilon(t - t_1)] = kl[(x - x_1) + \varepsilon r], \\ y' - y'_1 &= l(y - y_1), \\ z' - z'_1 &= l(z - z_1). \end{aligned} \tag{7}$$

Above, in §3, we found:

$$l^4\left(\sum f'^2 - \sum \alpha'^2\right) = \sum f^2 - \sum \alpha^2.$$

Therefore $\sum f^2 = \sum \alpha^2$ leads to $\sum f'^2 = \sum \alpha'^2$.

On the other hand, by starting from equations (9) from §1, it is found:

$$l^4 \sum f'\alpha' = \sum f\alpha,$$

which shows that $\sum f\alpha = 0$ leads to $\sum f'\alpha' = 0$.

I now state that

$$\sum f'(x' - x'_1) = 0, \quad \sum \alpha'(x' - x'_1) = 0. \tag{8}$$

In fact, because of equations (7) (and also equations 9 from §1) the left-hand sides of the two equations (8) are written respectively:

$$\begin{aligned} &\frac{k}{l}\sum f(x - x_1) + \frac{k\varepsilon}{l}[fr + \gamma(y - y_1) - \beta(z - z_1)], \\ &\frac{k}{l}\sum \alpha(x - x_1) + \frac{k\varepsilon}{l}[\alpha r - h(y - y_1) + g(z - z_1)], \end{aligned}$$

They therefore become zero because of the equations $\sum f(x - x_1) = \sum \alpha(x - x_1) = 0$ and because equations (6). This is precisely what it was a matter of proving.

It is also possible to arrive at the same result by simple considerations of homogeneity.

In fact, ψ, F, G, H are homogeneous functions of $(x - x_1)$, $(y - y_1)$, $(z - z_1)$, $\xi_1 = \mathrm{d}x_1/\mathrm{d}t_1$, $\eta_1 = \mathrm{d}y_1/\mathrm{d}t_1$, $\zeta_1 = \mathrm{d}z_1/\mathrm{d}t_1$, of degree -1 in $x, y, z, t, x_1, y_1, z_1, t_1$ and their derivatives.

The derivatives of, ψ, F, G, H with respect to x, y, z, t (and consequently also to the two fields f, g, h; α, β, γ) will be homogeneous of degree -2 in the same quantities if we additionally recall that the relation

$$t - t_1 = r = \sqrt{\sum (x - x_1)^2}$$

is homogeneous in these quantities.

Now these derivatives or these fields depend on $x - x_1$, speeds dx_1/dt_1 and accelerations $d^2x_1/dt_1{}^2$; they are made up of a term independent of the accelerations (velocity wave) and a linear term in the accelerations (acceleration wave). Hence dx_1/dt_1 is homogeneous of degree 0 and $d^2x_1/dt_1{}^2$ is homogeneous of degree -1; from this it follows that the velocity wave is homogeneous of degree -2 in $(x - x_1)$, $(y - y_1)$, $(z - z_1)$, and the acceleration wave is homogeneous of degree -1. Therefore, at a very distant point, the acceleration wave dominates and can consequently be regarded as being the same as the total wave. Additionally, the law of homogeneity shows us that the acceleration wave is self-similar at a distant point and at an arbitrary point. It is therefore, at an arbitrary point, similar to the total wave at a distant point. Hence at a distant point the perturbation can only propagate by plane waves such that the two fields must be equal, perpendicular to each other and perpendicular to the direction of propagation.

I will limit myself to referring to the article by Langevin in the *Journal de Physique* (1905)[9] for more details.

§6 – Contraction of Electrons

We assume a single electron driven in a motion of straight and uniform translation. Based on what we just saw, the study of the field created by this electron in the case where the electron is immobile can be determined using the Lorentz transformation; the Lorentz transformation therefore replaces the real moving electron by an ideal immobile electron.

Let $\alpha, \beta, \gamma, f, g, h$; be the real field; let $\alpha', \beta', \gamma', f', g', h'$ be what the field becomes after the Lorentz transformation, such that the ideal field α', f' corresponds to the case of an immobile electron; it follows:

$$\alpha' = \beta' = \gamma' = 0, \quad f' = -\frac{d\psi'}{dx'}, \quad g' = -\frac{d\psi'}{dy'}, \quad h' = -\frac{d\psi'}{dz'};$$

and for the real field (because of formulas 9 from §1):

$$\begin{aligned} \alpha &= 0, & \beta &= \varepsilon h, & \gamma &= -\varepsilon g, \\ f &= l^2 f', & g &= kl^2 g', & h &= kl^2 h'. \end{aligned} \tag{1}$$

It is now a matter of determining the total energy due to the motion of the electron, the corresponding action and the electromagnetic moment in order to be able to calculate the electromagnetic masses of the electron. For a distant point, it is sufficient to consider the electron as reduced to a single point; one is then back at the formulas (4) from the previous section which are generally suitable. But here they would not be sufficient, because the energy is principally located in the parts of the ether closest to the electron.

Several hypotheses can be made on this subject.

According to Abraham's hypothesis, the electrons would be spherical and undeformable.

Then, when the Lorentz transformation would be applied, since the real electron would be spherical, the ideal electron would become an ellipsoid. Following §1, the equation of this ellipsoid would be:

$$k^2(x' - \varepsilon t - \xi t' + \varepsilon\xi x')^2 + (y' - \eta k t' + \eta k\varepsilon x')^2 + (z' - \zeta k t' + \zeta k\varepsilon x')^2 = l^2 r^2.$$

But here we have:

$$\xi + \varepsilon = \eta = \zeta = 0, \quad 1 + \varepsilon\xi = 1 - \varepsilon^2 = \frac{1}{k^2},$$

such that the equation of the ellipsoid becomes:

$$\frac{x'^2}{k^2} + y'^2 + z'^2 = l^2 r^2.$$

If the radius of the real electron is r, the axes of the ideal electron would therefore be:

$$klr, \quad lr, \quad lr.$$

In contrast, in Lorentz's hypothesis, the moving electrons would be deformed such that it would be the real electron which would be an ellipsoid whereas the immobile ideal electron would always be a sphere of radius r; the axes of the real electron will then be:

$$\frac{r}{lk}, \quad \frac{r}{l}, \quad \frac{r}{l}.$$

Call

$$A = \frac{1}{2}\int f^2 \mathrm{d}\tau$$

the *longitudinal electrical energy*;

$$B = \frac{1}{2}\int (g^2 + h^2)\mathrm{d}\tau$$

the *transverse electric energy*; and

$$C = \frac{1}{2}\int (\beta^2 + \gamma^2)\mathrm{d}\tau$$

the *transverse magnetic energy*. There is no longitudinal magnetic energy because $\alpha = \alpha' = 0$. Designate the corresponding quantities in the ideal system by A', B', C'. It is first found that:

$$C' = 0, \quad C = \varepsilon^2 B.$$

Additionally, we can observe that the real field depends only on $x + \varepsilon t$, y and z, and write:

$$\begin{aligned}\mathrm{d}\tau &= \mathrm{d}(x + \varepsilon t)\mathrm{d}y\mathrm{d}z,\\ \mathrm{d}\tau' &= \mathrm{d}x'\mathrm{d}y'\mathrm{d}z' = kl^3\mathrm{d}\tau;\end{aligned}$$

hence

$$A' = kl^{-1}A, \quad B' = k^{-1}l^{-1}B, \quad A = \frac{lA'}{k}, \quad B = klB'.$$

In Lorentz's hypothesis, $B' = 2A'$, and A', which is inversely proportional to the radius of the electron, is a constant independent of the speed of the real electron; in this way, it is possible to find the total energy:

$$A + B + C = A'lk\left(3 + \varepsilon^2\right)$$

and the action (per unit time):

$$A + B - C = \frac{3A'l}{k}.$$

We now calculate the electromagnetic momentum; we will find:

$$D = \int (g\gamma - h\beta)\mathrm{d}\tau = -\varepsilon\int \left(g^2 + h^2\right)\mathrm{d}\tau = -2\varepsilon B = -4\varepsilon klA'.$$

But there must be some relations between the energy $E = A + B + C$, the action per unit time $H = A + B - C$ and the momentum D. The first of these relations is:

$$E = H - \varepsilon \frac{dH}{d\varepsilon},$$

the second is:

$$\frac{dD}{d\varepsilon} = -\frac{1}{\varepsilon}\frac{dE}{d\varepsilon};$$

hence:

$$D = \frac{dH}{d\varepsilon}, \quad E = H - \varepsilon D. \tag{2}$$

The second of equations (2) is always satisfied; but the first is satisfied only if

$$l = \left(1 - \varepsilon^2\right)^{\frac{1}{6}} = k^{-\frac{1}{3}},$$

meaning if the volume of the ideal electron is equal to that of the real electron, or also if the volume of the electron is constant; that is Langevin's hypothesis.

This stands in contradiction with the result from §4 and with the results obtained by Lorentz in another way. This contradiction is what needs to be explained.

Before bringing up this explanation, I observe that, whatever the hypothesis adopted we will have:

$$H = A + B - C = \frac{l}{k}(A' + B'),$$

or, because $C' = 0$,

$$H = \frac{l}{k} H'. \tag{3}$$

We can compare this result for the equation $J = J'$ obtained in §3.

We have in fact:

$$J = \int H dt, \quad J' = \int H' dt'.$$

We will observe the state of the system depends only on $x + \varepsilon t$, y and z, meaning x', y', z', and that we have:

$$t' = \frac{l}{k} t + \varepsilon x', \quad dt' = \frac{l}{k} dt. \tag{4}$$

By combining equations (3) and (4), it is found that $J = J'$.

We place ourselves in an arbitrary hypothesis which could be either that of Lorentz, Abraham, or Langevin, or an intermediate hypothesis.

Let

$$r, \quad \theta r, \quad \theta r$$

be the three axes of the real electron; those of the ideal electron will be:

$$klr, \quad \theta lr, \quad \theta lr.$$

Then $A' + B'$ will be the electrostatic energy due to an ellipsoid having axes klr, θlr, θlr.

Let us assume that the electricity spreads over the surface of the electron like that of a conductor or spreads uniformly inside this electron. This energy will be of the form:

$$A' + B' = \frac{\varphi\left(\frac{\theta}{k}\right)}{klr},$$

where φ is a known function.

Abraham's hypothesis consists of assuming:

$$r = \text{const.} \quad \theta = 1\,.$$

Lorentz's hypothesis:

$$l = 1, \quad kr = \text{const.} \quad \theta = k.$$

Langevin's hypothesis:

$$l = k^{-1/3}, \quad k = \theta, \quad klr = \text{const.}$$

Next find:

$$H = \frac{\varphi\left(\frac{\theta}{k}\right)}{k^2 r}.$$

Abraham found, up to differences of notation (Göttinger Nachrichten, 1902, p. 37):

$$H = \frac{a}{r}\,\frac{1-\varepsilon^2}{\varepsilon}\log\frac{1+\varepsilon}{1-\varepsilon},$$

where a is a constant. Now, in Abraham's hypothesis, $\theta = 1$; therefore:

$$\varphi\left(\frac{1}{k}\right) = ak^2 \frac{1-\varepsilon^2}{\varepsilon} \log \frac{1+\varepsilon}{1-\varepsilon} = \frac{a}{\varepsilon} \log \frac{1+\varepsilon}{1-\varepsilon} \tag{5}$$

which defines the function φ.

Having laid that out, imagine that the electron is subject to a binding force, such that there is a relation between r and θ; under Lorentz's hypothesis, this relation would be $\theta r = \text{const.}$, in Langevin's $\theta^2 r^3 = \text{const.}$. We will assume more generally:

$$r = b\theta^m,$$

where b is a constant; hence:

$$H = \frac{1}{bk^2}\theta^{-m}\varphi\left(\frac{\theta}{k}\right).$$

What shape will the electron take when the velocity becomes $-\varepsilon t$,[10] *if it is assumed that the only forces involved are binding forces?* That shape will be defined by the equality:

$$\frac{\partial H}{\partial \theta} = 0, \tag{6}$$

or

$$-m\theta^{-m-1}\varphi + \theta^{-m}k^{-1}\varphi' = 0,$$

or

$$\frac{\varphi'}{\varphi} = \frac{mk}{\theta}.$$

If we want there to be a balance such that $\theta = k$, it must be that for $\theta/k = 1$, the logarithmic derivative of φ is equal to m.

If we expand $1/k$ and the right-hand side of (5) in powers of ε, equation (5) becomes:

$$\varphi\left(1 - \frac{\varepsilon^2}{2}\right) = a\left(1 + \frac{\varepsilon^2}{3}\right),$$

by neglecting higher powers of ε.

By differentiating, it follows that:

$$-\varepsilon\varphi'\left(1-\frac{\varepsilon^2}{2}\right)=\frac{2}{3}\varepsilon a.$$

For $\varepsilon = 0$, meaning when the argument of φ is equal to 1, these equations become:

$$\varphi = a, \quad \varphi' = -\frac{2}{3}a, \quad \frac{\varphi'}{\varphi} = -\frac{2}{3}. \tag{7}$$

Therefore it must be that $m = -2/3$ as in Langevin's hypothesis.

This result must be compared with the result concerning the first equation (2) and from which, in reality, it is not different. In fact, let us assume that any element $\mathrm{d}\tau$ of the electron is subject to a force $X\mathrm{d}\tau$ parallel to the x-axis, where X is the same for all elements; we will then have, conforming to the definition of the momentum:

$$\frac{\mathrm{d}D}{\mathrm{d}t} = \int X\mathrm{d}\tau.$$

Additionally, the principle of least action gives us:

$$\delta J = \int X\delta U\mathrm{d}\tau\mathrm{d}t, \quad J = \int H\mathrm{d}t, \quad \delta J = \int D\delta U\mathrm{d}t,$$

where δU is the displacement of the center of gravity of the electron; H depends on θ and ε, if it is accepted that r is related to θ by the binding equation; it then follows:

$$\delta J = \int\left(\frac{\partial H}{\partial\varepsilon}\delta\varepsilon + \frac{\partial H}{\partial\theta}\delta\theta\right)\mathrm{d}t.$$

Additionally, $\delta\varepsilon = -\frac{\mathrm{d}\delta U}{\mathrm{d}t}$; hence, by integrating by parts:

$$\int D\delta\varepsilon\mathrm{d}t = \int D\delta u\mathrm{d}t,$$

or

$$\int\left(\frac{\partial H}{\partial\varepsilon}\delta\varepsilon + \frac{\partial H}{\partial\theta}\delta\theta\right)\mathrm{d}t = \int D\delta\varepsilon\mathrm{d}t;$$

hence

$$D = \frac{\partial H}{\partial\varepsilon}, \quad \frac{\partial H}{\partial\theta} = 0.$$

But the derivative $dH/d\varepsilon$, which appears in the right-hand side of the first equation (2) is the derivative taken by assuming θ is expressed as a function of ε, such that

$$\frac{dH}{d\varepsilon} = \frac{\partial H}{\partial \varepsilon} + \frac{\partial H}{\partial \theta}\frac{d\theta}{d\varepsilon}.$$

Equation (2) is therefore equivalent to equation (6).

The conclusion is that if the electron is subject to a binding between its three axes, *and if no other force is involved apart from the binding forces*, the shape that this electron will take, when driven at a uniform speed, can only be that of the ideal electron corresponding to a sphere, or that in the case where the binding will be such that the volume is constant, as assumed in Langevin's hypothesis.

In that way we are led to state the following problem: what additional forces, other than the binding forces, would need to be involved to incorporate Lorentz's law or, more generally, any law other than that of Langevin?

The simplest hypothesis, and the first that we needed to examine, is that these additional forces derive from a special potential deriving from the three axes of the ellipsoid and consequently from θ and r; let $F(\theta, r)$ be that potential; in that case the expression for the action will be:

$$J = \int [H + F(\theta, r)]dt$$

and the equilibrium conditions will be written:

$$\frac{dH}{d\theta} + \frac{dF}{d\theta} = 0, \quad \frac{dH}{dr} + \frac{dF}{dr} = 0. \tag{8}$$

If we assume that r and θ are linked by the relationship $r = b\theta^m$, we will be able to regard r as a function of θ, consider F as only depending on θ and retain only the first equation (8) with:

$$H = \frac{\varphi}{bk^2\theta^m}, \quad \frac{dH}{d\theta} = \frac{-m\varphi}{bk^2\theta^{m+1}} + \frac{\varphi'}{bk^3\theta^m}.$$

It must be, for $k = \theta$, that equation (8) is satisfied, which gives, in light of equations (7):

$$\frac{dF}{d\theta} = \frac{ma}{b\theta^{m+3}} + \frac{2}{3}\frac{a}{b\theta^{m+3}},$$

hence:

$$F = \frac{-a}{b\theta^{m+2}} \frac{m + \frac{2}{3}}{m + 2}$$

and in the Lorentz hypothesis, where $m = -1$:

$$F = \frac{a}{3b\theta}.$$

Now let us assume that there is *no* binding and, regarding r and θ as two independent variables, we retain the two equations (8); it will follow:

$$H = \frac{\varphi}{k^2 r}, \quad \frac{\mathrm{d}H}{\mathrm{d}\theta} = \frac{\varphi'}{k^3 r}, \quad \frac{\mathrm{d}H}{\mathrm{d}r} = \frac{-\varphi}{k^3 r^2}.$$

Equations (8) will have to be satisfied for $k = \theta$, $r = b\theta^m$; which gives:

$$\frac{\mathrm{d}F}{\mathrm{d}r} = \frac{a}{b^2\theta^{2m+2}}, \quad \frac{\mathrm{d}F}{\mathrm{d}\theta} = \frac{2}{3}\frac{a}{b\theta^{m+3}}. \tag{9}$$

One of the ways to satisfy these conditions is to set:

$$F = Ar^\alpha \theta^\beta, \tag{10}$$

where A, α and β are constants; equations (9) must be satisfied for $k = \theta$ and $r = b\theta^m$, which gives:

$$A\alpha b^{\alpha-1}\theta^{m\alpha-m+\beta} = \frac{a}{b^2\theta^{2m+2}}, \quad A\beta b^\alpha \theta^{m\alpha+\beta-1} = \frac{2}{3}\frac{a}{b\theta^{m+3}}.$$

By identification, it follows:

$$\alpha = 3\gamma, \quad \beta = 2\gamma, \quad \gamma = -\frac{m+2}{3m+2}, \quad A = \frac{a}{\alpha b^{\alpha+1}}. \tag{11}$$

But the volume of the ellipsoid is proportional to $r^3\theta^2$, such that the additional potential is proportional of the volume of the electron to the power γ.

In the Lorentz hypothesis, $m = -1$ and $\gamma = 1$.

This is therefore the Lorentz hypothesis on the condition of adding an additional potential proportional to the volume of the electron.

Langevin's hypothesis corresponds to $\gamma = \infty$.

§7 – Quasi-Stationary Motion

It remains to be seen whether this hypothesis about the contraction of electrons reflects the impossibility of showing absolute motion; I will start by studying quasi-stationary motion of an electron which is isolated or only subject to the action of other distant electrons.

It is known that motion is called quasi-stationary motion when the changes in velocity are sufficiently slow that the magnetic and electrical energies due to the motion of the electron differ slightly from what they would be in uniform motion; it is also known that it is by starting from this concept of quasi-stationary motion that Abraham arrived at the concept of transverse and longitudinal electromagnetic masses.

I believe I have to be more specific. Let H be our action per unit time:

$$H = \frac{1}{2}\int\left(\sum f^2 - \sum \alpha^2\right)\mathrm{d}\tau$$

where for the moment we only consider the electric and magnetic fields due to the motion of an isolated electron. In the previous section, considering the motion to be uniform, we regarded H as dependent on the speed ξ, η, ζ of the center of gravity of the electron (in the previous section, these three components had values $-\varepsilon, 0, 0$) and the parameters r and θ which define the shape of the electron.

But, if the motion is no longer uniform, H will depend not only on the values ξ, η, ζ, r, θ at the moment being considered, but on the values of the same quantities at other moments which will be different from them by quantities of the same order as the time taken by light to go from one point of the electron to another; in other words, H will depend not only on ξ, η, ζ, r, θ but also on their derivatives with respect to time of all orders.

Hence, the motion will be called quasi-stationary when the partial derivatives of H with respect to the successive derivatives of ξ, η, ζ, r, θ will be negligible compared to the partial derivatives of H with respect to the quantities ξ, η, ζ, r, θ themselves.

The equations of a similar motion could be written:

$$\begin{aligned}&\frac{\mathrm{d}H}{\mathrm{d}\theta}+\frac{\mathrm{d}F}{\mathrm{d}\theta}=\frac{\mathrm{d}H}{\mathrm{d}r}+\frac{\mathrm{d}F}{\mathrm{d}r}=0,\\ &\frac{\mathrm{d}}{\mathrm{d}t}\frac{\mathrm{d}H}{\mathrm{d}\xi}=-\int X\mathrm{d}\tau,\quad \frac{\mathrm{d}}{\mathrm{d}t}\frac{\mathrm{d}H}{\mathrm{d}\eta}=-\int Y\mathrm{d}\tau,\quad \frac{\mathrm{d}}{\mathrm{d}t}\frac{\mathrm{d}H}{\mathrm{d}\zeta}=-\int Z\mathrm{d}\tau.\end{aligned} \tag{1}$$

In these equations, F has the same meaning as in the previous section; X, Y, Z are the components of the force which acts on the electron: this force is solely due to the electric and magnetic fields produced by *other* electrons.

We observed that H depends on ξ, η, ζ only through the combination

$$V = \sqrt{\xi^2 + \eta^2 + \zeta^2},$$

meaning the magnitude of the velocity; by again calling D the momentum, it follows:

$$\frac{dH}{d\xi} = \frac{dH}{dV}\frac{\xi}{V} = -D\frac{\xi}{V}$$

hence:

$$-\frac{d}{dt}\frac{dH}{d\xi} = \frac{D}{V}\frac{d\xi}{dt} - D\frac{\xi}{V^2}\frac{dV}{dt} + \frac{dD}{dV}\frac{\xi}{V}\frac{dV}{dt}, \tag{2}$$

$$-\frac{d}{dt}\frac{dH}{d\eta} = \frac{D}{V}\frac{d\eta}{dt} - D\frac{\eta}{V^2}\frac{dV}{dt} + \frac{dD}{dV}\frac{\eta}{V}\frac{dV}{dt}, \tag{2'}$$

with

$$V\frac{dV}{dt} = \sum \xi\frac{d\xi}{dt}. \tag{3}$$

If we take the x-axis as the current direction of the velocity, it follows:

$$\xi = V, \quad \eta = \zeta = 0, \quad \frac{d\xi}{dt} = \frac{dV}{dt};$$

equations (2) and (2′) become:

$$-\frac{d}{dt}\frac{dH}{d\xi} = \frac{dD}{dV}\frac{d\xi}{dt}, \qquad -\frac{d}{dt}\frac{dH}{d\eta} = \frac{D}{V}\frac{d\eta}{dt}$$

and the three equations (1) become:

$$\frac{dD}{dV}\frac{d\xi}{dt} = \int X d\tau, \quad \frac{D}{V}\frac{d\eta}{dt} = \int Y d\tau, \quad \frac{D}{V}\frac{d\zeta}{dt} = \int Z d\tau. \tag{4}$$

This is why Abraham gave dD/dV the name *longitudinal mass* and D/V the name *transverse mass*; recall that $D = dH/dV$.

In Lorentz's hypothesis, we have:

$$D = -\frac{dH}{dV} = -\frac{\partial H}{\partial V},$$

where $\partial H/\partial V$ represents the derivative with respect V, after r and θ have been replaced by their values as a function of V drawn from the first two equations (1); and additionally it follows after this substitution,

$$H = +A\sqrt{1 - V^2}.$$

We will now choose the units such that the constant factor A is equal to 1, and I set $\sqrt{1 - V^2} = h$, hence:

$$H = +h, \quad D = \frac{V}{h}, \quad \frac{\mathrm{d}D}{\mathrm{d}V} = h^{-3}, \quad \frac{\mathrm{d}D}{\mathrm{d}V}\frac{1}{V^2} - \frac{D}{V^3} = h^{-3}.$$

We will also set:

$$M = V\frac{\mathrm{d}V}{\mathrm{d}t} = \sum \xi \frac{\mathrm{d}\xi}{\mathrm{d}t}, \quad X_1 = \int X\mathrm{d}\tau$$

and we will find for the equation of quasi-stationary motion:

$$h^{-1}\frac{\mathrm{d}\xi}{\mathrm{d}t} + h^{-3}\xi M = X_1. \tag{5}$$

Let's look at what becomes of these equations under the Lorentz transformation. We will set $1 + \xi\varepsilon = \mu$, and we will first have:

$$\mu\xi' = \xi + \varepsilon, \quad \mu\eta' = \frac{\eta}{k}, \quad \mu\zeta' = \frac{\zeta}{k},$$

from which it is easy to find

$$\mu h' = \frac{h}{k}.$$

We also have

$$\mathrm{d}t' = k\mu\mathrm{d}t,$$

hence:

$$\frac{\mathrm{d}\xi'}{\mathrm{d}t'} = \frac{\mathrm{d}\xi}{\mathrm{d}t}\frac{1}{k^3\mu^3}, \quad \frac{\mathrm{d}\eta'}{\mathrm{d}t'} = \frac{\mathrm{d}\eta}{\mathrm{d}t}\frac{1}{k^2\mu^2} - \frac{\mathrm{d}\xi}{\mathrm{d}t}\frac{\eta\varepsilon}{k^2\mu^3}, \quad \frac{\mathrm{d}\zeta'}{\mathrm{d}t'} = \frac{\mathrm{d}\zeta}{\mathrm{d}t}\frac{1}{k^2\mu^2} - \frac{\mathrm{d}\xi}{\mathrm{d}t}\frac{\zeta\varepsilon}{k^2\mu^3},$$

and again:

$$M' = \frac{\mathrm{d}\xi}{\mathrm{d}t}\frac{\varepsilon h^2}{k^3\mu^4} + \frac{M}{k^3\mu^3}$$

and

$$h'^{-1}\frac{\mathrm{d}\xi'}{\mathrm{d}t'}+h'^{-3}\xi' M'=\left[h^{-1}\frac{\mathrm{d}\xi}{\mathrm{d}t}+h^{-3}(\xi+\varepsilon)M\right]\mu^{-1}, \tag{6}$$

$$h'^{-1}\frac{\mathrm{d}\eta'}{\mathrm{d}t'}+h'^{-3}\eta' M'=\left(h^{-1}\frac{\mathrm{d}\eta}{\mathrm{d}t}+h^{-3}\eta M\right)\mu^{-1}h^{-1}. \tag{7}$$

Let us now refer to equations (11′) from §1; there X_1, Y_1, Z_1 can be regarded as having the same meaning as in equations (5). Also, we have $l=1$ and $\rho'/\rho=k\mu$; these equations therefore become:

$$\begin{aligned} X_1' &= \mu^{-1}\left(X_1+\varepsilon\sum X_1\xi\right),\\ Y_1' &= k^{-1}\mu^{-1}Y_1. \end{aligned} \tag{8}$$

When we calculate $\sum X_1\xi$ using equations (5), we will find:

$$\sum X_1\xi=h^{-3}M,$$

hence:

$$\begin{aligned} X_1' &= \mu^{-1}\left(X_1+\varepsilon h^{-3}M\right),\\ Y_1' &= k^{-1}\mu^{-1}Y_1. \end{aligned} \tag{9}$$

By comparing equations (5), (6), (7) and (9), we finally find:

$$\begin{aligned} h'^{-1}\frac{\mathrm{d}\xi'}{\mathrm{d}t'}+h'^{-3}\xi' M' &= X_1',\\ h'^{-1}\frac{\mathrm{d}\eta'}{\mathrm{d}t'}+h'^{-3}\eta' M' &= Y_1', \end{aligned} \tag{10}$$

which shows that the equations of quasi-stationary motion are unaltered by the Lorentz transformation, but that does not yet prove that Lorentz's hypothesis is the only one which leads to this result.

To establish that point, we are going to restrict ourselves, as Lorentz did, to some specific cases which will obviously be sufficient for us to prove a negative proposition.

How are we first going to extend the hypothesis on which the previous calculation rests:

1) Instead of assuming $l=1$ in the Lorentz transformation, we will assume that l is arbitrary.
2) Instead of assuming that F is proportional to the volume, and consequently that H is proportional to h, we are going to assume that F is an arbitrary function of θ and r, such that (after having replaced θ and r by their values as functions of V, drawn from the first two equations (1)) H is an arbitrary function of V.

I first note that, if it is assumed that $H = h$, one should in fact have $l = 1$; and in fact equations (6) and (7) will remain, except that the left-hand side will be multiplied by $1/l$; equations (9) also, except that the right-hand sides will be multiplied by $1/l^2$; and finally equations (10) except that the right-hand side will be multiplied by $1/l$. If one wants the equations of motion to be unaltered by the Lorentz transformation meaning that equations (10) are not different from equations (5) except for the accenting of the letters, it must be assumed that:

$$l = 1.$$

Now assume that $\eta = \zeta = 0$, hence $\xi = V$ and $\frac{\mathrm{d}\xi}{\mathrm{d}t} = \frac{\mathrm{d}V}{\mathrm{d}t}$; equations (5) will take the form:

$$-\frac{\mathrm{d}}{\mathrm{d}t}\frac{\mathrm{d}H}{\mathrm{d}\xi} = \frac{\mathrm{d}D}{\mathrm{d}V}\frac{\mathrm{d}\xi}{\mathrm{d}t} = X_1, \quad -\frac{\mathrm{d}}{\mathrm{d}t}\frac{\mathrm{d}H}{\mathrm{d}\eta} = \frac{D}{V}\frac{\mathrm{d}\eta}{\mathrm{d}t} = Y_1\,. \tag{5'}$$

We can additionally set:

$$\frac{\mathrm{d}D}{\mathrm{d}V} = f(V) = f(\xi), \quad \frac{D}{V} = \varphi(V) = \varphi(\xi)\,.$$

If the equations of motion are unaltered by the Lorentz transformation, it should be that:

$$\begin{aligned}
&f(\xi)\frac{\mathrm{d}\xi}{\mathrm{d}t} = X_1,\\
&\varphi(\xi)\frac{\mathrm{d}\eta}{\mathrm{d}t} = Y_1,\\
&f(\xi')\frac{\mathrm{d}\xi'}{\mathrm{d}t'} = X_1' = l^{-2}\mu^{-1}\left(X_1 + \varepsilon\sum X_1\xi\right) = l^{-2}\mu^{-1}X_1(1+\varepsilon\xi) = l^{-2}X_1,\\
&\varphi(\xi')\frac{\mathrm{d}\eta'}{\mathrm{d}t'} = Y_1' = l^{-2}k^{-1}\mu^{-1}Y_1.
\end{aligned}$$

and consequently:

$$\begin{aligned}
f(\xi)\frac{\mathrm{d}\xi}{\mathrm{d}t} &= l^2 f(\xi')\frac{\mathrm{d}\xi'}{\mathrm{d}t'}\\
\varphi(\xi)\frac{\mathrm{d}\eta}{\mathrm{d}t} &= l^2 k\mu\varphi(\xi')\frac{\mathrm{d}\eta'}{\mathrm{d}t'}
\end{aligned} \tag{11}$$

But we have:

$$\frac{d\xi'}{dt'} = \frac{d\xi}{dt}\frac{1}{k^3\mu^3}, \quad \frac{d\eta'}{dt'} = \frac{d\eta}{dt}\frac{1}{k^2\mu^2},$$

hence:

$$f(\xi') = f\left(\frac{\xi+\varepsilon}{1+\xi\varepsilon}\right) = f(\xi)\frac{k^3\mu^3}{l^2},$$
$$\varphi(\xi') = \varphi\left(\frac{\xi+\varepsilon}{1+\xi\varepsilon}\right) = \varphi(\xi)\frac{k\mu}{l^2};$$

hence, by eliminating l^2, we find the functional equation:

$$k^2\mu^2\frac{\varphi\left(\frac{\xi+\varepsilon}{1+\xi\varepsilon}\right)}{\varphi(\xi)} = \frac{f\left(\frac{\xi+\varepsilon}{1+\xi\varepsilon}\right)}{f(\xi)},$$

or, by setting

$$\frac{\varphi(\xi)}{f(\xi)} = \Omega(\xi) = \frac{D}{V\frac{dD}{dV}},$$

this:

$$\Omega\left(\frac{\xi+\varepsilon}{1+\xi\varepsilon}\right) = \Omega(\xi)\frac{1+\varepsilon^2}{(1+\xi\varepsilon)^2}$$

equation which must be satisfied for all values of ξ and ε. For $\zeta = 0$ one finds:

$$\Omega(\varepsilon) = \Omega(0)\left(1-\varepsilon^2\right),$$

hence:

$$D = A\left(\frac{V}{\sqrt{1-V^2}}\right)^m,$$

where A is a constant, and where I made $\Omega(0) = 1/m$.

One then finds:

$$\varphi(\xi) = \frac{A}{\xi}\left(\frac{\xi}{\sqrt{1-\xi^2}}\right)^m, \quad \varphi(\xi') = \frac{A}{\xi}\left(\frac{\xi+\varepsilon}{\sqrt{1-\xi^2}\sqrt{1-\varepsilon^2}}\right)^m.$$

However $\varphi(\xi') = \varphi(\xi)k\mu/l^2$, so it follows:[11]

$$(\xi+\varepsilon)^{m-1}\left(1-\varepsilon^2\right)^{-\frac{m}{2}}=-\xi^{m-1}\left(1-\varepsilon^2\right)^{-\frac{1}{2}}l^{-2}.$$

As l must only depend on ε (because, if there are several electrons, l must be the same for all electrons whose velocities ξ can be different), this identity can only hold if one has:

$$m=1, \quad l=1\,.$$

Thus Lorentz's hypothesis is the only one which is compatible with the impossibility of showing absolute motion; if this impossibility is accepted, it must be accepted that moving electrons contract so as to become ellipsoids of revolution two axes of which remain constant; the existence of an additional potential proportional to the volume of the electron also has to be accepted, as we showed in the previous section.

Lorentz's analysis is therefore found to be fully confirmed, but we can do better by observing the true reason for the fact we are dealing with; this reason must be sought in the considerations from §4. *The transformations which do not change the equations of motion must form a group and that can occur only if* $l=1$. Since we must not be able to recognize whether an electron is at rest or in absolute motion, it must be that when it is in motion it experiences a deformation which must be precisely that which the corresponding transformation of the group demands of it.

§8 – Arbitrary Motion

The previous results only apply to quasi-stationary motion, but it is easy to extend them to the general case; it is sufficient to apply the principles from §3, meaning to start with the principle of least action.

It is appropriate to add to the expression for the action:

$$J=\int\left(\frac{\sum f^2}{2}-\frac{\sum\alpha^2}{2}\right)\mathrm{d}t\mathrm{d}\tau,$$

a term representing the additional potential F from §6; this term will obviously take the form:

$$J_1=\int\sum(F)\mathrm{d}t,$$

where $\sum(F)$ represents the sum of the additional potentials due to the various electrons, where each of them is proportional to the volume of the corresponding electron.

I'm writing (F) between parentheses so as not to confuse it with the vector F, G, H.

The total action is then $J + J_1$. We saw in §3 that J is unchanged by the Lorentz transformation; it now needs to be shown that the same is true of J_1.

For one of the electrons, it holds that:

$$(F) = \omega_0 \tau,$$

where ω_0 is a coefficient specific to the electron and τ is its volume; I can then write:

$$\sum(F) = \int \omega_0 \mathrm{d}\tau,$$

where the integral has to extend to all space, but does so in a way that the coefficient ω_0 is zero outside of the electrons and that inside of each electron it is equal to the special coefficient for that electron. It then follows:

$$J_1 = \int \omega_0 \mathrm{d}\tau \mathrm{d}t,$$

and after the Lorentz transformation:

$$J_1' = \int \omega_0' \mathrm{d}\tau' \mathrm{d}t',$$

Hence $\omega_0 = \omega_0'$; because if the point belongs to an electron, the corresponding point after the Lorentz transformation still belongs to the same electron. Further, we found in §3:

$$\mathrm{d}\tau' \mathrm{d}t' = l^4 \mathrm{d}\tau \mathrm{d}t$$

and, because we now assume $l = 1$,

$$\mathrm{d}\tau' \mathrm{d}t' = \mathrm{d}\tau \mathrm{d}t.$$

We then have:

$$J_1 = J_1'.$$

Which was to be proven

The theorem is therefore general; at the same time, it gives us a solution to the question that we asked at the end of §1: to find additional forces unchanged by the Lorentz transformation. The additional potential (F) satisfies that condition.

We can therefore generalize the results stated at the end of §1 and write:

If the inertia of the electrons is exclusively of electromagnetic origin, if they are only subject to forces of electromagnetic origin, or to forces which give rise to the additional potential (F), *no experiment will be able to show absolute motion.*

What then are these forces which give rise to the potential (F)? They can obviously be compared to a pressure which governs inside the electron; everything happens as if each electron had a hollow capacitor subject to a constant internal pressure (independent of the volume); the work due to such a pressure would obviously be proportional to changes in the volume.

I must again observe that this pressure is negative. Let's go back to equation (10) from §6, which in Lorentz's hypothesis is written:

$$F = Ar^3\theta^2;$$

equations (11) from §6 will give us:

$$A = \frac{a}{3b^4}.$$

Our pressure is equal to A, up to a constant coefficient, which furthermore is negative.

We now evaluate the electron mass; I want to speak of the "experimental mass", meaning the mass for small velocities; we have (see §6):

$$H = \frac{\varphi\left(\frac{\theta}{k}\right)}{k^2 r}, \quad \theta = k, \quad \varphi = a, \quad \theta r = b;$$

hence

$$H = \frac{a}{bk} = \frac{a}{b}\sqrt{1 - V^2}.$$

For very small V, I may write:

$$H = \frac{a}{b}\left(1 - \frac{1}{2}V^2\right),$$

such that the mass, both longitudinal and transverse, will be a/b.

However, a is a numeric constant; this shows that: *the pressure to which our additional potential gives rise is proportional to the fourth power of the experimental mass of the electron.*

Since Newtonian attraction is proportional to this experimental mass, one is tempted to conclude that there is some relation between the cause which gives rise to gravitation and that which gives rise to this additional potential.

§9 – Hypotheses on Gravitation

Thus Lorentz's theory would fully explain the impossibility of showing absolute motion, if all the forces were of electromagnetic origin.

But there are other forces to which an electromagnetic origin cannot be attributed, such as gravitation for example. It can in fact happen that two systems of bodies produce equivalent electromagnetic fields, meaning exerting the same action on charged bodies and on currents and that however these two systems do not exert the same gravitational action on Newtonian masses. The gravitational field is therefore distinct from the electromagnetic field. Lorentz was therefore compelled to extend his hypothesis by assuming that *forces of any origin, and in particular gravitation, are affected by a translation* (or, if you prefer, by the Lorentz transformation) *in the same way as the electromagnetic forces.*

It is now appropriate to go into the details and examine more closely this hypothesis. If we want the Newtonian force to be affected in the same way by the Lorentz transformation, we can no longer allow that this force depends solely on the relative position of the attracting body and the attracted body at the moment under consideration. It will have to additionally depend on the velocity of both bodies. And that is not all: it will be natural to assume that the force which acts on the attracted body at the moment t depends on the position and velocity of this body at this moment t; but it will additionally depend on the position and velocity of the *attracting* body, not at the moment t, but at *an earlier moment*, as if the gravitation had taken some time to propagate.

We therefore consider the position of the attracted body at the moment t_0 and let, at this moment, x_0, y_0, z_0 be its coordinates, and ξ, η, ζ be the components of its velocity; we will additionally consider the attracting body at the corresponding moment $t_0 + t$ and let, at that moment, $x_0 + x$, $y_0 + y$, $z_0 + z$ be its coordinates; and ξ_1, η_1, ζ_1 be the components of its velocity.

We will first have to have a relation

$$\varphi(t, x, y, z, \xi, \eta, \zeta, \xi_1, \eta_1, \zeta_1) = 0 \tag{1}$$

in order to define the time t. This relation will define the laws of propagation of the gravitational action (I am in no way imposing the condition that the propagation occurs with the same speed in all directions).

Now let X_1, Y_1, Z_1 be the three components of the action exerted at the moment t on the attracted body; it is a matter of expressing X_1, Y_1, Z_1 as functions of

$$t, x, y, z, \xi, \eta, \zeta, \xi_1, \eta_1, \zeta_1. \tag{2}$$

What are the conditions to be satisfied?

1) The condition (1) must not be changed by transformations from the Lorentz group.

2) The components X_1, Y_1, Z_1 will have to be affected by the Lorentz transformations in the same way as the electromagnetic forces designated by the same letters, meaning according to equations (11′) from §1.
3) When the two bodies are at rest, the ordinary law of attraction must be restored.

It needs to be remarked that in this last case, the relation (1) would disappear, because time t doesn't play any role if the two bodies are at rest.

The problem thus stated is obviously indeterminate. We will therefore seek to satisfy other additional conditions as much as possible:

4) Since astronomical observations do not seem to show any meaningful deviation from Newton's law, we will choose the solution which deviates the least from this law, for small velocities of the two bodies.
5) We will make every effort to situate ourselves such that t is always negative; if in fact it is thought that the effect of gravitation requires some time to propagate, it would be more difficult to understand how this effect could depend on the position *not yet reached* by the attracting body.

There is a case where the indeterminacy of this problem disappears; it is the one where both bodies are it relative rest with each other, meaning where:

$$\xi = \xi_1, \quad \eta = \eta_1, \quad \zeta = \zeta_1;$$

this is therefore the case that we are going to examine first, by assuming that these velocities are constant, such that both bodies are driven in a shared, straight and uniform translational motion.

We can assume that the x-axis was taken parallel to this translation such that $\eta = \zeta = 0$; and we will take $\varepsilon = -\xi$.

If we apply the Lorentz transformation under these conditions, then after the transformation both bodies will be at rest and it will be that:

$$\xi' = \eta' = \zeta' = 0.$$

Then the components will X'_1, Y'_1, Z'_1 will have to be determined by Newton's law and it will hold, up to a constant factor, that:

$$X'_1 = -\frac{x'}{r'^3}, \quad Y'_1 = -\frac{y'}{r'^3}, \quad Z'_1 = -\frac{z'}{r'^3}, \tag{3}$$
$$r'^2 = x'^2 + y'^2 + z'^2.$$

But, according to §1, we have:

$$\begin{aligned}
&x' = k(x + \varepsilon t), \quad y' = y, \quad z' = z \quad t' = k(t + \varepsilon x),\\
&\frac{\rho'}{\rho} = k(1 + \xi\varepsilon) = k\left(1 - \varepsilon^2\right) = \frac{1}{k}, \quad \sum X_1\xi = -X_1\varepsilon,\\
&X_1' = k\frac{\rho}{\rho'}\left(X_1 + \varepsilon\sum X_1\xi\right) = k^2X_1\left(1 - \varepsilon^2\right) = X_1,\\
&Y_1' = \frac{\rho}{\rho'}Y_1 = kY_1,\\
&Z_1' = kZ_1.
\end{aligned}$$

Additionally

$$x + \varepsilon t = x - \xi t, \quad r'^2 = k^2(x - \xi t)^2 + y^2 + z^2$$

and

$$X_1 = \frac{-k(x - \xi t)}{r'^3}, \quad Y_1 = \frac{-y}{kr'^3}, \quad Z_1 = \frac{-z}{kr'^3}; \tag{4}$$

which can be written:

$$X_1 = \frac{\mathrm{d}V}{\mathrm{d}x}, \quad Y_1 = \frac{\mathrm{d}V}{\mathrm{d}y}, \quad Z_1 = \frac{\mathrm{d}V}{\mathrm{d}z}, \quad V = \frac{1}{kr'}. \tag{4$'$}$$

At first it seems that the indeterminacy remains, because we have made no assumption about the value of t, meaning the speed of transmission and additionally that x is a function of t, but it is easy to see that $x - \xi t$, y, z, which alone appear in our formulas, do not depend on t.

It can be seen that if the two bodies are simply driven in a shared translation, the force which acts are the attracted body is normal to an ellipsoid that has the attracting body at its center.

To go farther, we need to seek the *invariants of the Lorentz group.*

We know that the substitutions from this group (with the assumption $l = 1$) are linear substitutions which do not change the quadratic form:

$$x^2 + y^2 + z^2 - t^2.$$

Let us also set:

$$\begin{aligned}
\xi &= \frac{\delta x}{\delta t}, \quad & \eta &= \frac{\delta y}{\delta t}, \quad & \zeta &= \frac{\delta z}{\delta t},\\
\xi_1 &= \frac{\delta_1 x}{\delta_1 t}, \quad & \eta_1 &= \frac{\delta_1 y}{\delta_1 t} \quad & \zeta_1 &= \frac{\delta_1 z}{\delta_1 t}
\end{aligned}$$

we see that the Lorentz transformation will have the effect of making δx, δy, δz, δt and $\delta_1 x$, $\delta_1 y$, $\delta_1 z$, $\delta_1 t$ undergo the same linear substitutions as x, y, z, t.

Let us regard

$$\begin{array}{cccc} x, & y, & z, & t\sqrt{-1}, \\ \delta x, & \delta y, & \delta z, & \delta t\sqrt{-1}, \\ \delta_1 x, & \delta_1 y, & \delta_1 z, & t\sqrt{-1}, \end{array}$$

as the coordinates of three points P, P', P'' in four-dimensional space. We have seen that the Lorentz transformation is solely a rotation of this space around the origin, which is regarded as fixed. We will have no other distinct invariants besides the six distances of the three points P, P', P'' between each other and the origin, or, if one prefers besides the two expressions:

$$x^2 + y^2 + z^2 - t^2, \quad x\delta x + y\delta y + z\delta z - t\delta t\,,$$

or the four expressions of the same form that are deduced by arbitrarily permuting the three points P, P', P''.

But what we are looking for are functions of 10 variables (2) which are invariants; we therefore need to look among the combinations of our six invariants for those which only depend on these 10 variables, meaning those which are homogeneous of zeroth degree both in δx, δy, δz, δt and in $\delta_1 x$, $\delta_1 y$, $\delta_1 z$, $\delta_1 t$. Thus we will be left with four distinct invariants which are:

$$\sum x^2 - t^2, \quad \frac{t - \sum x\xi}{\sqrt{1 - \sum \xi^2}}, \quad \frac{t - \sum x\xi}{\sqrt{1 - \sum \xi_1^2}}, \quad \frac{1 - \sum \xi\xi_1}{\sqrt{(1 - \sum \xi^2)(1 - \sum \xi_1^2)}}. \tag{5}$$

Let us now work on the transformations undergone by the components of the force; let us take up equations (11) from §1 which refer not to the force X_1, Y_1, Z_1, that we are considering here, but to the force X, Y, Z referred to the unit volume. Let us additionally set:

$$T = \sum X\xi;$$

we will see that these equations (11) can be written (with $l = 1$) as:

$$\begin{aligned} X' &= k(X + \varepsilon T), & T' &= k(T + \varepsilon X), \\ Y' &= Y, & Z' &= Z; \end{aligned} \tag{6}$$

such that X, Y, Z, T undergo the same transformation as x, y, z, t. The invariants of the group will therefore be:

$$\sum X^2 - T^2, \quad \sum Xx - Tt, \quad \sum X\delta x - T\delta t, \quad \sum X\delta_1 x - T\delta_1 t$$

But these are not the X, Y, Z that we need those are X_1, Y_1, Z_1 with

$$T_1 = \sum X_1 \xi.$$

We see that

$$\frac{X_1}{X} = \frac{Y_1}{Y} = \frac{Z_1}{Z} = \frac{1}{\rho}.$$

Therefore the Lorentz transformation will act on X_1, Y_1, Z_1, T_1, in the same way as on X, Y, Z, T with the difference that these expressions will be additionally multiplied by

$$\frac{\rho}{\rho'} = \frac{1}{k(1+\xi\varepsilon)} = \frac{\delta t}{\delta t'}.$$

Similarly, the transformation will act on ξ, η, ζ, 1, in the same way as on δx, δy, δz, δt with the difference that these expressions will be additionally multiplied by the *same* factor:

$$\frac{\delta t}{\delta t'} = \frac{1}{k(1+\xi\varepsilon)}.$$

Let us then consider X, Y, Z, $T\sqrt{-1}$ as coordinates of a fourth point Q;
then the invariants will be functions of the mutual distances between five points

$$0, P, P', P'', Q$$

and among these functions we will have to keep only those which are homogeneous of zeroth degree both in

$$X, Y, Z, T, \delta x, \delta y, \delta z, \delta t$$

(variables that can next be replaced with X_1, Y_1, Z_1, T_1, ξ, η, ζ, 1), and also in

$$\delta_1 x, \delta_1 y, \delta_1 z, 1$$

(variables that can next be replaced by ξ_1, η_1, ζ_1, 1).

We will thus find, beyond the four invariants (5), for new distinct invariants, which are:

$$\frac{\sum X_1^2 - T_1^2}{1 - \sum \xi^2}, \quad \frac{\sum X_1 x - T_1 t}{\sqrt{1 - \sum \xi^2}}, \quad \frac{\sum X_1 \xi_1 - T_1}{\sqrt{1 - \sum \xi^2}\sqrt{1 - \sum \xi_1^2}}, \quad \frac{\sum X_1 \xi - T_1}{1 - \sum \xi^2}. \tag{7}$$

The last invariant is always zero according to the definition of T_1.
Having set that, what are the conditions to be satisfied?

1) The left-hand side of the relation (1), which defines the propagation velocity, must be a function of the four invariants (5).

One can obviously make a load of hypothesis; we will consider only two of them:

A) One could have:

$$\sum x^2 - t^2 = r^2 - t^2 = 0,$$

where $t = \pm r$, and, because t must be negative, $t = -r$. This is the same as saying the speed of propagation is equal to that of light. At first it seems this hypothesis must be rejected without examination. Laplace in fact showed that the propagation is either instantaneous or much faster than that of light. But Laplace had examined the hypothesis of the finite propagation speed, everything else remaining the same; here, in contrast, this hypothesis is complicated by many others and it could happen that there might be a more or less perfect compensation between them, like those for which the applications of the Lorentz transformation have already given us many examples.

B) One could have:

$$\frac{t - \sum x\xi_1}{\sqrt{1 - \sum \xi_1^2}} = 0 \quad t = \sum x\xi_1 .$$

The speed of propagation is then much faster than that of light; but in some cases t could be negative, which, as we already stated, hardly seems admissible. *We will therefore keep hypothesis A.*

2) The four invariants (7) must be a function of the invariants (5).
3) When both bodies are at absolute rest, X, Y, Z must have the value deduced from Newton's law and when they are at relative rest, the value deduced from equations (4).

In the scenario of absolute rest, the first two invariants (7) must reduce to

$$\sum X_1^2, \quad \sum X_1 x,$$

or, by Newton's law, to

$$\frac{1}{r^4}, \quad -\frac{1}{r};$$

on the other hand, in scenario (A), the second and third invariants (5) become:

$$\frac{-r - \sum x\xi}{\sqrt{1 - \sum \xi^2}}, \quad \frac{-r - \sum x\xi_1}{\sqrt{1 - \sum \xi_1^2}},$$

meaning, for absolute rest:

$$-r, \quad -r.$$

We can therefore allow, *for example*, that the first two invariants (4) reduce to

$$\frac{\left(1 - \sum \xi_1^2\right)^2}{\left(r + \sum x\xi_1\right)^4}, \quad -\frac{\sqrt{1 - \sum \xi^2}}{r + \sum x\xi_1};$$

but other combinations are possible.

The choice must be made between these combinations, and, additionally, in order to define X_1, Y_1, Z_1 a third equation is needed. For such a choice, we need to make an effort to come as close as possible to Newton's law. Let us therefore look at what happens when (still keeping $t = -r$) the squares of the velocities ξ, η, etc. are neglected. The four invariants (5) then become:

$$0, \quad -r - \sum x\xi, \quad -r - \sum x\xi_1, \quad 1$$

and the four invariants (7):

$$\sum X_1^2, \quad \sum X_1(x + \xi r), \quad \sum X_1(\xi_1 - \xi), \quad 0.$$

But to be able to make a comparison with Newton's law, another transformation is necessary; here $x_0 + x$, $y_0 + y$, $z_0 + z$ represent the coordinates of the attracting body at the moment $t_0 + t$, and $r = \sqrt{\sum x^2}$; in Newton's law, $x_0 + x_1$, $y_0 + y_1$, $z_0 + z_1$ of the attracting body at the moment t_0, and the distance $r = \sqrt{\sum x_1^2}$ need to be considered.

We can neglect the square of the time t needed for propagation and consequently proceed as if the motion were uniform; we then have:

$$x = x_1 + \xi_1 t, \quad y = y_1 + \eta_1 t, \quad z = z_1 + \zeta_1 t,$$
$$r(r - r_1) = \sum x\xi_1 t;$$

or, because $t = -r$,

$$x = x_1 - \xi_1 r, \quad y = y_1 - \eta_1 r, \quad z = z_1 - \zeta_1 r, \quad r = r_1 - \sum x\xi_1\,;$$

such that our four invariants (5) become:

$$0, \quad -r_1 + \sum x(\xi_1 - \xi), \quad -r_1, \quad 1$$

and our four invariants (7) become:

$$\sum X_1^2, \quad \sum X_1[x_1 + (\xi - \xi_1)r_1], \quad \sum X_1(\xi_1 - \xi) \quad 0.$$

In the second of these expressions, I wrote r_1 instead of r, because r is multiplied by $\xi - \xi_1$ and because I neglected the square of ξ.

On the other hand, Newton's law would give us, for these four invariants (7),

$$\frac{1}{r_1^4}, \quad -\frac{1}{r_1} - \frac{\sum x_1(\xi - \xi_1)}{r_1^2}, \quad \frac{\sum x_1(\xi - \xi_1)}{r_1^3}, \quad 0\,.$$

If therefore we call A and B the second and third of the invariants (5) and M, N, P the first three invariants (7), we will satisfy Newton's law, up to terms of order of the square of the velocities, by making:

$$M = \frac{1}{B^4}, \quad N = \frac{+A}{B^2}, \quad P = \frac{A - B}{B^3}. \tag{8}$$

This solution is not unique. In fact, let C be the fourth invariant (5), $C - 1$ is of order of the square of ξ, and it is the same for $(A - B)^2$.

We could therefore add a term formed from $C - 1$ multiplied by an arbitrary function of A, B, C and a term formed from $(A - B)^2$ also multiplied by a function of A, B, C to the right hand side of each of equations (8).

At first sight, the solution (8) seems the simplest; it cannot however be adopted. In fact, since M, N, P are functions of X_1, Y_1, Z_1 and of $T_1 = \sum X_1\xi$, the values for X_1, Y_1, Z_1 can be drawn from these three equations (8), but in some cases these values could become imaginary.

To avoid this disadvantage, we will work in another way. Let us set:

$$k_0 = \frac{1}{\sqrt{1 - \sum \xi^2}}, \quad k_1 = \frac{1}{\sqrt{1 - \sum \xi_1^2}},$$

which is justified by analogy with the notation

$$k = \frac{1}{\sqrt{1-\varepsilon^2}}$$

which appears in Lorentz's substitution.

In this case and because of the condition $-r = t$, the invariants (5) become:

$$0, \quad A = -k_0\left(r + \sum x\xi\right), \quad B = -k_1\left(r + \sum x\xi_1\right), \quad C = k_0 k_1\left(1 - \sum \xi\xi_1\right).$$

On the other hand, we see that the following system of quantities:

$$\begin{array}{cccc} x, & y, & z, & -r = t \\ k_0 X_1, & k_0 Y_1, & k_0 Z_1, & k_0 T_1 \\ k_0\xi, & k_0\eta, & k_0\zeta, & k_0 \\ k_1\xi_1, & k_1\eta_1, & k_1\zeta_1, & k_1 \end{array}$$

undergo the *same* linear substitutions when the transformations of the Lorentz group are applied to them. This leads us to set:

$$\begin{aligned} X_1 &= x\frac{\alpha}{k_0} + \xi\beta + \xi_1\frac{k_1}{k_0}\gamma, \\ Y_1 &= y\frac{\alpha}{k_0} + \eta\beta + \eta_1\frac{k_1}{k_0}\gamma, \\ Z_1 &= z\frac{\alpha}{k_0} + \zeta\beta + \zeta_1\frac{k_1}{k_0}\gamma, \\ T_1 &= -r\frac{\alpha}{k_0} + \beta + \frac{k_1}{k_0}\gamma. \end{aligned} \tag{9}$$

It is clear that if α, β, γ are invariants, X_1, Y_1, Z_1, T_1 will satisfy the fundamental condition, meaning will undergo an appropriate linear substitution under the effect of the Lorentz transformation.

But in order for these equations (9) to be compatible, it needs to be that we have:

$$\sum X_1\xi - T_1 = 0,$$

which, by replacing X_1, Y_1, Z_1, T_1 by their values (9) and by multiplying by k_0^2, becomes:

$$-A\alpha - \beta - C\gamma = 0. \tag{10}$$

What we want is that if the squares of the velocities ξ, etc. and also the products of the accelerations by distances are neglected compared to the square the speed of light as we have done above, then the values X_1, Y_1, Z_1 continue to satisfy Newton's laws.

We can take:

$$\beta = 0, \quad \gamma = -\frac{A\alpha}{C}.$$

With the adopted order of approximation, it follows:

$$k_0 = k_1 = 1, \quad C = 1, \quad A = -r_1 + \sum x(\xi_1 - \xi), \quad B = -r_1.$$
$$x = x_1 + \xi_1 t = x_1 - \xi_1 r.$$

The first equation (9) then becomes:

$$X_1 = \alpha(x - A\xi_1)$$

But if the square of ξ is neglected, $A\xi_1$ can be replaced by $-r_1\xi_1$ or even.by – r, which gives:

$$X_1 = \alpha(x + \xi_1 r) = \alpha x_1.$$

Newton's law would give:

$$X_1 = -\frac{x_1}{r_1^3}.$$

For the invariant α, we need to choose the value which reduces to $-1/r_1^3$ for the chosen order of approximation, meaning $1/B^3$. Equations (9) will become:

$$\begin{aligned}
X_1 &= \frac{x}{k_0 B^3} - \xi_1 \frac{k_1}{k_0} \frac{A}{B^3 C}, \\
Y_1 &= \frac{y}{k_0 B^3} - \eta_1 \frac{k_1}{k_0} \frac{A}{B^3 C}, \\
Z_1 &= \frac{z}{k_0 B^3} - \zeta_1 \frac{k_1}{k_0} \frac{A}{B^3 C}, \\
T_1 &= -\frac{r}{k_0 B^3} - \frac{k_1}{k_0} \frac{A}{B^3 C}.
\end{aligned} \tag{11}$$

We can first see that the corrected attraction is made up of two components: one parallel to the vector which joins the positions of the two bodies and the other parallel to the velocity of the attracting body.

We recall that when we speak of the position or the velocity of the attracting body, it is about its position or its velocity at the moment when the gravitational wave leaves it; for the attracted body it is instead about its position or its velocity at the moment when the gravitational wave reaches it, with the assumption that this wave propagates at the speed of light.

I think that it would be premature to try to move the discussion of these formulas farther; I will therefore limit myself to a few remarks.

1) The solutions (11) are not unique; $1/B^3$, which enters as a factor throughout, can in fact be replaced by

$$\frac{1}{B^3} + (C-1)f_1(A,B,C) + (A-B)^2 f_2(A,B,C),$$

where f_1 and f_2 are arbitrary functions of A, B, C or even β can now be taken as non-zero but some arbitrary terms can be added to α, β, γ provided that they satisfy the condition (10) and that they be of second-order in ξ, as it relates to α, and first-order as it relates to β and γ.

2) The first equation (11) can be written:

$$X_1 = \frac{k_1}{B^3 C}\left[x\left(1 - \sum \xi\xi_1\right) + \xi_1\left(r + \sum x\xi\right)\right] \tag{11'}$$

and the quantity between square brackets can itself be written:

$$(x + r\xi_1) + \eta(\xi_1 y - x\eta_1) + \zeta(\xi_1 z - x\zeta_1), \tag{12}$$

such that the total force can be broken down into three components corresponding to the three parentheses from the expression (12); the first component has a vague analogy with mechanical force due to the electric field and the two others with the mechanical force due to the magnetic field. To complete the analogy, I can, because of the first remark, replace $1/B^3$ in equations (11) with C/B^3, such that X_1, Y_1, Z_1 now only depend linearly on the velocity ξ, η, $\zeta$of the attracted body because C has disappeared from the denominator of (11′).

Then set:

$$\begin{array}{lll} k_1(x + r\xi_1) = \lambda, & k_1(y + r\eta_1) = \mu, & k_1(z + r\zeta_1) = \nu, \\ k_1(\eta_1 z - \zeta_1 y) = \lambda', & k_1(\zeta_1 x - \xi_1 z) = \mu', & k_1(\xi_1 y - \eta_1 x) = \nu', \end{array} \tag{13}$$

it follows, since C has disappeared from the denominators of (11′), that:

$$\begin{aligned} X_1 &= \frac{\lambda}{B^3} + \frac{\eta\nu' - \zeta\mu'}{B^3}, \\ Y_1 &= \frac{\mu}{B^3} + \frac{\zeta\lambda' - \xi\nu'}{B^3}, \\ Z_1 &= \frac{\nu}{B^3} + \frac{\xi\mu' - \eta\lambda'}{B^3}, \end{aligned} \tag{14}$$

and we will additionally have:

$$B^2 = \sum \lambda^2 - \sum \lambda'^2 \tag{15}$$

Then λ, μ, ν or λ/B^3, μ/B^3, ν/B^3, is a kind of electric field while λ', μ', ν' or instead λ'/B^3, μ'/B^3, ν'/B^3 is kind of magnetic field.

3) The relativity postulate would compel us to adopt solution (11) or solution (14) or any one of the solutions which could be deduced from them using the first remark. But, the first question which comes up is that of knowing whether they are compatible with astronomical observations. The divergence from Newton's law is of order ξ^2, meaning 10,000 times smaller than if it were of order ξ, meaning if the propagation occurs with the speed of light, everything else being equal; one could therefore hope that it will not be too large. But we will only be able to learn that from an in-depth discussion.

Paris, July 1905.

H. Poincaré

Translator's Notes

1. See Part II, Chapter 13, p. 251–252 for discussion of the pronunciation of this name.
2. For convenience, the content of this article is reformatted and provided in Part III, Chapter 14.
3. The first reference appears to be to Kaufmann, W. (1901). Die magnetische und electrische Ablenkbarkeit der Becquerelstrahlen und die scheinbare Masse der Elektronen. *Nachrichten von der Königl. Gesellschaft der Wissenschaften zu Göttingen, 2*, 143–155; the second reference could be to Abraham, M. (1902). Dynamik des Electrons. *Nachrichten von der Gesellschaft der Wissenschaften zu Göttingen*, 20–41; or to Abraham, M. (1903). Prinzipien der Dynamik des Eleckrons. *Annalen der Physik, Ser. 4 vol. 10 supplement*, 105–179.
4. Part II, Chapter 12 discusses this choice of notation and on page 228 shows how this half-page would look with our vector formalism.
5. These equations are in Part III, page 263.
6. See for example, Lorentz, H. A. (1902) Contributions to the theory of electrons. I, *Proceedings of the KNAW*, vol. 5 (1902), 608–628.
7. Presumably Poincaré meant "charge" not "mass."
8. Presumably this is a reference to Langevin, P. (1905). Sur l'origine des radiations et l'inertie électromagnétique. J. Phys. Theor. Appl., 4(1), 156–183. Poincaré provides a more complete citation on p. 46.
9. Langevin, P. (1905). Sur l'origine des radiations et l'inertie électromagnétique. *J. Phys. Theor. Appl., 4*(1), 156–183.
10. Presumably Poincaré meant "$-\varepsilon$", not "$-\varepsilon t$".
11. In this equation, note that there should not be a minus sign immediately after the equal sign.

Chapter 6
Dynamics of the Electron

I. Introduction

Are the general principles of dynamics—which since Newton have served as the foundation of physical science and which seem unshakable—on the point of being abandoned or at least profoundly changed? Many people have been wondering that for a few years now. According to them, the discovery of radium could lead to overturning scientific dogmas that were thought to be the most solid: both the impossibility of transmutation of metals and also the fundamental postulates of mechanics. Maybe it's too hasty to consider these discoveries as firmly established and to break our idols from the past; maybe it would be better, before taking sides, to wait for more numerous and more probative experiments. It is no less necessary today to understand the new doctrines and the already very serious arguments on which they rest.

Let us first review in a few words what these principles consist of:

A) The motion of an isolated material point without any external force is straight and uniform; this is the principle of inertia: no acceleration without force.

B) The acceleration of a moving point has the same direction as the resultant of all the forces to which this point is subject; it is equal to the quotient of this resultant and the coefficient of the mobile point called *mass*.

The mass of a mobile point, defined in that way, is a constant; it does not depend on the velocity acquired by this point; it is the same whether the force, being parallel to this velocity, only tends to accelerate or slow the motion of the point or, in contrast, being perpendicular to this velocity, it deflects this motion towards the right or the left, meaning to curve the trajectory.

Poincaré, H. (1908). La dynamique de l'électron. *Revue générale des sciences pures et appliquées, 19*, 386–402.

B. D. Popp, *Henri Poincaré: Electrons to Special Relativity*,
https://doi.org/10.1007/978-3-030-48039-4_6

C) All forces experienced by a material point arise from the action of other material points; they depend only on the *relative* positions and velocities of these various material points.
   By combining the two principles B and C, one arrives at the *principle of relative motion*, according to which the laws of motion of the system are the same whether this system is referred to fixed axes or referred to axes moving with a straight and uniform motion, such that it is impossible to distinguish absolute motion from relative motion with respect to similar mobile axes.
D) If a material point A acts on another material point B, the body B reacts on A and these two actions are two equal and oppositely directed forces. This is the *principle of the equality of action and reaction*, or more briefly, the *conservation of momentum.*

Astronomical observations, the most familiar physical phenomenon, seem to have provided these principles a complete, unchanging and very precise confirmation. We now say, that's true, but it's because we only ever dealt with small velocities; for example, Mercury, which is the fastest planet, scarcely goes 100 km/s. Would this body behave in the same way if it went 1000 times faster? Here we can see that there's still no reason to be concerned: however fast the progress with cars, it will still be a very long time before we need to give up applying the classical principles of dynamics to our machines.

With that in mind, how would it be possible to achieve speeds a thousand times larger than that of Mercury, equal for example to one-tenth or one-third of the speed of light or getting even closer to the speed? It would be possible using cathode rays or radium rays.

It is known that radium emits three kinds of rays that are designated by the three Greek letters α, β and γ. In the following, unless expressly indicated otherwise, it always involves β rays which are analogous to cathode rays.

After the discovery of cathode rays, two theories took hold: Crookes attributed the phenomena to a genuine molecular bombardment and Hertz to specific waves of the ether. It was a renewal of the debate concerning light that had divided physicists a century earlier. Crookes represented the particle theory, that had been abandoned for light and Hertz held for the wave theory. The facts seem to favor Crookes.

It was recognized first the cathode rays transported a negative electric charge with them. They are deflected by a magnetic field and by an electric field, and these deflections are precisely what would be produced by these same fields on projectiles that are driven at a very high speed and are highly electrically charged. These two deflections depend on two quantities: first, the velocity and, second, the ratio of the electric charge of the particle to its mass; it is not possible to know the absolute value of this mass, nor that of the charge, but only the ratio. It is in fact clear that if both the charge and the mass are doubled, without changing the speed, the force which tends to deflect the projectile will double, but since its mass is also doubled, the acceleration and the observed deflection will be unchanged. The observation of the two deflections will therefore provide two equations for determining these two unknowns. A speed of 10,000 to 30,000 km/s is found and the charge-to-mass ratio is very large. It can be compared to the corresponding ratio for the hydrogen ion in electrolysis. It is then found that the cathode projectile transports about 1000 times more electricity than an equal mass of hydrogen would transport in an electrolyte.

To confirm these views, a direct measurement of this velocity would be needed so that it could be compared with the velocity calculated that way. Old experiments by J. J. Thomson had given results more than 100 times too weak; but they were subject to certain sources of error. The question was taken up again by Wiechert in a device where Hertzian oscillations were used and results agreeing with theory, at least within an order of magnitude, were found; it would be very interesting to repeat these experiments. However it may be, wave theory seems powerless to account for this set of facts.

The same calculations, done on beta rays from radium, have given even larger speeds: 100,000 or 200,000 km/s or even more. These speeds greatly exceed all others that we know of. It is true, and it has been known for a long time, that light travels 300,000 km/s, but it does not transport matter; whereas, according to the theory of the emission of cathode rays, they would have material molecules actually driven at the speeds in question and it is appropriate to see whether the ordinary laws of mechanics are still applicable to them.

II. Longitudinal and Transverse Mass

It is known that electric currents give rise to induction phenomena and in particular to self-induction. When a current increases, it develops an electromotive force of self-induction which tends to oppose the current; in contrast, when the current decreases, the electromotive force of self-induction tends to maintain the current. The self-induction therefore opposes any variation in the current intensity just as the inertia of a body in mechanics opposes any variation of the body's velocity. *Self-induction is a true inertia.* Everything happens as if the current could not become established without moving the surrounding ether and as if the inertia of this ether tended, as a consequence, to maintain the constant intensity of this current. To establish the current it would be necessary to overcome this inertia; to stop the current it would also be necessary to overcome it.

A cathode ray, which is a rain of charged projectiles with negative electricity, can be compared to a current; undoubtedly, this current differs, on first consideration at least, from ordinary conduction currents where the matter is immobile and where the electricity moves through the matter. It is a *convection current*, where electricity, attached to a material vehicle, is carried along by the motion of this vehicle. But Rowland showed that the convection currents produce the same magnetic effects as the conduction currents; they must also produce the same induction effects. First, if it weren't that way, then the principle of energy conservation would be violated; additionally, Crémieu and Pender had used a method where these induction effects were shown *directly*.

If the velocity of a cathodic corpuscle varies, the intensity of the corresponding current will also vary, and it will develop self-induction effects which will tend to oppose this variation. These corpuscles must therefore have a double inertia: first their own inertia and the apparent inertia due to self-induction which produces the same effects. They will therefore have a total apparent mass composed of their real mass and a fictional mass of electromagnetic origin. Calculation shows that this fictional mass varies with the velocity and that the inertial force of self-induction is not the same when the velocity of the projectile accelerates or slows, or even when it is deflected; it is therefore the same way as for the total apparent inertial force.

The total apparent mass is therefore not the same when the real force applied to the corpuscle is parallel to its velocity and tends to vary its magnitude, and when this force is perpendicular to the velocity and tends to vary the direction. It is therefore necessary to distinguish between the *total longitudinal mass* and the *total transverse mass*. These two total masses additionally depend on the velocity. This is what results from the theoretical work of Abraham.

In the measurements that we were talking about it in the previous section, what is determined by measuring the two deflections? It is both the velocity and also the ratio of the charge to the total transverse mass. How, under these conditions, to tell the portion of the actual mass and that of the fictive electromagnetic mass in this total mass? If there were only cathode rays themselves, it would not be necessary to think about it; but, fortunately, there are radium rays, which, as we have seen, are distinctly faster. These rays are not all identical and do not behave in the same way under the action of an electric and magnetic field. It is found that the electric deflection is a function of the magnetic deflection, and one can photograph the curve which represents the relation between these two deflections by receiving them on a plate sensitive to the radium rays which have undergone the action of the two fields. This is what Kaufmann did; he deduced the relationship between the velocity and the charge to apparent total mass ratio from it. We will call this ratio ε.

One could assume that there are several types of rays, each characterized by a set velocity, a set charge and a set mass. But this hypothesis is unlikely; for what reason, in fact, would all corpuscles with the same mass always have the same velocity. It is more natural to assume that the charge and also the actual mass are the same for all projectiles and that only their velocity is different. If the ratio ε is a function of the velocity, it is not because the actual mass varies with this velocity; but, as the fictional electromagnetic mass depends on this velocity, the total apparent mass, which alone is observable, must depend on it even though the actual mass would not depend on it and would be constant.

Abraham's calculations let us know the law under which the *fictional* mass varies depending on the velocity; Kaufmann's experiment let us know the law of variation of the *total* mass. Comparison of these two laws therefore allows us to determine the ratio of the *real* mass to the total mass.

This is the method which Kaufmann used to determine this ratio. The result is very surprising: *the real mass is zero*.

In this way we are led to concepts which are entirely unexpected. What had only been shown for cathode corpuscles was extended to all bodies. What we call mass would only be an appearance; all inertia would be of electromagnetic origin. But then the mass would no longer be constant, it would increase with velocity; substantially constant for velocities which could range up to 1000 km/s, it would next increase and become infinite for the speed of light. The transverse mass would no longer be equal to the longitudinal mass: the masses would only be approximately equal if the velocity is not too large. Principal B of mechanics would no longer be true.

III. Channel Rays

At the point where we are, this conclusion may seem premature. Is it possible to apply to all matter what has only been established for these corpuscles, which are so light that they are only an emanation of matter and not true matter itself? But, before taking up this question, it is necessary to say a word about another type of ray. I want to first talk about channel rays, Goldstein's *kanalstrahlen*. At the same time as negatively electrically charged cathode rays, the cathode emits positively electrically charged channel rays. In general, since these channel rays are repelled by the cathode, they remain confined to the immediate neighborhood of this cathode, where they constitute the "chamois layer," which is not very easy to see; but, if the cathode is pierced with holes, and if it nearly completely blocks the tube, the channel rays are going to propagate *behind* the cathode in the opposite direction from that of the cathode rays and it will become possible to study them. This is the way that it can be demonstrated that their charge is positive and shown that magnetic and electrical deflections still exist, as for cathode rays, but are much weaker.

Radium also emits rays that are analogous to channel rays, and relatively very absorbable, which are called α rays.

As for the cathode rays, it is possible to measure both deflections and from that deduce the speed and the ratio ε. The results are less constant than for the cathode rays, but the speed is smaller and also the ratio ε; the positive corpuscles are less charged than the negative corpuscles; or if, which is more natural, it is assumed that the charges are equal and of opposite sign, the positive corpuscles are much heavier. These corpuscles—ones charged positively and the others negatively—have received the name of *electrons*.

IV. Lorentz's Theory

But the electrons do not manifest their existence solely in these rays, where they seem to us driven at enormous speeds. We are going to see them in very different roles, and they are the ones which are going to make us aware of the principal phenomena of Optics and Electricity. We are going to say a word about the brilliant synthesis that comes from Lorentz.

Matter is entirely formed from electrons bearing enormous charges and, if they appear neutral to us, it is because the charges of opposite sign of these electrons balance out. For example, one could represent a sort of solar system formed from a large positive electron around which would gravitate a number of small planets which would be negative electrons attracted by the electricity of the opposite name which charges the central electron. The negative charges of these planets would compensate the positive charge of this Sun such that the algebraic sum of all these charges would be zero.

All these electrons would be bathed in the ether. The ether would be identical to itself everywhere, and disturbances would propagate in it according to the same laws as light or the Hertzian oscillations *in vacuum*. Apart from electrons and ether, there would be nothing. When a light wave entered into a part of the ether where electrons were numerous, these electrons would enter into motion under the influence of the disturbance of the ether and they would next react on the ether. That is the way in which refraction, dispersion, double refraction and absorption would be explained. Similarly if an electron entered into motion for an arbitrary cause, it would disturb the ether around it and give birth to light waves which would explain the emission of light by incandescent bodies.

In some bodies, metals for example, we would have immobile electrons between which mobile electrons would move enjoying a complete freedom, except that of leaving the metal body and crossing the surface which separates it from the outside vacuum, or air, or any other nonmetallic body. These mobile electrons then behave inside the metallic body as do gas molecules, according to the kinetic theory of gases, inside the container in which this gas is enclosed. But, under the influence of a potential difference, the negative mobile electrons would tend to all go to one side, and the mobile positive electrons to the other. This is what would produce electric currents and it is why these bodies would be conductors. Additionally, the speeds of our electrons would be that much larger as the temperature is higher, if the comparison with the kinetic theory of gases is accepted. When one of these mobile electrons would encounter the metal body's surface, a surface that it cannot cross, it would be reflected, like a billiard ball touching the bumper, and its velocity would undergo an abrupt change of direction. But, when an electron changes direction, as we will see later, it becomes the source of a light wave, and that is why hot metals are incandescent.

In other bodies, dielectrics and transparent bodies, mobile electrons enjoy a much smaller freedom. They remain as attached to the fixed electrons which attract them. The more separated they become; the larger this attraction becomes and it tends to bring them back. They can therefore only undergo small separations; they cannot travel but only oscillate around their average position. This is why these bodies would not be conductors; they would additionally most often be transparent and they would be refracting because the light vibrations would communicate to the mobile electrons, subject oscillation, and a disturbance would result from that.

I cannot give the details of the calculations here; I limit myself to stating that this theory considers all known facts and that it has predicted new ones, such as the Zeeman Effect.

V. Mechanical Consequences

We can now consider two hypotheses: 1) The positive electrons have a real mass, much larger than their fictive electromagnetic mass; the negative electrons alone are without real mass. It could even be assumed that apart from electrons of these two signs, there are neutral atoms which have no other mass than their real mass. In this case, mechanics is not threatened; we do not need to touch its laws: real mass is constant. Only motions are disturbed by the effects of self-induction, which was always known. These disruptions are additionally roughly negligible, except for the negative electrons, which, since they don't have real mass, are not true matter.

2) But, there is another perspective: it can be assumed that there are no neutral atoms, and that the positive electrons are without real mass in the same way as the negative electrons. But then, the real mass goes away, or else the word *mass* no longer has any meaning, or else it will have to be that it designates the fictive electromagnetic mass. In this case, the mass will no longer be constant, the transverse *mass* will no longer be equal to the longitudinal mass, and the principles of mechanics will be overturned.

First, a word of explanation is necessary. We stated that, for an equal charge, the *total* mass of a positive electron is much larger than that of a negative electron. And then, it is natural to think that this difference is explained because the positive electron has, beyond its fictive mass, a considerable real mass—this would bring us back to the first hypothesis. But, it can also be allowed that the real mass is zero for both of them, but that the fictive mass of the positive electron is much larger because this electron is much smaller. I definitely mean much smaller. And, in fact, in this hypothesis, the inertia is almost exclusively of electromagnetic origin, it reduces to the inertia of the ether. Electrons are no longer anything on their own; they are only holes in the ether, and around which the ether moves. The smaller these holes the more ether there will be and consequently the inertia of the ether will be large.

How to decide between these two hypotheses? By working with the channel rays as Kaufmann did with β rays? It is impossible; the speed of these rays is much too small. Will everyone have to decide according to their own temperament: the conservatives going to one side and the friends of the new going to the other? Maybe. But, to make the arguments of the innovators better understood, other considerations need to be brought into play.

VI. Aberration

The phenomenon of aberration, discovered by Bradley, is understood. Light emanating from a star takes some time to pass through a telescope; during this time, the telescope, carried along by the motion of the Earth, is displaced. If the telescope is therefore turned in the *true* direction of the star, the image would be formed at the

Fig. 1

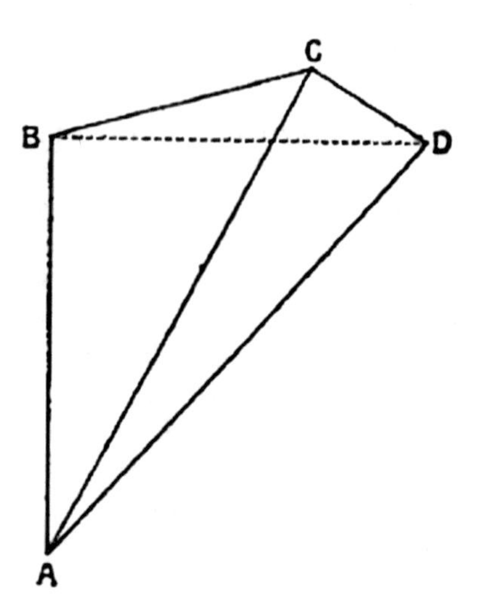

point occupied by the crossed wires of the reticle when the light reached the objective. And this cross would no longer be in this same point when the light reached the plane of the reticle. One would therefore be led to de-point the telescope to return the image onto the crossed wires. The result of this is that the astronomer will not point the telescope in the direction of the absolute velocity of the light, meaning at the true position of the star, but in the direction of the relative velocity of the light with respect to the Earth, that is what is called the apparent position of the star. In Figure 1, we have shown the absolute velocity of the light with AB (changed in direction, because the observer is at A and the star at a very large distance in the direction AB), the velocity of the Earth with BD and the *relative* velocity of the light with AD (changed in direction). The astronomer should point their instrument to the direction AB; the astronomer does point it in the direction AD.

The magnitude of AB, meaning the speed of light, is known. One might think that we have the means to calculate BD, meaning the *absolute* velocity of the Earth. (I will soon explain this word absolute.) That is not how it is. We do know the apparent position of the star—meaning the direction AD that we observe—but we do not know its true position. We only know the magnitude of AB and not its direction.

If therefore the absolute velocity of the Earth were straight and uniform, we would have never suspected the phenomena of aberration, but it is variable; and made up of two parts: the velocity of the solar system which is straight and uniform and what I represent by BC, and the velocity of the Earth with respect to the sun, which is variable and which I represent by CD such that the resultant is represented by BD.

Since BC is constant, the direction AC does not vary: it defines the *mean* apparent position of the star. In contrast the direction AD, which is variable, defines the actual apparent position, which describes a small ellipse around the average apparent position and it is this ellipse that is observed.

We know CD in magnitude and direction from Kepler's laws and our knowledge of the distance to the sun. We know AC and AD in direction and we can consequently construct the triangle ACD; knowing AC, we will have the speed of light (shown by AB), because, since BC is assumed very small with respect to AB AC differs very little from AB. Only the *relative* velocity of the Earth compared to the sun is involved.

Stop right there. We have considered AC as equal to AB; that is not rigorous—it is only approximate. Let us push the approximation a little farther. The dimensions of the ellipse described during one year by the apparent position of a star depend on the ratio of CD (which is known) to the length AC. This length is known to us from observation. We compared the major axis of the ellipse for different stars: for each of them, we will have the means of determining the magnitude and direction of AC. The length AB is constant (it is the speed of light), such that the points B corresponding to various stars will all be on a sphere with center A. Since BC is constant in magnitude and direction, the points B corresponding to various stars will all be on a sphere of radius AB and centered on A′, with the vector AA′ equal to and parallel to BC. If it were then possible to determine, as we just stated, the various points C, then this sphere, its center A′ and consequently the magnitude and direction of the absolute velocity BC would be known.

There would then be a means for determining the absolute velocity of the Earth. That might be less shocking than it might seem at first. It would not in fact be about the velocity relative to an absolutely empty space, but the velocity relative to the ether, that is considered *by definition* as being an absolute rest.

Furthermore, this means is purely theoretical. In fact, aberration is very small; the possible variations of the aberration ellipse are even much smaller, and, if the aberration is regarded of as being of first-order, the variations must therefore be regarded as second-order: about 1000th of a second—that absolutely cannot be assessed with our instruments. We will see later why the preceding theory must be rejected and why we would be unable to determine BC even when our instruments would be 10,000 times more precise!

One could hope for another means and one has in fact hoped for it. The speed of light is not the same in water and air; could one compare the two apparent positions of the stars seen through both a telescope full of air and a telescope full of water? The results were negative; the apparent laws of reflection and refraction are unaltered by the Earth's motion. There are two explanations for this phenomenon:

1) It could be assumed that the ether is not at rest, but that it is dragged along by moving bodies. It would then not be surprising that the phenomena of refraction might have been unchanged by the motion of the Earth, because everything—prisms, lenses and ether—are dragged along at the same time by the same motion. As for the operation itself, it could be explained by a sort of refraction which could be produced at the surface of separation between the ether at rest in interstellar space and the ether dragged along by the motion of the Earth. Hertz's theory on the electrodynamics of moving bodies is based on this hypothesis (complete dragging of the ether).

2) In contrast, Fresnel assumed that the ether is at absolute rest in the vacuum and at nearly absolute rest in air, whatever the velocity of this air, and that it is partially dragged along by refracting media. Lorentz gave this theory a more satisfactory form. For him, the ether is at rest, only the electrons are moving. In the vacuum, where only the ether is in play, or in the air where nearly it alone is in play, the drag is zero or nearly zero. In refracting media, where the disturbance is produced both by vibrations of the ether and by vibrations of electron shaken by the motion of the ether, the waves are *partially* dragged.

To decide between these two hypotheses, we have Fizeau's experiment. He compared, by measurements of interference fringes, the speed of light in air at rest or in motion, and also in water at rest or in motion. These experiments confirmed Fresnel's hypothesis of partial dragging. They were repeated with the same result by Michelson. *Hertz's theory must therefore be rejected.*

VII. The Principle of Relativity

But, if the ether is not dragged by the motion of the Earth, is it possible to show, with optical phenomena, the absolute motion of the Earth, or stated another way its velocity relative to the fixed ether? The experimental response is negative even though the experimental procedures were varied in all possible ways. Whatever the means used, only relative velocities—by which I mean the velocities of some material bodies relative to other material bodies—could ever be detected. In fact, if the light source and the observation equipment are on the Earth and participate in the Earth's motion, the experimental results are always the same whatever the orientation of the equipment relative to the direction of the orbital motion of the Earth. If astronomical aberration occurs, it is because the source, which is a star, is moving relative to the observer.

The hypotheses made right here fully reflect this general result, *if very small quantities of order the square of the aberration are neglected.* The explanation rests on the concept of *local time*, which we are going to try to understand, and which was introduced by Lorentz. We are going to assume that two observers positioned at A and B want to set their watches by means of optical signals. They agree that B will send a signal to A when his watch shows a set time and that A will set his watch to the time at the instant that he sees the signal. If they were only to operate in this way, there would be a systematic error, because, since light takes some time t to go from B to A, the watch of A is going to be slow by the time t compared to the watch of B. This error is easily corrected. It is sufficient to cross the signals. A in turn has to send signals to B, and, after this new adjustment, the watch B will be the one slow by the time t compared A. It will then be sufficient to take the arithmetic average between the two settings.

This operating procedure assumes that light takes the same time to go from A to B and to return from B to A. That is true if the observers are stationary; it is no longer true if they are carried with a shared motion, because then A, for example, will go ahead of the light coming from B, and B will instead fall behind the light coming from A. If the observers are therefore driven in shared motion, and if they are unaware of it, their adjustment will be defective; their watches will not indicate the same time, each of them will indicate the *local time*, appropriate to the point where they are located.

The two observers will not have any means of becoming aware of it, if the motionless ether can only send them light signals travelling with the same speed, and if the other signals that they could send are sent by media dragged with them in their motion. The phenomenon that each of them will observe will either be early or late; it will only occur at the same time if there were no motion; but, since the observations are done with an incorrectly set watch, it will go unnoticed and the appearances will be unchanged.

The result of this is that the compensation is easy to explain so long as the square of the aberration is neglected, and for a long time experiments were too imprecise for there to be away to include them. Then, one day Michelson imagined a much more delicate method: he produced interference from light rays which it traveled different trajectories after being reflected on mirrors; each of the trajectories was about one meter long and with the interference fringes it was possible to assess differences of a fraction of a thousandth of a millimeter; the square of the aberration could no longer be neglected and all the same *the results were still negative*. Something had to be added to the theory and it was provided by *the hypothesis of Lorentz and Fitz Gerald*.

These two physicists assumed that all bodies in motion underwent a contraction in the direction of this motion, whereas the dimensions perpendicular to this motion remained invariant. *This contraction is the same for all bodies*; it is additionally very small, about 200 millionths for a velocity like that of the Earth. Our instruments would additionally be unable to detect it, even if they were much more precise; the meter sticks with which we measure undergo the same contraction in fact as the objects to be measured. If a body lines up exactly with the meter stick when the body and also the meter stick are oriented in the direction of the Earth's motion, it will still line up exactly on the meter stick in another orientation and does so because the body and the meter stick have changed length at the same time as the orientation, and precisely because the change is the same for each of them. But it is not the same if instead of measuring a length with a meter stick, it is measured with the time the light takes to travel the distance, and this is precisely what Michelson did.

A body that is spherical when it is at rest will thus take on the form of a flattened ellipsoid of revolution when it is in motion; but the observer will still believe it to be spherical, because the observer has also undergone an analogous deformation along with all the objects used as reference points. On the other hand, the wave surfaces of light, which have remained rigorously spherical, will seem to the observer to be elongated ellipsoids.

Then what is going to happen? Let us assume an observer and a source carried along together in motion, the wave surfaces emanating from the source will be spheres centered on the successive positions of the source; the distance from the center to the current position of the source will be proportional to the time passed

Fig. 2

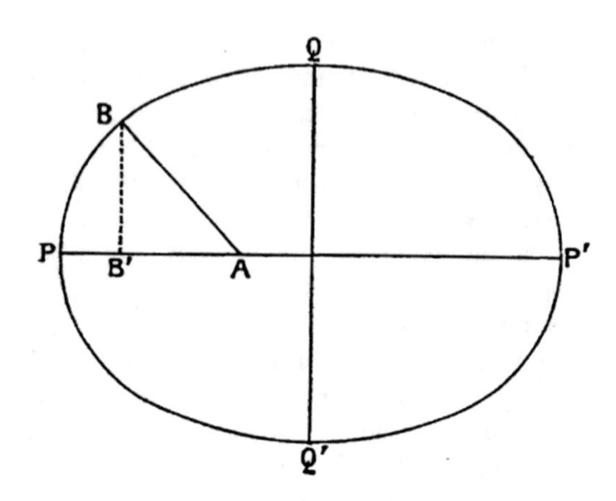

since the emission, meaning proportional to the radius of the sphere. All these spheres will therefore each be shifted and scaled relative to the current position S of the source. But, for our observer, because of the contraction, all the spheres will appear to be elongated ellipsoids; and all these ellipsoids will again be shifted and scaled relative to the point S; the eccentricity of all these ellipsoids is the same and depends only on the velocity of the Earth. *We will choose the law of contraction such that the point S is at the focus of the meridional section of the ellipsoid.*

What are we then going to do to evaluate the time it takes light to go from B to A? At A and B (Figure 2) I show the *apparent* positions of these two points. I construct an ellipsoid similar to the ellipsoids from the waves that we just defined and having its major axis in the direction of the Earth's motion. I construct this ellipsoid such that it passes through B and has its focus at A.

According to a well-known property of ellipsoids, there is a relation between the apparent distance AB between two points and its projection AB'; this relationship is:

$$AB + e \cdot AB' = OQ\sqrt{1 - e^2}.$$

But the semi-minor axis of the ellipsoid, which is unchanged in shape, is equal to Vt, where V is the speed of light and t the length of the transmission; hence:

$$AB + e \cdot AB' = Vt\sqrt{1 - e^2}.$$

The eccentricity e is a constant that only depends on the speed of the Earth; we therefore have a linear relationship between AB, AB' and t. But AB' is the difference between the abscissas of the points A and B. Assume that the difference between the real time and the local time at an arbitrary point is equal to the abscissa of this point multiplied by the constant:

$$\frac{e}{V\sqrt{1-e^2}};$$

the *apparent* transmission time will be:

$$\tau = t - AB' \frac{e}{V\sqrt{1-e^2}}$$

hence:

$$AB = Vt\sqrt{1-e^2}$$

That means that the *apparent* length of the transmission is proportional to the *apparent* distance. This time, the compensation is *rigorous* and is what explains the Michelson experiment.

Above I stated that, according to ordinary theories, observations of astronomical aberration could let us know the absolute velocity of the Earth, if our instruments were 1000 times more precise. I have to change that conclusion. Yes, the observed angles would be changed by the effect of this absolute velocity, but the divided circles which we used for measuring the angles would be deformed by the translation; they would become ellipses. The result would be an error in the angle measured, and *this second error would exactly compensate the first.*

On first appearance, this hypothesis by Lorentz and Fitz Gerald would seem to be very extraordinary; for now all that we can say in its favor is that it is the immediate manifestation of Michelson's experimental result, if lengths are defined by the time that light takes to traverse them.

However it may be, it is impossible to escape this impression that the principle of relativity is a general law of nature and that one cannot by any imaginable means show anything other than relative speeds and by that I understand not only the speeds of bodies relative to the ether, but also the speeds of bodies relative to each other. It is in any case appropriate to look at the consequences this perspective would lead us to and then put these consequences to experimental verification.

VIII. The Conservation of Momentum

Now let us look at what happens in Lorentz's theory to the principal that action and reaction are equal and opposite. Look at an electron A which undergoes motion because of an arbitrary cause. It produces a disturbance in the ether; at the end of some time, this perturbation reaches another electron B which will be disturbed from its equilibrium position. Under these conditions, the action and reaction cannot be equal, unless the ether is not considered but only the electrons *which alone are observable*, because our matter is made of electrons.

In fact, it is the electron A which disturbs the electron B. Even though the electron B could react on A, this reaction could be equal to the action but it could not in any case be simultaneous because the electron B could only enter into motion after some time, necessary for the propagation. If we subject the problem to a very precise calculation, we arrive at the following result: Let us assume a Hertz emitter placed at the focus of a parabolic mirror to which it is mechanically connected; this emitter emits electromagnetic waves and the mirror returns all these waves in the same direction. The emitter is therefore going to radiate energy in a set direction. Now, the calculation shows that the *emitter is going to recoil* like a cannon which shoots a projectile. In the case of the canon, the recoil is the natural result of the quality of action and reaction. The canon recoils, because the projectile on which it acted reacts on it.

But now it is no longer the same. What we have sent far away is no longer a material projectile: it is energy and energy does not have mass, there is no exchange. And, instead of an emitter, we could have simply considered a lamp with a reflector concentrating its rays in a single direction.

It is true that, if the energy emanating from the emitter or the lamp comes to reach a material object, that object is going to experience a mechanical thrust as if it'd been reached by a real projectile and this thrust will be equal to the recoil of the emitter and the lamp, if it has not lost energy along the way and if the object completely absorbs this energy. One would therefore be tempted to state that there is still balance between action and reaction. But this balance, even though it is complete is still delayed. It can never happen if the light, after having left the source, wanders in interstellar space without ever encountering a material body; it is incomplete, if the body that it strikes is not completely absorbing.

Are these mechanical actions too small to be measured, or else are they accessible to experiment? These actions are none other than those which are due to the *Maxwell-Bartoli* pressures; Maxwell had called for these pressures by calculations relating to electrostatics and magnetism; Bartoli had arrived at the same result by thermodynamic considerations.

This is the way in which *comet tails* are explained. Small particles detach from the core of the comet and they are struck by sunlight which drives them back as would a rain of projectiles coming from the Sun. The mass of these particles is so small that this repulsion is stronger than gravitational attraction. They are therefore going to form the tail by moving away from the Sun.

Direct experimental verification was not easy to get. The first attempt led to the construction of the *radiometer*. But this device *turns backwards*—in the direction opposite to the theoretical direction—and the explanation for its rotation, since discovered, is completely different. It was finally successful, both by making a better vacuum and by blackening one of the surfaces of the blades and directing a light beam on one of the surfaces. The radiometric effects and the other causes of disturbance are eliminated by series of meticulous precautions and a deviation results which is very small, but which is, it seems, in keeping with theory.

The same effects of the Maxwell-Bartoli pressure are also provided by Hertz's theory, which we spoke about above, and by Lorentz's theory. But there is a difference. Assume that the energy, in light form for example, goes from a light source to an arbitrary body through a transparent medium. The Maxwell-Bartoli pressure will act, not only on the originating source, and on the body illuminated at arrival, but on the matter of the transparent medium that it traverses. At the moment when the light wave reaches a new region of this material, this pressure will push the matter forward which becomes spread and will gather it behind when the wave leaves this region. Such that the recoil of the source is compensated by the forward motion of the transparent matter which is in contact with this source; a little later, the recoil of this same matter is compensated by the forward motion of the transparent matter which is located a little farther on, and so on.

Only, is the compensation perfect? Is the action of the Maxwell-Bartoli pressure on the matter of the transparent medium equal to its reaction on the source, and is it so whatever this matter is? Or does this action become that much smaller as the medium is less refracting and more rarefied, for finally becoming zero in vacuum? If Hertz's theory is accepted, which regards matter as mechanically linked to the ether, such that the ether is completely dragged by the matter, then the answer to the first question would have to be yes and to the second question no.

There would then be perfect compensation, as the principle of the equality of action and reaction requires, even in less refracting media, even in air, even in interplanetary vacuum, where it would suffice to assume residual matter, however slight it might be. If, instead, Lorentz's theory is accepted, the compensation, always imperfect, is undetectable in air and become zero in vacuum.

But we saw above that Fizeau's experiment does not allow retaining Hertz's theory; Lorentz's theory has to be adopted and, consequently, *the conservation of momentum renounced.*

IX. Consequences the Principle of Relativity

We have seen above the reasons which support regarding the Principle of Relativity as a general law of nature. Let's look at what consequences this principle would lead us to if we were to regard it as definitively proven.

First, it would force us to generalize the Lorentz and Fitz Gerald hypothesis on the contraction of all bodies in the direction of translation. In particular, we will have to extend this hypothesis to electrons themselves. Abraham considered these electrons as spherical and undeformable; we will have to accept that these electrons, while spherical when they are at rest, undergo the Lorentz contraction when they are moving and therefore take the shape of flattened ellipsoids.

This deformation of the electrons is going to have an influence on their mechanical properties. In fact, I state that the motion of these charged electrons is a true convection current and that their apparent inertia is due to the self-induction of this current, exclusively as it pertains to negative electrons, exclusively or not, we don't know anything about it yet, for positive electrons. Such that, the deformation of electrons, a deformation which depends on their velocity, is going to change the charge distribution on their surface and consequently the intensity of the convection current that they produce and as a further consequence the laws according to which the self-induction of this current will vary as a function of velocity.

For this price, the compensation will be perfect and conform to the requirements of the Principle of Relativity, but under two conditions:

1) That the positive electrons not have real mass, but only a fictive electromagnetic mass; or at least that their real mass, if it exists, is not constant and varies with velocity according to the same laws as their fictive mass;
2) That all forces are of electromagnetic origin, or at least that they vary with velocity according to the same laws as the forces of electromagnetic origin.

Lorentz is again the one who made this remarkable synthesis; let us stop here for a moment and look at what results from it. First, there is no longer matter, because positive electrons no longer have real mass, or at least no longer have constant real mass. The actual principles of our mechanics, based on the constancy of mass, must therefore be modified.

Next, an electromagnetic explanation has to be sought for all known forces, in particular gravitation, or at least the law of gravitation must be modified such that this force is altered by the velocity in the same way as the electromagnetic forces. We will return to this point.

At first glance, all this seems a little artificial. In particular, this deformation of the electrons seems thoroughly hypothetical. But it can be presented differently, so as to avoid putting this hypothesis of deformation at the base of our reasoning. We will consider electrons as material points and we will ask how their mass must vary as a function of the velocity so as to not contradict the principle of relativity. Or, rather again we will ask what their acceleration must be under the influence of an electric or magnetic field so that this principle is not violated and we return to the ordinary laws by assuming very small velocity. We will find that the variations of this mass, or these accelerations, must occur as if the electron underwent the Lorentz deformation.

X. Kaufmann's Experiment

And so we're faced with two theories: one where electrons are undeformable (Abraham's theory); the other where they undergo the Lorentz deformation. In both cases, their mass increases with velocity, for finally becoming infinite when this velocity is equal to that of light, but the law of variation is not the same. The method used by Kaufmann for showing the law of variation of the mass seems to therefore provide us an experimental means for deciding between the two theories.

Unfortunately, his first experiments were not sufficiently precise to do that. He thought that he had to resume with greater precautions and by measuring the field intensities with greater care. In their new form, *the experiments supported Abraham's theory*. The Principle of Relativity would therefore not have the rigorous value that one was tempted to give to it; there were no longer be any reason to believe the positive electrons are without real mass like negative electrons.

However, before permanently adopting this conclusion, some thought is necessary. The question is of such importance that it would be desirable for Kaufmann's experiment to be repeated by another experimenter. Unfortunately, this experiment is very delicate and it will only be possible to profitably perform it by a physicist of the same caliber as Kaufmann. All precautions were suitably taken and it's not clear what objection could be made.

There is however one point to which I would like to draw attention: it is the measurement of the electrostatic field; everything depends on this measurement. This field was produced between the two plates of a capacitor, and, between these plates, a very high vacuum had to be made in order to obtain complete insulation. The potential difference was then measured between the two plates and the field resulted from dividing this difference by the separation of the plates. That assumes that the field is uniform; are we sure of that? Couldn't there have been a sudden drop of potential in the neighborhood of one of the plates, the negative plate, for example? There could have been a potential difference at the contact between the metal and the vacuum, and it could be that this difference is not the same on the positive and negative side; it is the electric valve effects between mercury and vacuum which leads me to believe it. However small the probability for it to be that way, it seems that there is reason to consider it.

XI. The Principle of Inertia

In the new Dynamics, the Principle of Inertia is still true, meaning that an *isolated* electron will have a straight and uniform motion. At least, it is generally agreed to allow it; however, Lindemann raised objections to this outlook; I don't want to take part in that discussion, which I'm not able to present here because of its overly arduous nature. Small modifications to the theory would in any case be sufficient to shelter it from Lindemann's objections.

We know that a body immersed in water experiences, when it is moving, a significant resistance, but this is because our fluids are viscous. In an ideal fluid, completely free of viscosity, the body would leave a liquid stern behind it, a kind of wake; at the start, it would take a large force to get it to move, because it would need to stir up not only the body itself, but the liquid in its wake. But, once the motion started, it would perpetuate itself without resistance, because the body, as it advanced, would simply transport the disturbance of the liquid with it without the total energy of this liquid increasing. Everything would therefore happen as if its inertia had increased. An electron moving forward in the ether would behave in the same way: around it, the ether would be stirred, but this disturbance would accompany the body in its motion; in that way, for an observer carried along with the electron, the electric and magnetic fields accompanying this electron would appear unchanging; and could only change if the speed of the electron should vary. A force would therefore be needed to start the electron moving, because it would create the energy of these fields; in contrast, once moving, no force would be necessary to maintain it, because the energy created would have nothing more than to follow along behind the electron like a wake. This energy can therefore only increase the inertia of the electron, like the agitation of the liquid increases that of the body immersed in a perfect fluid. And even negative electrons, at least, have no other inertia than that.

In Lorentz's hypothesis, the total energy, which is nothing other than the energy of the ether, is not proportional to the v^2, but to:

$$\frac{V - \sqrt{V^2 - v^2}}{\sqrt{V^2 - v^2}},$$

where V represents the speed of light; the momentum is no longer proportional to v, but to:

$$\frac{v}{\sqrt{V^2 - v^2}};$$

the transverse mass is in inverse proportion to $\sqrt{V^2 - v^2}$, and the longitudinal mass is in inverse proportion to the cube of this quantity.

It can be seen that, if v is very small, the energy is substantially proportional to v^2, the momentum substantially proportional to v and the two masses substantially constant and equal to each other. But, *when the velocity approaches the speed of light, the energy, momentum and both masses increase without limit.*

In Abraham's hypothesis, the expressions are a little more complicated, but what we just stated continues in its main features.

Thus the mass, the momentum and the energy become infinite when the velocity is equal to that of light. The result of this is that *no body can reach a velocity greater than that of light by any means.* And, in fact, as its velocity increases, its mass increases, such that its inertia opposes any new increase of velocity, becoming a larger and larger obstacle.

Authors who have written on the Dynamics of the Electron speak, it's true, of bodies which go faster than light. But, it is to ask how a body would behave whose *initial* velocity is greater than that of light and which consequently would have already gone past the limit before being considered.

The question then comes up: let us accept the Principle of Relativity; an observer in motion must not have any means of perceiving its own motion. If therefore no body in its own absolute motion can exceed the speed of light, but can approach it arbitrarily closely, it must be the same as it relates to its motion relative to our observer. And then one could attempt to reason as follows: the observer can reach a velocity of 200,000 km/s; the body, in its relative motion compared to the observer can reach the same velocity; its absolute velocity will then be 400,000 km/s which is impossible, because it is a figure greater than speed of light.

The way in which relative velocity should be evaluated must be considered; they must be counted not with real time, but with the *local* time. Let A and B be two points unchangeably linked to the observer; let t and $t + h$ be the moments when the body passes by A and B, moments evaluated in real time; let αt and $\alpha(t + h)$ be the same moments evaluated in local time for A, let $\alpha(t + \varepsilon)$ and $\alpha(t + h + \varepsilon)$ be the same moments evaluated in local time for B. If the length of the trip was evaluated in real time, this length would be h and the relative velocity AB/h; but we will have to evaluate it in local time, meaning noting the moment of the passage at A in local time for A, and that of the passage at B in local time at B, such that the length of the trip will be $\alpha(\varepsilon + h)$ and the relative velocity:

$$\frac{AB}{\alpha(\varepsilon + h)}$$

And that is how the compensation happens.

XII. Acceleration Wave

When an electron is moving, it produces a disturbance in the ether surrounding it; if its motion is straight and uniform, this disturbance reduces to the wake which we talked about in the previous section. But it is no longer the same if the motion is curved or varied. The disturbance can then be regarded as the superposition of two others, to which Langevin gave the names *velocity wave* and *acceleration wave*.

The velocity wave is nothing other than the wake which is produced during uniform motion. Let me clarify: let M be an arbitrary point of the ether, considered at an instant t; let P be the position which the electron occupied at an earlier moment $t - h$, such that h is precisely the time that the light would take to travel from P to M. Let v be the velocity that the electron had at that moment $t - h$. Such that, if we consider only the velocity wave, the disturbance at the point M will be the same as if the electron had continued its trajectory since the moment $t - h$, by keeping the velocity v and with a straight and uniform motion.

As for the acceleration wave, it is a disturbance, entirely analogous to light waves, which leaves the electron at the moment when it is accelerated, and which then propagates at the speed of light with successive spherical waves.

This consequence follows: in straight and uniform motion, energy is entirely conserved; but, once there is acceleration, there is a loss of energy, which dissipates in the form of light waves and goes out to infinity through the ether.

However, the effects of this acceleration wave, in particular the corresponding energy loss, are negligible in most cases, meaning not only in ordinary mechanics and in the motions of celestial bodies, but even in radium waves, where the velocity is very large without the acceleration being large. We can then limit ourselves to applying the laws of mechanics, by writing that the force is equal to the mass times acceleration, where this mass, however, varies with the speed according to the laws presented above. We then say that the motion is *quasi-stationary*.

It would no longer be the same in all the cases where the acceleration is large, and for which the principles are the following: 1) In incandescent gases, some electrons take on a very high frequency oscillatory motion; the motions are very small, the velocities are finite, and the accelerations very large. The energy is then communicated to the ether and that is why these gases radiate light with the same period as the oscillations of the electron. 2) Inversely, when a gas receives light, these same electrons are shaken with large accelerations and they absorb the light. 3) In the Hertz emitter, the electrons which move in the metal mass experience, at the moment of the discharge, a sudden acceleration and then take on a high frequency oscillatory motion. The result of this is that a part of the energy is radiated in the form of Hertz waves. 4) In incandescent metal, the electrons enclosed in this metal are driven at large velocities, when they arrive at the surface of the metal, which they cannot cross, they are reflected and thus experience a considerable acceleration. This is why the metal emits light. It is what I already explained in section IV. The details of the laws of emission of light by blackbodies are perfectly explained by this hypothesis. 5) Finally, when cathode rays strike the anode, the negative electrons which form these rays and which are driven at very large speeds are abruptly stopped. As a result of the accelerations that they thus experience, they produce waves in the ether. According to some physicists, this would be the origin of Röntgen waves, which would be nothing other than very short wavelength light waves.

XIII. Gravitation

Mass can be defined in two ways: First, by the ratio of force to acceleration; this is the real definition of mass which measures the inertia of the body; and second, by the attraction that the body exerts on an external body according to Newton's law of gravitation. We therefore have to distinguish the mass coefficient of inertia and the mass coefficient of attraction. According to Newton's law there is rigorous proportionality between these two coefficients. But that is only proven for the velocities at which the general principles of dynamics are applicable. Now, we've seen that the mass coefficient of inertia increases with velocity; do we have to conclude that the mass coefficient of attraction also increases with velocity and remains proportional to the inertial coefficient, or, in contrast, that this coefficient of attraction remains constant? That is a question where we have no means of deciding.

On the other hand, if the coefficient of attraction depends on the velocity, as the velocities of the two mutually attracting bodies are generally not the same, how does this coefficient depend on these two velocities?

We can only form a hypothesis on this subject, but we are naturally led to seek which of these hypotheses would be compatible with the Principle of Relativity. They are many; the only one which I will talk about here is that of Lorentz; I will present it briefly.

First consider some electrons at rest. Two electrons of the same sign repel and two electrons of opposite sign attract; in the ordinary theory, their mutual actions are proportional to their electric charges; if therefore we have four electrons, two positive A and A', and two negative B and B' and if the charges of these four electrons are the same in absolute value, the repulsion of A and A' will be, at the same distance, equal to the repulsion of B and B' and also equal to the attraction of A on B' or A' on B. Therefore if A and B are very close to each other and likewise A' and B', and if we examine the action of the system $A + B$ on the system $A' + B'$, we will have two repulsions and two attractions which will exactly balance and the resulting action will be zero.

Hence, molecules of matter must be precisely regarded as kinds of solar systems where electrons, some positive, others negative, and *such that the algebraic sum of all the charges is zero*. A molecule of matter is therefore in all aspects comparable to the system $A + B$ which we just spoke of, such that the total mutual electric action of two molecules on each other should be zero.

But experience shows us that these molecules later attract by Newtonian gravitation; and we can then make two hypotheses: we can assume the gravitation has no connection with electrostatic attraction, which is due to a completely different cause and that it simply superimposes with it; or else we can allow that there is no proportionality of the attractions of the charges and that the attraction exerted by a charge $+1$ on a charge -1 is larger than the mutual repulsion of two charges $+1$, or that of two charges -1.

In other words, the electric field produced by positive electrons and that which negative electrons produce could be superposable while remaining distinct. The positive electrons would be more sensitive to the field produced by negative electrons than to the field produced by positive electrons; it would be the opposite for the negative electrons. It is clear that this hypothesis complicates electrostatics a bit, but the hypothesis brings back gravitation into it. It is, on the whole, Franklin's hypothesis.

What would happen now if the electrons are moving? The positive electrons are going to lead to a disturbance in the ether and will make it give rise to an electric field and a magnetic field. It will be the same for the negative electrons. Electrons, both positive and negative, will next experience a mechanical impulse from the action of these various fields. In ordinary theory, the electromagnetic field due to the motion of the positive electrons, exerts, on two electrons of opposite sign and same absolute charge, actions that are equal and of opposite sign. Without disadvantage, one can then not distinguish the field due to the motion of positive electrons and the field due to negative electrons and only consider the algebraic sum of these two fields, meaning the resultant.

In the new theory, in contrast, the action on positive electrons of the electromagnetic field due to the positive electrons is done according to ordinary laws; it is the same for the action on negative electrons of the field due to negative electrons. Now consider the action of the field due to the positive electrons on the negative electrons (or vice versa); it will again follow the same laws, but *with a different coefficient*. Each electron is more sensitive to the field created by electrons of opposite name than to the field created by electrons the same name.

Such is Lorentz's hypothesis, which reduces the Franklin's hypothesis for small velocities; it will therefore cover, for these small velocities, Newton's law. Additionally, as gravitation arises from forces of electrodynamic origin, the general theory of Lorentz will apply to it and consequently the Principal Relativity will not be violated.

It is seen that Newton's law is no longer applicable at large velocities and that it must be modified, for moving bodies, precisely in the same way as the laws of electrostatics for moving electricity.

It is known that electromagnetic disturbances propagate with the speed of light. One will therefore be tempted to reject the preceding theory, recalling the gravitation propagates, according to Laplace's calculations, at least 10 million times faster than light, and that, consequently, it cannot be of electrodynamic origin. Laplace's result is well known, but we are generally unaware of the meaning. Laplace assumed that, if the propagation of gravitation is not instantaneous, its propagation velocity combines with that of the attracted body (as it happens for light in the phenomenon of astronomical aberration) in such a way that the effective force is not directed along the line which joins the two bodies, but makes a small angle with that line. That is a very specific hypothesis, rather poorly justified, and in a case entirely different from that of Lorentz. Laplace's result does not prove anything against Lorentz's theory.

XIV.Comparison with Astronomical Observations

Can the preceding theories be reconciled with astronomical observations? First, if they are adopted, the energy of the planetary motions will be continuously dissipated by the effect of the *acceleration wave*. From this, it would result that the mean motions of the bodies would continuously be accelerating, as if these bodies were moving in a resistant medium. But this effect is extremely small, much too small for being detected by the most precise observations. The acceleration of celestial bodies is relatively weak, such that the effects of the acceleration wave are negligible and the motion can be regarded as *quasi-stationary*. It is true that the effects of the acceleration wave are going to constantly accumulate, but this accumulation itself is so slow that it would take many thousands of years of observation for it to become noticeable.

Let us therefore do the calculation by considering the motion as quasi-stationary and do so under the following three hypotheses:

A. Allow Abraham's hypothesis (electrons are undeformable) and keep Newton's law in its usual form;
B. Allow Lorentz's hypothesis on the deformation of electrons and keep Newton's law as usual;
C. Allow Lorentz's hypothesis on electrons and modify Newton's law, as we did in Chapter XIII, so as to make it compatible with the Principle of Relativity.

The motion of Mercury is where the effect will be the most sensitive, because this planet has the largest velocity. Tisserand had done an analogous calculation earlier, by accepting Weber's law; recall that Weber had sought to explain both electrostatic and electrodynamic phenomena by assuming that electrons (whose name had not yet been invented) exerted attractions and repulsions among them directed along the straight line which joins them and depending not only on their separation, but the first and second derivatives of these distances, consequently meaning their speeds and their accelerations. This law by Weber, fairly different from those which seem to prevail today, nonetheless has some analogy with them.

Tisserand found that, if Newtonian attraction occurred in keeping with Weber's law, the result would be a secular variation of the perihelion of Mercury of 14″, *in the same direction as that which was observed and could not be explained*, but smaller, because the variation is 38″.

Let us go back to hypotheses A, B and C, and first study the motion of a planet attracted by a fixed center. The hypotheses B and C are no longer distinguishable because, if the attracting point is fixed, the field that it produces is a purely electrostatic field; hence the attraction varies according to the inverse square of the distances, following the electrostatic law of Coulomb, identical to that of Newton.

The conservation of energy equation remains, by taking the new definition for the total energy; likewise, the equation of the areas (Kepler's second law) is replaced by another equivalent law; the moment of the momentum is a constant, but the momentum must be defined as was done in the New Dynamics.

The only detectable effect will be a secular motion of the perihelion. With Lorentz's theory, we will find for this motion half of what Weber's law would give; with the theory of Abraham, two fifths.

If we now assume two mobile bodies gravitating around their shared center of gravity, the results are hardly different, although the calculations are a little more complicated. The motion of the perihelion of Mercury would therefore be 7″ in Lorentz's theory and 5.6″ in that of Abraham.

Additionally, the effect is proportional to n^3a^3, where n is the mean motion of the body and a the radius of its orbit. For planets, because of Kepler's third law, the effect varies according to $1/\sqrt{a^5}$; it is therefore insignificant except for Mercury.

It is also insignificant for the Moon, even though n is large, because a is extremely small; in brief, it is five times smaller than for Venus and 600 times smaller for the moon than for Mercury. For Venus and the Earth, we should also add that the motion of the perihelion (for the same angular velocity for this motion) would be much more difficult to detect by astronomical observations because the eccentricity of their orbits is much smaller than for Mercury.

In brief, the only notable effect on astronomical observations would be a motion of Mercury's perihelion, in the same direction as that which was observed without explanation, but in particular much smaller.

That could be regarded as an argument in favor of the New Dynamics, because another explanation must still be sought for the largest portion of the anomaly of Mercury; but, it could still, to a lesser extent, be regarded as an argument against it.

XV. The Theory of Lesage

It is appropriate to confront these considerations with a theory proposed long ago for explaining universal gravitation. Let us assume that, in interplanetary space, very small corpuscles move in all directions, at very high speeds. A spatially isolated body would not appear to be affected by the impact of these corpuscles, because these impacts are equally distributed in all directions. But, if two bodies A and B are present, the body B will play the role of a screen and will intercept a portion of the corpuscles which, without it, would have struck A. Then, the impacts received by A in the direction opposite that of B will no longer be counterbalanced or will only be imperfectly balanced and they will push A towards B.

Such is the theory of Lesage; and we are going to discuss it by placing ourselves first in the perspective of Ordinary Mechanics. First, how must the impacts called for by this theory take place? Is it according to the laws of perfectly elastic bodies, or according to those of inelastic bodies, or according to an intermediate law? The Lesage corpuscles cannot behave like perfectly elastic bodies; without that, there would be no effect, corpuscles intersected by the body B would be replaced by others which would have bounced off B and calculation shows that the compensation would be perfect.

The impact must therefore drain energy from the corpuscles and this energy should reappear as heat. But how much heat would be produced that way? We observe that the attraction passes through the bodies; the Earth therefore has to be represented, for example, not as a solid screen, but as formed of a very large number of very small spherical molecules which individually play the role of small screens, but between which the Lesage corpuscles can move freely. Thus, not only is the Earth not a solid screen, but it is not even a sieve, because the spaces in it take much more space than the solids. So that we can be aware of it, recall that Laplace proved that the attraction, when passing through the Earth is weakened by at most one ten-millionth, and his proof leaves nothing to be desired: if, in fact, the attraction was absorbed by the bodies that it passes through, it would no longer be proportional to the masses; it would be *relatively* weaker for large bodies than for the small, because it would have a greater thickness to pass through. The attraction of the Sun on the Earth would therefore be *relatively* weaker than that of the Sun on the Moon and its result would be a very noticeable inequality in the motion of the moon. We will therefore have to conclude (if we adopt the theory of Lesage) that the total surface of the spherical molecules which make up the Earth is at most one ten-millionth of the total surface of the Earth.

Darwin proved that the theory of Lesage only led exactly to Newton's law by assuming the corpuscles were entirely without elasticity. The attraction exerted by the Earth on a mass 1 at a distance 1 will then be proportional both to the total surface S of the spherical molecules which make it up, the velocity v of the corpuscles, and the square root of the density ρ of the medium formed by the corpuscles. The heat produced will be proportional to S, the density ρ and the cube of the velocity v.

But, the resistance experienced by a body which moves in such a medium has to be considered; it cannot move, in fact, without going in front of some impacts and by fleeing, in contrast, before those coming in the opposite direction, such that the balance achieved in the state of rest can no longer continue. The calculated resistance is proportional to S, ρ and v; now, it is known that celestial bodies move as if they did not experience any resistance, and the precision of the observations allows us to set a limit on the resistance of the medium.

Since this resistance varies with $S\rho v$, whereas the attraction varies as $S\sqrt{\rho}v$, we see that the ratio or the resistance to the square the attraction is the inverse of the product Sv.

We therefore have a lower limit on the product. We already had an upper limit of S (by the absorption of the attraction by the bodies that it traverses); we therefore have a lower limit on the velocity v, which must be at least equal to 24×10^{17} times that of light.

From that, we can deduce ρ and the quantity of heat produced; this quantity would be sufficient for raising the temperature by 10^{26} degrees per second; the Earth would therefore receive in a given time 10^{20} times more heat than the Sun emits in the same time; I don't mean the heat from the sun reaching the Earth, but the heat that the Sun radiates in all directions.

It is obvious that the Earth would not withstand such a situation for long.

The results would be no less fantastic if, in contrast to the views of Darwin, the corpuscles of Lesage were given an imperfect and nonzero elasticity. In truth, the energies of these corpuscles would not be completely converted to heat, but the attraction produced would also be less, such that it would only be the portion of this energy converted into heat which would contribute to producing the attraction and that would result in the same thing; this could be remarked with a judicious use of the Virial Theorem.

The theory of Lesage can be transformed; let us eliminate the corpuscles and imagine that the ether is traversed in all directions by light waves coming from all points in space. When a material object receives a light wave, this wave exerts a mechanical action on it due to the Maxwell-Bartoli pressure, just as if it had received the impact of a material projectile. The waves in question can therefore play the role of Lesage corpuscles. That is what, for example, Tommasina accepts.

This does not remove the difficulties; the speed of propagation cannot be that of light and one is thus led, by the resistance of the medium, to an unacceptable figure. Additionally, if the light were completely reflected, the effect is null, just as with the hypothesis of perfectly elastic corpuscles. In order to have attraction, the light must be partially absorbed; but then heat would be produced. The calculations are essentially unchanged from those done in the ordinary theory of Lesage, and the results in that way maintain an aura of fantasy.

From a different perspective, the attraction is not absorbed by the bodies that it passes through or it is hardly absorbed; it is not the same with the light that we know. Light which produces Newtonian attraction would have to be considerably different from ordinary light and have, for example, a very short wavelength. Not to mention that, if our eyes were sensitive to this light, the entire sky would have to appear to us much brighter than the sun, such that the sun would seem to stain it black, without which the sun would repel us instead of attracting us. For all these reasons, the light which would serve to explain the attraction would have to be much more like Röntgen's x-rays than ordinary light.

Yet still, x-rays would not be sufficient; however penetrating they may seem to us, they would not be able to pass through the entire Earth; one would have to imagine x′-rays that are much more penetrating than ordinary x-rays. Next, a portion of the energy from these x′-rays would have to be destroyed, since without it there would be no attraction. If one doesn't want it transformed into heat, which would lead to an enormous production of heat, it would have to be accepted that it is radiated in all directions in the form of secondary rays, which could be called x″-rays and which would have to be much more penetrating still than x′-rays, without which they would in turn disturb the phenomenon of attraction.

Such are the complicated hypotheses to which one is led when one wants to make the theory of Lesage viable.

But, everything that we just stated assumes the ordinary laws of Mechanics. Will things work out better if we accept the New Dynamics? And first, can we keep the Principle of Relativity? Let's first give the theory of Lesage its most basic form and assume space is plowed by material corpuscles; if these corpuscles were perfectly elastic, the laws for their impact would satisfy this Principle of Relativity, but we know that then they would have no effect. It has to be assumed that these corpuscles are not elastic, and then it is difficult to imagine a law for their impact compatible with the Principle of Relativity. Further, a considerable production of heat would again be found and again a very sizable resistance of the medium.

If we eliminate the corpuscles and if we return to the hypothesis of Maxwell-Bartoli pressure, the difficulties will not be any less. That is what Lorentz tried in his Article for the Amsterdam Academy of Sciences, April 25, 1900.

Let us consider a system of electrons immersed in an ether traversed in all directions by light waves; one of these electrons, struck by one of these waves is going to go into vibration; its vibration will be synchronous with that of the light; but it can have a different phase, if the electron absorbs part of the incident energy. If, in fact, it absorbs energy, it is because it is the vibration of the ether which *drives* the electron; the electron must therefore lag behind the ether. A moving electron is comparable to a convection current; therefore any magnetic field, in particular the one which is due to the disruption from the light itself, must exert a mechanical action on this electron. This action is very small; further, it changes sign in the course of the period; nonetheless, the mean action is not zero if there is a phase difference between the vibrations of the electron and that of the ether. The average action is proportional to this difference and consequently to the energy absorbed by the electron.

I'm not able to go into the calculations in detail here; let us just say that the final result is an attraction between two arbitrary electrons, equal to:

$$\frac{EE_1}{4\pi E' r^2}.$$

In this formula, r is the distance between the two electrons, E and E_1 are the energy absorbed by the two electrons during unit time, and E' is the energy of the incident wave per unit volume.

Therefore, there cannot be attraction without absorption of light and consequently without production of heat and that is what compelled Lorentz to abandon this theory which in the end is not different from that of Lesage-Maxwell-Bartoli. It would have been much scarier still if he had carried the calculation to the end. He would have found that the temperature of the Earth would have to increase 10^{13} degrees per second.

XVI. Conclusions

I have made an effort in just a few words to give as complete an idea as possible of these new doctrines; I sought to explain how they had come to be, without giving the reader room to be scared by their difficulty. The new theories are not yet proven, much is still needed; they are based only on a sufficiently serious set of probabilities that one mustn't treat them with disdain.

Most likely new experiments will teach us what we must finally think of it. The crux of the question is in Kaufmann's experiment and those that we can try to verify it.

Allow me to end with a wish. Let us assume that in a few years these theories undergo new tests and that they win out; our secondary instruction then runs a serious danger: some teachers are most likely going to want to make a place for the new theories. New things are so attractive and it is so hard to appear left behind! At least, one wants to give glimpses to the children and, before teaching them Ordinary Mechanics, one will warn them that its time has passed and that it was good at most for that old jewel from Laplace. And then, they won't take on the habit of Ordinary Mechanics.

Is it good to warn them that Ordinary Mechanics is only approximate? Yes. But later, when they will have gotten all the way to the marrow, when they will have formed the habit of only thinking with it, when they are no longer be at risk of forgetting it, then one will be able to show them its limitations without harm.

They will have to live with ordinary mechanics; it is the only one that they will always have to apply; whatever the progress of the automobile, our cars will never reach speeds where it is no longer true. The other is just a luxury, and one must only think of the luxury when there is no risk of harming the essential.

Henri Poincaré,

of the *Académie des Sciences* and the *Académie française*.

Part II
Discussion

Chapter 7
Discovery of the Electron: Cathode Rays

J. J. Thomson

J. J. Thomson was awarded the Nobel Prize in Physics in 1906 "in recognition of the great merits of his theoretical and experimental investigations on the conduction of electricity by gases." (Nobel Media AB 2019, n.d.). The citation does not mention electrons or cathode rays, even though it is commonly stated that he discovered the electron.

Between 1897 and 1899, J. J. Thomson published three papers in the *Philosophical Magazine* about his work on "conduction of electricity by gases"; they are (Thomson, Cathode Rays, 1897a), (Thomson, On the Charge of Electricity Carried by the Ions Produced by Röntgen Rays, 1898) (Thomson, On the Masses of Ions and Gases at Low Pressures, 1899). In this series of three papers, J. J. Thomson respectively measures the charge-to-mass ratio and velocity of the particles in cathode rays, the charge on ions produced in gas irradiated with X-rays, and the charge-to-mass ratio and mass of the negative ions released from a polished zinc surface by the photoelectric effect. George Smith (Smith, 2001) rightly points out that they need to be considered together.

Before these, J. J. Thomson also published two preliminary papers early in 1897 in the Proceedings of the *Cambridge Philosophical Society* and *The Electrician*. The first is a discussion of the work he had just started on cathode rays and has no description of equipment or results (Thomson, On the cathode rays, 1897c), and the second in contrast is clearly worthy of note—it has a discussion of several experiments, introduces corpuscles as a component of atoms and identifies the negative corpuscles as particles making up cathode rays, and ends with a value of the charge-to-mass ratio near the bottom of the last column (Thomson, Cathode Rays, 1897b). This measurement was done using magnetic deflection and the method is described in the first paper from the series (Thomson, Cathode Rays, 1897a, pp. 300–307).

Continuing with the first in the series of three papers (Thomson, Cathode Rays, 1897a), this would seem to be a good point to ask, what was J. J. Thomson trying to

B. D. Popp, *Henri Poincaré: Electrons to Special Relativity*,
https://doi.org/10.1007/978-3-030-48039-4_7

accomplish? After this series, J. J. Thomson did not return to publish subsequent articles on this subject. In contrast, Walter Kaufmann published several papers with improved measurements of the charge-to-mass ratio. This suggests that J. J. Thomson's main interest was not obtaining values (and increasingly accurate values) for properties of the electron. If it weren't measuring properties of electrons, what was he trying to do. Referring to the first paper in the series (Thomson, Cathode Rays, 1897a), the second sentence states there are "diverse opinions" about cathode rays and contrasts the view that they are wholly processes in the ether against the view that they are "wholly material, and ... mark the paths of particles of matter charged with negative electricity." Two paragraphs later he also writes, "The following experiments were made to test some of the consequences of the electrified-particle theory." With this series of experiments, J. J. Thomson is looking for an experiment to decide between an ether-vortex theory of subatomic processes and an electrified-particle theory. After the first paper on cathode rays, the series continues with other electrified particles. In hindsight, the experiments are recognized as decisive evidence of negatively charged particles of very small mass compared to positive ions encountered in electrolysis.

Jean Perrin

The introduction to the experiment described in the first section of (Thomson, Cathode Rays, 1897a), "Charge Carried by the Cathode Rays," reinforces this perspective. The description starts with a reference to an experiment done by Jean Perrin. J. J. Thomson does not provide a reference; however, it seems unambiguous that he is referring to the experiment reported in (Perrin, 1895). It is relevant for us to look at it.

Jean Perrin—working in Paris—conducted an experiment showing that cathode rays transported a negative charge[1]. It appears to be the first published work by the winner of the 1926 Nobel Prize in physics, famous for his study of Brownian motion. He starts by stating, "Two hypotheses have been imagined for explaining the properties of cathode rays." Like J. J. Thomson's introduction described above this introduction establishes the contrast between the two theories. Jean Perrin continues, "Some ... think that this phenomenon, like light, is due to vibrations of the ether." and then "Others,"—and here he includes J. J. Thomson—"think that these rays are formed by negatively charged matter." In these terms, Jean Perrin, sets up the same contrast as J. J. Thomson between the competing explanations for cathode rays. Jean Perrin goes on to describe his experimental apparatus,

[1] Another aspect of Jean Perrin's paper is worth noting. He wrote that he looked for positive charges corresponding to the negatively charged cathode rays and writes, "I think I found them in the same region where the cathode rays form." (Perrin, 1895) These could be the channel rays (kanalstrahlen) discussed by Poincaré in (Poincaré, La dynamique de l'électron, 1908) section III and translated on page 107 of this book.

corresponding to the description provided by J. J. Thomson, and concludes, "*the cathode rays are therefore charged with negative electricity*." He is providing experimental evidence for particles over ether. This work is also described in the first part of his doctoral thesis at the *École Normale Supérieure*.

While J. J. Thomson and physicists in France and Germany cited this article by Perrin and I have suggested it was a motivation for their work on cathode rays, I have seen no indication why Perrin chose to work on this subject. There is no suggestion in the article itself or in his thesis (including the forward) to indicate who or even whether someone suggested this topic to him. The footnote at the very end of the article states, "This work was done at the laboratory of the *École Normale* and at the laboratory of Mr. Pellat, at the Sorbonne." (Perrin, 1895) The forward to Jean Perrin's thesis indicates that his work in 1895 was done in Mr. Pellat's laboratory. Henri Pellat (b. 1850, d. 1909) had been named adjunct professor with his own laboratory in 1893. After looking at his bibliography of published books[2] and looking at some of his notes published a few years earlier in *Comptes rendus hebdomadaires de l'Académie des Sciences*, including a note in the same volume from 1895 as Jean Perrin's article, nothing would suggest Henri Pellat had a research interest that would lead him to suggest such a topic to a student. The selection of this topic by a beginning graduate student remains mysterious.

The second part of Perrin's thesis involved Röntgen rays.

J. J. Thomson continues this first section noting that the experimental apparatus of Jean Perrin criticized for not excluding the possibility that negative charges incidentally followed the same straight-line path as the fundamental ether phenomenon. To address this objection, J. J. Thomson modifies the configuration used by Jean Perrin by adding a bend in the glasswork and providing a magnet to bend both the cathode rays (visible by their discharge) and the electric charges (measured by an electroscope) onto the target.

J. Perrin's experiment and results alone might be sufficient and J. J. Thomson's results with the improved design of the layout of the path for the cathode rays are sufficient for showing that the cathode rays are particles. J. Perrin stopped at that point; J. J. Thomson continued.

Cathode Rays

J. J. Thomson (Cathode Rays, 1897a) continues by describing the effect of an electric field on the rays. When the rays are made to pass between two parallel aluminum plates connected to a storage battery, the rays were deflected towards the plate connected to the positive pole of the battery if the vacuum was good enough. Previously H. Hertz had used this method to measure the electrostatic deflection of cathode rays, to determine their charge, and did not measure any deflection. His

[2]https://data.bnf.fr/fr/12744499/henri_pellat/

negative result was seen as supporting an ether-disturbance interpretation of cathode rays, but the negative result had come to be viewed as suspect by the time of J. J. Thomson's experimental work. J. J. Thomson provides an explanation: H. Hertz did not have an adequate vacuum. J. J. Thomson notes, "it was only when the vacuum was a good one that the deflexion took place, but . . . the absence of deflexion is due to the conductivity of the medium." Here we have an indication of the importance of vacuum pumping technology.

This is an opportune point to digress and indicate the importance of mastery of laboratory skill and technology. One aspect of skill involved glass blowing to produce glass tubes with metal structures inside and appropriate electrodes passing through the walls in the right locations. The second aspect involves producing vacuums—with the associated technology of pumps and processes of pumping, waiting for outgassing and repeated pumping—with which to achieve low pressures corresponding to greatly reduced quantities of residual gases. With the need for skill, also came risks. J. J. Thomson had a reputation as a klutz when it came to handling laboratory equipment and was generally steered away. Jean Perrin, in a footnote, indicates that he could not continue a particular line of measurements because his glasswork broke.[3]

At this point J. J. Thomson (Cathode Rays, 1897a) had established that cathode rays are particles and plausibly explained the contradicting results of the earlier experiment by H. Hertz. Additional experiments to determine particular properties of the particles were then well justified. The important experimental results are the mass-to-charge ratio (*m*/*e*) obtained by two methods involving magnetic and electrostatic deflection and the velocity of cathode ray particles under different conditions. In presenting his results, J. J. Thomson first notes that the value of the mass-to-charge ratio is independent of the nature of the residual gas and metal of the electrodes and is about 10^{-7} cgs. Then, interpreting the newly measured mass-to-charge ratio in this context, J. J. Thomson writes that this value "is very small compared with the value 10^{-4}, which is the smallest value of this quantity previously known, and which is a value for the hydrogen ion in electrolysis." J. J. Thomson argues that the ratio is smaller because of the size but accepts that it could also be combined with a larger charge. It would seem plausible for him to argue that the charge on the hydrogen ion (in electrolysis) and on the electron (in cathode rays) need to be equal in magnitude and opposite in sign to produce a net neutrality; he does not. Interestingly he does speculate, "If, in the very intense electric field in the neighborhood of the cathode, the molecules of the gas are dissociated and are split up, not into the ordinary chemical atoms but into these primordial atoms, which we shall call for brevity call corpuscles; and if these corpuscles are charged with electricity and projected from the cathode by the electric field, they would behave exactly like cathode rays." In this sentence, we can read J. J. Thomson defining what

[3]When I, in an undergraduate physics laboratory many years ago, measured the mass-to-charge ratio of an electron, glasswork and vacuum pumps were not involved; I was shown an instrument on a laboratory table, told "do this," "measure that," and given a handout with more information.

he meant by "corpuscle": a building block of atoms. And we can add, knowing the comparatively small mass-to-charge ratio for the corpuscles in cathode rays, that they must be quite small, subatomic in fact.

At the end of the first article in the series, J. J. Thomson has the mass-to-charge ratio of his negatively charged corpuscle, plausible speculation that it is very small compared to the positive ions in electrolysis, a theory that it is a primordial component of all atoms and an expectation, consequently, that corpuscles may be ubiquitous where atoms form positive ions. Trusting and following his experimental results, he has gone much farther than the project set out in the first two paragraphs of the paper.

In the next two papers in the series, J. J. Thomson looks at two other configurations were corpuscles may be found in order to measure the charge and confirm that corpuscles have a very small mass compared atoms.

In the second article (Thomson, On the Charge of Electricity Carried by the Ions Produced by Röntgen Rays, 1898), the charge on gas ions produced by X-rays is measured but without information on the mass of the ions whose charge is being measured in the experiments. And in the third paper (Thomson, On the Masses of Ions and Gases at Low Pressures, 1899) the mass-to-charge ratio and mass of the ions released from a negatively-charged, polished, zinc plate exposed to ultraviolet radiation, the photoelectric effect, are measured. The experiment in the second article is not able to confirm that the mass of the electron is very small compared to positive ions. The third article does this and identifies the electron in a different context than cathode rays. That is, for us, the essential point of the series of papers by J. J. Thomson; he has identified cathode rays as a particle with a distinctive property and shown that the same particle is found in an experiment entirely unrelated to cathode rays.

Ether

To understand the phrase "processes in the ether" in the sentence from J. J. Thomson quoted above and therefore to understand a distinction between an ether-vortex or continuous charge model of atomic or subatomic phenomena (exemplified by (Larmor, 1893), for example), and a material charge or particle model (hard sphere, in theoretical work by Lorentz and other work discussed in Chapter 9) it is necessary to consider the role of ether in the earlier conception of charge. The ether formerly had a role both as a medium for the propagation of light waves (luminiferous ether) and also as a component of electrodynamics.

The aspect of ether as an absolute medium for the propagation of light waves is probably more familiar to most readers because it is connected with special relativity dispensing with the possibility of any absolute frame of reference.

However, from the publication of Maxwell's treatise in 1873 until the time of J. J. Thomson's work, a larger role was cut out for the ether than just a medium for the propagation of light. As understood by Maxwell and the Maxwellians[4], charge was a continuous fluid and analogies with fluid flow were prominent. Terminology suggestive of fluid analogies is still used like continuity equation, flux and field lines. In this fluid view, the appearance of charge on a surface of an object represents the boundary between a dielectric and matter. Further, charge appears discrete because it only occurs at the end of Faraday tubes and, for their part, Faraday tubes are a vortex (or other phenomenon) of the fluid ether.

Because of investigations, for example, with electrolytic solutions, which suggested the presence of small charged ions in the solution, it became necessary to have a theory that could account for these ions. This is where the work on an ether-vortex theory by (Larmor, 1893) mentioned above fits in.

Several points need to be noted when referring to discussions of the ether or singling out Larmor's work as an example, as I just did. Over the 20 years following the publication of Maxwell's treatise in 1873, different approaches to understanding and interpreting what Maxwell had written developed. One school of thought, in England, was represented by the Maxwellians, already mentioned; other schools, in Germany, developed from work by H. von Helmholtz and by H. Hertz. None of these theories, including Larmor's, were entirely satisfactory in addressing the demands placed on them, and this was especially true in relation to interactions between magnetic fields, light and matter.

Radiation and optical phenomena provide the connection between ether as charged fluid and ether as a medium for the propagation of light. Accelerated charge is associated with an acceleration of the ether resulting in a disturbance propagating as a transverse wave. The need for a continuous medium for the propagation of light provides its own set of constraints on the properties of the ether. The interactions between ether and matter, notably the motion of the Earth through the ether, impose further constraints and raises questions such as: whether the ether is dragged by the annual motion of the Earth. Experiments first with aberration and then by Michelson and Morley looked for effects of this drag to first-order and then to second-order in the ratio of the velocity of the observer to the speed of light, respectively; no effect was found.

While J. J. Thomson's work led to acceptance of the concept of discrete, material particle-based charge over the course of a few years it did not have the same effect on the aspect of ether relating to a medium for the propagation of light. In 1905 eight years after J. J. Thomson's work, Poincaré and Einstein did not need in their papers to argue for the discrete, material nature of electrical charge; they did need to use the 18-year-old results from Michelson and Morley to argue that there was no medium for the propagation of light and no absolute frame of reference. The aspect of the

[4]For information on who the Maxwellians were and what they did, see (Hunt, 1991). Notably, Oliver Heaviside, whose contribution to mathematical notation is discussed in Chap. 6, is among them.

ether as an absolute reference frame is considered in Chapters 9 and 10. Poincaré presents other arguments against the ether as a medium for propagation of light in a later article (Poincaré, La dynamique de l'électron, 1908). These arguments are largely based on the absurdities that would result from these constraints imposed on a theory of the ether.

J. J. Thomson's Motivation

The timing of J. J. Thomson's work, that is to say his motivation for starting the work and bring it to a conclusion, need to be considered. Isobel Falconer persuasively shows that there was a burst of work on the nature of cathode rays starting in 1896 and 1897 (Falconer, 1987). Figure 1 on page 246 is particularly effective in this respect. In a later subsection of this chapter, a similar table will be built with a list of references provided in a review article (Kaufmann, Die Entwicklung des Electronenbegriffs, 1901a). That table also shows a significant jump in articles published on cathode rays in 1896 and 1897. Falconer (Corpuscles, Electrons and Cathode Rays: J. J. Thomson and the 'Discovery of the Electron', 1987, p. 249) writes, "The effect of the discovery of X-rays on cathode ray research was clearly dramatic." Did the interest in these new rays motivate the research on cathode rays seen in the jump in publications or is the timing fortuitous? W. Röntgen famously took an X-ray image of the bones in his wife's hand in December 1895 and word quickly spread in January 1896. Röntgen's discovery was directly connected with H. Becquerel's decision to look for radiation directly from natural sources, specifically uranium. (This is discussed more in Chapter 8.) This led to his discovery in the following months of what were called, for a while, Becquerel rays. Becquerel rays promptly took their place alongside Röntgen rays (X-rays) and Lenard rays as a topic of interest, discussion and research. In Germany in the review article by Walter Kaufmann (Die Entwicklung des Electronenbegriffs, 1901a), he clearly indicates that X-rays were important to their motivation. In contrast to Becquerel and Kaufman, Falconer's argument for what caused this jump in interest is less persuasive when applied to J. J. Thomson. As discussed at the beginning of this paragraph J. J. Thomson stated that his motivation was providing an experimental determination between competing understandings of cathode rays. He explicitly refers to Jean Perrin's experiment (Perrin, 1895) reported within a week or two of Röntgen's X-ray image of his wife's hand. J. J. Thomson does not refer to Becquerel, Lenard or Röntgen or their discoveries in his three papers. I take at face value J. J. Thomson's statement of his intent in (Thomson, Cathode Rays, 1897a) discussed above. I also see the timing as following from the work by Jean Perrin which coincidentally appeared at the same time is Röntgen's discovery.

Naming the Electron

J. J. Thomson in describing the subject of his work did not use the term "electron," even many years later; he preferred "corpuscle." In the first full-length article (Thomson, Cathode Rays, 1897a), indicates what he means by corpuscles as discussed above.

The origin of the term electron for a charge-bearing atomic component is commonly credited to George Johnstone Stoney who used it in 1891 in a paper on atomic spectra. (Stoney, On the Cause of Double Lines and of Equidistant Satellites in the Spectra of Gases, 1891, p. 585). Stoney also defends his claim of priority for the use of this term and meaning in 1894 in a letter to the editor (Stoney, Of the "Electron" or Atom of Electricity, 1894). It is also suggested that the physicist George Fitz Gerald, an editor of the Philosophical Magazine where Stoney and Thomson published and, notably, a nephew of Stoney, encouraged the use of the term electron to refer to the ion or particle investigated by Thomson. In a conference paper presented by W. Kaufmann (Kaufmann, Die Entwicklung des Electronenbegriffs, 1901a) September 25, 1901, he cited Stoney's 1891 paper (cited above) and credits him with deciding on the now commonly accepted name, electron.

By 1906, the term electron was in common use by H. Poincaré (in particular in the works translated in Part I of this book), H. A. Lorentz, and others whose work is mentioned in other chapters. This can be seen as a sign that they, between five and 10 years after J. J. Thomson's experiments, were not concerned about making a strong distinction between a particle view and an ether-vortex view for the explanation of subatomic phenomena. In the preface to (Poincaré, Sur la dynamique de l'électron, 1906), Poincaré indicates the Michelson Morley experiment as grounds for justifying the reworking of electrodynamics and electron theory without an ether.

Meaning of Discovery

If we view J. J. Thomson's series of three papers as an experimental determination of the particle nature of cathode rays (in contrast to an ether-vortex origin) and further of the mass-to-charge ratio and mass of these particles showing that they were much smaller than the positively charged ions, then what does it mean to say that "J. J. Thomson discovered the electron." Before approaching this question directly, it is useful to consider some examples of discoveries in physics (and the sciences).

The discovery of radiation from natural sources and of polonium and radium by Henri Becquerel, and Pierre and Marie Curie (discussed in the next chapter), and the discovery of the Zeeman effect by Pieter Zeeman all involve easily observed and reproduced phenomena. In the first case, at the most basic level, Henri Becquerel observed that photographic plates stored in the dark could be exposed by a rock of uranium ore placed nearby and that a metal object placed between the ore and the plate even cast a shadow. Pieter Zeeman showed that when the source (typically a flame) of some emission spectral lines was placed in a strong static magnetic field, certain lines became split into several components. At their core, both involve a previously unknown natural phenomenon that, once discovered, can be investigated and studied. Some component of what J. J. Thomson did can be understood in this sense: even under high vacuum an electric current could still pass from cathode to anode.

Some discoveries involve the recognition of patterns (empirical laws) in experimental data. In his introduction (Poincaré, Sur la dynamique de l'électron, 1906), Poincaré mentions Johannes Kepler's work that led to his three laws of planetary motion. By analyzing Tycho Brahe's observations of the position of Mars on the sky, Kepler was able to establish his laws notably including that planetary orbits are elliptical.

Later, Isaac Newton was able to establish that Kepler's laws of planetary motion could be confirmed as a mathematical consequence of Newton's laws of motion and gravitation. This involves taking a new discovery (Newton's laws of motion and gravitation) and looking back for confirmation or explanation of earlier discoveries (or laws) or at least consistency with them. Newton's laws, together with observations, were also used to look forward and make predictions resulting in discoveries.

Edmund Halley in 1705 published a book (Halley, 1705) containing a list of the orbital elements (including the ascending node, orbital inclination and perihelion) for 24 comets observed between 1337 and 1698. He had calculated the orbital elements using a method described by Newton based on the laws of motion and gravitation applied to observations of the comets' positions on the sky over the days each was visible. The observations he used included his own observations of a comet in 1682 and observations by Kepler and Longomontanus of a comet that appeared in 1607. Halley indicates that the main use of the table with the orbital elements for 24 comets is "that whenever a new comet shall appear, we may be able to know, by comparing together the elements, whether it be any of those which has appeared before … and to foretell its return." Even though the appearance of the comet (for example its peak brightness, or the shape or length of its tail) may vary from one return to another, it could be identified as the same comet if the orbital elements were the same. And indeed, Halley notes the similarity of the orbital elements for the comet that appeared in 1531, 1607 and 1682. He concluded that it was the same comet, and inferred an orbital period of about 76 years. He then predicted, "Hence I dare venture to foretell, that it will return again in the year 1758." Halley died in 1742, so he did not see his prediction confirmed in 1758. In honor of its discoverer, this comet is named after Edmund Halley.

This gives us a way to look again at what J. J. Thomson did. He provides us with a description of what he found ("a building block of atoms"), and a property and value with which to distinguish it. His measurement of the mass-to-charge ratio of cathode ray particles is key to distinguishing the electron as a particle and further to identify it as the same particle when it appears in a different context. Independent of the intent and context within which he made the measurement, the measurement itself does provide a distinctive property with which to confirm the presence of the electron in other contexts. Measuring the mass-to-charge ratio of the ions released by the photoelectric effect in his third paper (Thomson, On the Masses of Ions and Gases at Low Pressures, 1899) allows J. J. Thomson to confirm that he is working with the same particle as he did in his first paper and therefore to associate the mass measurement also in this third paper with both the negative ions in cathode rays and the photoelectric effect. This property was also useful at that time for identifying the β ray produced in radioactive decay of uranium as an electron, as discussed in Chapter 8.

Priority

In discussing the history of physics in the decade between 1895 and 1905, some English authors seem to write about work at Cavendish Laboratories in Cambridge, and about *continental* ideas or physics[5]. Work on the continent in this timeframe was not monolithic and did follow a variety of directions including cathode rays. Awarding of the Nobel Prize to J. J. Thomson suggests he deserves priority for discovery of the electron. The situation is less clear cut and we need to look at work on cathode rays in Germany that has not been discussed so far in this section.

In Germany starting in 1897, so roughly at the same time as J. J. Thomson in England, several people started measurements of the mass-to-charge ratio of cathode ray particles. In a review paper presented by W. Kaufmann (Die Entwicklung des Electronenbegriffs, 1901a) at the 73. Naturforscherversammlung zu Hamburg conference on September 25, 1901[6], he lists seven authors of 13 papers published in 1897 and 1898 measuring properties (including the mass-to-charge ratio) of the electron. His list of authors and the references to their published work is provided in Table 7.1, sorted by submission date.

Like Figure 1 in (Falconer, 1987, p. 246), Kaufmann's list of references prepared in 1901 shows the number of people working on the subject and with the dates added it shows the concentration of the work in just two years. Just as Falconer's Figure 1 reflects her choice of journals to survey, this list reflects Kaufmann's selection of articles that were worth referencing. Together, the conclusion is stronger: cathode rays attracted the research interest of many people in a short time span.

Like the first paper in the series by J. J. Thomson discussed above, several of these papers reference Jean Perrin's work published in December 1895 and already discussed. In his conference review paper, Kaufmann does not mention Perrin's work but he does provide an illuminating statement, "The more facts were

[5]To provide a relevant example, but without suggesting that it is more or less deserving of critique, consider the first chapter provided by A. B. Pippard from the University of Cambridge for the book *Electron: a Centenary Volume* (Springford, 1997). The adjective continental is used three times in the first 10 pages.

[6]An English adaptation of this paper was published within two months in *The Electrician* (Kaufmann, The Development of the Electron Idea, 1901b). The adaptation does not include this list of references.

Table 7.1 Authors working on measuring properties of cathode rays from (Kaufmann, Die Entwicklung des Electronenbegriffs, 1901a)

Submitted	Published	Author	Citation	Title
January 7, 1897	January 7, 1897	E. Wiechert	Schriften der phys. ökon. Gesellsch. Königsberg 1897. S. 1	Über das Wesen der Elektricität
January 7, 1897	May 8, 1897	E. Wiechert	Naturwiss. Rundsch. Mai 1897	Über das Wesen der Elektricität
May 21, 1897	June 15, 1897	W. Kaufmann	Ann. der Phys. **61**, 544, 1897	Die magnetische Ablenkbarkeit der Kathodenstrahlen und ihre Abhängigkeit vom Entladungspotential
August 7, 1897	October 1897	J. J. Thomson	Phil. Mag. (5) **44**, 293, 1897	Cathode Rays
November 10, 1897	November 9, 1897	W. Wien	Verhdl. physilc. Ges. Berlin **16**, 165, 1897	Über die elektrostatischen Eigenschaften der Kathodenstrahlen
October 25, 1897	November 25, 1897	W. Kaufmann and E. Aschkinass	Ann. der Phys. **62**, 588, 1897	Über die Deflexion der Kathodenstrahlen
October 25, 1897	November 25, 1897	W. Kaufmann	Ann. der Phys. **62**, 596, 1897	Nachtrag zu der Abhandlung: „Die magnetische Ablenkbarkeit der Kathodenstrahlen etc.“
February 4, 1898	January 31, 1898	T. Des Coudres	Verhdl. physik. Ges. Berl. **17**, 17, 1898.	Ein neuer Versuch mit Lenhard'schen Strahlen
January 2, 1898	February 1, 1898	P. Lenard	Ann. der Phys. **64**, 279, 1898	Über die elektrostatischen Eigenschaften der Kathodenstrahlen
March 19, 1898		E. Wiechert	Gött. gel. Nachr. 1898. S. 260	Experimentelle Untersuchungen über die Geschwindigkeit und die magnetische Ablenkbarkeit der Kathodenstrahlen
March 19, 1898	May 15, 1898	W. Kaufmann	Ann. der Phys. **65**, 431, 1898	Die magnetische Ablenkbarkeit elektrostatisch beeinflusster Kathodenstrahlen
May 1, 1898	May 16, 1898	P. Lenard	Ann. der Phys. **65**, 504, 1898	Über das Verhalten den Kathodenstrahlen parallel zu elektrischer Kraft
October 3, 1898	November 11, 1898	W. Kaufmann	Ann. der Phys. **66**, 649, 1898	Bemerkungen zu der Mittheilung von A. Schuster: „Die magnetische Ablenkung der Kathodenstrahlen“

accumulated, the more puzzling became the cathode rays, and eventually matters went so far that it was almost considered unbecoming for a decent physicist to concern himself with phenomena so little amenable to theoretical and quantitative treatment. Then there came suddenly the most puzzling of all obscure phenomena—the X-rays discovered by Röntgen—and therewith a new stimulus for attacking the various riddles." (Kaufmann, The Development of the Electron Idea, 1901b). This is a clear statement of the motivation of the community by one of the key people involved with the measurement of the charge-to-mass ratio in Germany.

The place of J. J. Thomson's papers on this chronological list needs to be considered. The first of J. J. Thomson's series of three papers (Thomson, Cathode Rays, 1897a) is already fourth on the list. The third paper on the list (Kaufmann, Die magnetische Ablenkbarkeit der Kathodenstrahlen und ihre Abhängigkeit vom Entladungspotential, 1897) does cite a notice in *Nature* of publication of J. J. Thomson's first, preliminary announcement (Thomson, On the cathode rays, 1897c). As previously noted, this first announcement from J. J. Thomson does not include a value of the charge-to-mass ratio. J. J. Thomson's second announcement (Thomson, Cathode Rays, 1897a) with a value of the charge-to-mass ratio, as discussed in the first section of this chapter, slips in just before the paper by Kaufmann: Thomson's paper in *The Electrician* was published under a month before Kaufmann's paper in *Annalen der Physik*; and Thomson's paper is based on a "discourse delivered at the Royal Institution, Friday evening, April 30" (Thomson, Cathode Rays, 1897a) about three weeks before Kaufmann's paper was submitted.

Referring first to the two papers by Wiechert, although they have the same title, they do not have the same content. The second paper, published May 8, 1897 in *Naturwissenschaftliche Rundschau* (Wiechert, Ueber das Wesen der Elektricität, 1897b) does not have the second, experimental section that appears in the first paper in *Schriften der Physikalisch-Ökonomischen Gesellschaft zu Königsberg* (Wiechert, Über das Wesen der Elektricität, 1897a) and the second paper may also be further abridged. The second paper is not considered further.

The presentation by Wiechert of his experiment and results in (Wiechert, Über das Wesen der Elektricität, 1897a) is hard to follow. For example, the origin of the formula II with the ratio of the cathode ray charge-to-mass ratio to hydrogen ion charge-to-mass ratio, α, is unknown. Where do k and the numerical constants in this formula come from? It is hard to know whether to have any confidence in his results. In the conclusion he indicates that $1/\alpha$ is between 2000 and 4000. For comparison, the current accepted value of the ratio of the proton to electron mass is 1836 (National Institute of Standards and Technology, n.d.-b).

By contrast, Kaufmann's paper from the first half of 1897 (Kaufmann, Die magnetische Ablenkbarkeit der Kathodenstrahlen und ihre Abhängigkeit vom Entladungspotential, 1897) is much more satisfactory with a clear explanation of equipment, methodology and theory. Notably he has seven tables of measurements. Each table has measurements at 4 to 7 electrode voltage differences. For the results in the first table he uses aluminum electrodes and uses copper electrodes for the other tables. He also changes the distance between electrodes and in the last three tables uses illuminating gas[7], carbon dioxide and hydrogen in place of air. The breadth of the measurement conditions—metals for electrodes, residual gases in the tubes—is fully as broad as those used by J. J. Thomson in the first of his series of three papers.

Note that Kaufmann states that the value he determined for the charge-to-mass ratio of the electron is 10^7 cgs (Kaufmann, Die magnetische Ablenkbarkeit der Kathodenstrahlen und ihre Abhängigkeit vom Entladungspotential, 1897, p. 552). For comparison, J. J. Thomson summarizes his values for the charge-to-mass ratio (Thomson, On the Masses of Ions and Gases at Low Pressures, 1899, p. 554): 5×10^6 cgs for cathode rays and 7.3×10^6 cgs for electrons released by the photoelectric effect; he also quotes a value for cathode rays from Lenard of 6.4×10^6 cgs. The currently accepted value of the charge-to-mass ratio of electrons is 1.759×10^7 cgs (National Institute of Standards and Technology, n.d.-a)

Then Kaufmann stops without reducing the measurements of voltage, magnet current and deflection. He says he can't make sense of the results: they indicate that the charge-to-mass ratio is the same for all electrode materials and filling gases. Further, it looks like the charge-to-mass ratio is a thousand times larger than hydrogen ion. The paper ends with the statement that he doesn't have a satisfactory explanation of the observed results. It is a disappointing conclusion; in hindsight, it is easy to say that he should've followed where the facts pointed him. His conclusion however remains what it was and seems to knock him out of consideration for priority.

This leaves J. J. Thomson with priority for recognizing that cathode rays are streams of charged particles and measuring the charge-to-mass ratio of the particles. With this discovery, as with several others in this chapter, events are moving quickly and similar accomplishments occur with little time between. Here J. J. Thomson was first and his work that soon followed on the charge-to-mass ratio in ionized gases and in the particles freed from the metal surface in the photoelectric effect added weight to his finding by showing that a particle with this distinctive characteristic was found in different phenomena.

References

Falconer, I. (1987). Corpuscles, Electrons and Cathode Rays: J. J. Thomson and the 'Discovery of the Electron'. *British Journal for the History of Science*, 241–276.

Halley, E. (1705). *Synopsis of the Astronomy of Comets.* London.

Hunt, B. J. (1991). *The Maxwellians.* Ithaca, NY: Cornell University Press.

Kaufmann, W. (1897). Die magnetische Ablenkbarkeit der Kathodenstrahlen und ihre Abhängigkeit vom Entladungspotential. *Annalen der Physik, ser. 3, 61*, 544–552.

Kaufmann, W. (1901a). Die Entwicklung des Electronenbegriffs. *Physikalische Zeitschrift, 3*(1), 9–15.

Kaufmann, W. (1901b, November 8). The Development of the Electron Idea. *The Electrician, 48*, 95–97.

[7]A commercial gas mixture used for lighting.

Larmor, J. (1893). A Dynamical Theory of the Electric and Luminiferous Medium. *Proceedings of the Royal Society of London, 54*, 438–461.

National Institute of Standards and Technology. (n.d.-a). *CODATA Value: electron charge to mass quotient*. Retrieved November 12, 2019, from https://physics.nist.gov/cgi-bin/cuu/Value?esme

National Institute of Standards and Technology. (n.d.-b). *CODATA Value: proton-electron mass ratio*. Retrieved July 6, 2019, from https://physics.nist.gov/cgi-bin/cuu/Value?mpsme

Nobel Media AB 2019. (n.d.). *The Nobel Prize in Physics 1906*. Retrieved March 16, 2019, from NobelPrize.org: https://www.nobelprize.org/prizes/physics/1906/summary/

Perrin, J. (1895, December). Nouvelles propriétés des rayons cathodiques. *Comptes Rendus des Séances de l'Académie des Sciences, 121*, 1130–34.

Poincaré, H. (1906). Sur la dynamique de l'électron. *Rendiconti del circolo matematico di Palermo, 21*, 129–176.

Poincaré, H. (1908). La dynamique de l'électron. *Revue générale des sciences pures et appliquées, 19*, 386–402.

Smith, G. E. (2001). J. J. Thomson and the Electron. 1897–99. In J. Z. Buchwald, & J. Warwick, *Histories of the Electron: The Birth of Microphysics* (pp. 21–76). Cambridge, MA: Dibner Institute for the History of Science and Technology.

Springford, M. (Ed.). (1997). *Electron: a Centenary Volume*. Cambridge: Cambridge University Press.

Stoney, G. J. (1891). On the Cause of Double Lines and of Equidistant Satellites in the Spectra of Gases. *Scientific Transactions of the Royal Dublin Society, IV*, 563–680.

Stoney, G. J. (1894). Of the "Electron" or Atom of Electricity. *Philosophical Magazine and Journal of Science, Fifth Series, 38*, 418–420.

Thomson, J. J. (1897a, October). Cathode Rays. *Philosophical Magazine and Journal of Science, Fifth Series, 44*, 293–316.

Thomson, J. J. (1897b, May 21). Cathode Rays. *The Electrician, 39*, 104–9.

Thomson, J. J. (1897c, February). On the cathode rays. *Proceedings of the Cambridge Philosophical Society, 9*, 243–4.

Thomson, J. J. (1898). On the Charge of Electricity Carried by the Ions Produced by Röntgen Rays. *Philosophical Magazine and Journal of Science, Fifth Series, 46*, 528–45.

Thomson, J. J. (1899). On the Masses of Ions and Gases at Low Pressures. *Philosophical Magazine, Fifth Series, 48*, 547–567.

Wiechert, E. (1897a). Über das Wesen der Elektricität. *Schriften der Physikalisch-Ökonomischen Gesellschaft zu Königsberg., 38*, 3–20.

Wiechert, E. (1897b). Ueber das Wesen der Elektricität. *Naturwissenschaftliche Rundschau, 12*, 237–239.

Chapter 8
Discovery of the Electron: Radioactivity

Wilhelm Röntgen and X-Rays

In December 1895 at roughly the same time as Perrin's note on cathode rays appeared in the *Comptes rendus*, Wilhelm Röntgen famously took an x-ray, the first, of his wife's hand showing the bone structure. Word of the sensation spread quickly and a photograph was shared at the weekly meeting of the *Académie des sciences* on January 20, 1896; only a few lines in the minutes mark its discussion (Oudin & Barthélemy, 1896). In a memoir written in 1903, Henri Becquerel notes that, "at the moment when Mr. H. Poincaré had just shown the first radiographs sent by Mr. Röntgen, I asked my colleague whether it had been determined where the site of emission of these rays was in the empty tube producing the x-rays. He told me that the origin of the radiation was the bright spot on the wall receiving the cathode flux. Right away I thought of researching whether the new emission was a manifestation of vibratory movement which gave rise to the phosphorescence, and whether any phosphorescent body emitted similar radiation." (Becquerel, Recherche sur une propriété nouvelle de la matière, 1903, p. 3) Here we have a case of an interesting idea occurring to a person well prepared to pursue it. Becquerel had spent many years researching phosphorescence in crystals and minerals. Phosphorescence is a phenomenon in which a material absorbs light at one wavelength and reemits it at another wavelength. That means that in the above quote, Becquerel is suggesting that a phenomenon of phosphorescence may be occurring in the wall of the empty tube, with one kind of radiation striking the wall of the tube which then reemits the radiation as Röntgen rays. This would seem to implicitly assume that the cathode rays are radiation and not particles. Becquerel's response to this idea is to ask whether other fluorescent materials emit Röntgen rays. As shown in that session, the obvious property of Röntgen rays is to pass through seemingly opaque material, a hand, and expose a photographic plate. What Becquerel sets out to do is therefore to expose candidate, fluorescent materials to sunlight while they are on a photographic plate well sealed against light. If the photographic plates turn out to be

B. D. Popp, *Henri Poincaré: Electrons to Special Relativity*,
https://doi.org/10.1007/978-3-030-48039-4_8

exposed, it is not because of the sunlight, but because of the materials generating Röntgen rays by fluorescence, presumably.

By the end of the January 1896 many more lines would be written and photographs taken, including a noteworthy skeleton of a frog (Röntgen, 1896, p. 61) and a review by Poincaré (Poincaré, Les rayons cathodiques et les rayons Röntgen, 1896). In his review, Poincaré confirms that he had presented the pictures for Drs. Oudin and Barthelemy two weeks earlier and that he understood the phenomenon as fluorescence in the glass of the tube. He does not mention the article by Jean Perrin from one month earlier. He expresses uncertainty as to the nature of the cathode rays, while seeming to lean towards a wave interpretation.

Henri Becquerel and Radiation from Uranium

Those are the circumstances under which Becquerel started his project. After several attempts with fluorescent materials that provided negative results (i.e. the photographic plates with silver bromide suspended in gelatin were seen to be unexposed when developed), he returned to a material he had worked with previously. This is another aspect in which his previous work and experience prepared him for this project. Some 15 years earlier he had prepared "very beautiful lamellar crystals of uranyl and potassium double sulfate" and also had great expectations for working with uranium salts. (Becquerel, Recherche sur une propriété nouvelle de la matière, 1903, p. 8). As is now well known, Becquerel found in the spring of 1896 that the uranium salts were able to expose photographic plates sealed from light by wrapping in black paper, that metal objects placed between the uranium and the photographic plates cast shadows, and importantly that this happened even when the assembly of photographic plate, metal object and uranium salt was sealed in the dark in a metal enclosure. This last aspect is important because it shows that the phenomenon does not involve fluorescence which is dependent on incident light. Whatever is exposing the plates—causing the reaction of the suspended silver bromide—has its origin in the uranium salts and is not the result of the transformation of incident energy by the uranium salts. On March 3, 1896, Becquerel also used an assortment of salts showing that the effect was due to the uranium itself and not some feature of the chemical or crystalline configuration of the salt; this is a first step towards showing that the emission of the radiation is a property of the atom itself.

Within the week, he found another way to detect the radiation from the uranium salts. He observed that the uranium salts caused the charge on a gold-leaf electroscope to dissipate. In a gold leaf electroscope, two pieces of gold leaf are connected to each other and suspended from an electrode inside a grounded conducting container. When the gold leaf is charged through the electrode, the two pieces repel each other and the angular separation is proportional the net charge. As charge is conducted from the gold leaf to the conducting container, the angle of separation decreases. Becquerel first measured the background rate of change of the angular separation and then measured the rate of change of the separation with a uranyl and potassium double sulfate crystal inside the container with the gold leaf. The presence of the uranium salt caused the charge to dissipate over 60 times faster than the background rate. He also found that the discharge rate did not depend on whether the gold leaf was negatively or positively charged; this establishes that the discharge was not due to the photoelectric effect or to radiation from the uranium salt depositing net charge of either sign on the gold leaf. Contrary to the photographic technique, the gold leaf technique provided a quantitative measure of the intensity of the radiation from the uranium salt. Further experiments placing metal plates between the uranium salt and the gold leaf that blocked a direct path for the radiation from the uranium to the gold leaf but did allow circulation of air, also showed that the discharge of the gold leaf was due to the radiation from the uranium changing the conductivity of the air.

While Becquerel continued his research on the radiation from uranium salts—changing the configuration of the electroscope, looking for secondary radiation produced from exposed materials (consistent with his interest in fluorescence)—he notes that this line of research was not "at that time the most fertile for the progress of the question," (Becquerel, Recherche sur une propriété nouvelle de la matière, 1903, p. 102). In fact, a review of the entries for Becquerel in the author index for volumes 122 to 151 of the *Comptes rendus* (minutes) of the *Académie des Sciences* (Les Secrétaires Perpétuels, 1927, pp. 161–2) shows that from July 1897 through the beginning of 1899 he did not publish any notes on radioactivity; he did however present eight notes on electromagnetic radiation in magnetic fields (including the Zeeman effect).

I therefore want to take this narrative in a different direction.

Marie and Pierre Curie and Other Natural Sources of Radiation

My direction follows the search for other materials emitting radiation as uranium does. In April 1898 G. C. Schmidt in Erlangen (Schmidt, 1898) and Marie Curie in Paris (Sklodowska Curie M., 1898) almost simultaneously announced that thorium emits radiation capable of changing the conductivity of air[1]. Quantitatively, the leakage currents that flowed through the air—measured by a technique involving sensitive current measurements with innovative instruments developed by Pierre Curie—under the effect of radiation from thorium and uranium were nearly the same. Becquerel's work two years earlier showed that the radiation produced by uranium did not change with the chemical salt of uranium, the crystalline structure or over time elapsed since the discovery; this suggested that the radiation was an atomic property of uranium. The discovery of radiation from thorium by Schmidt and Curie

[1]Curie's paper was read in the April 12 weekly session of the Académie des Sciences; Schmidt's paper was submitted to the Annalen der Physik on March 24 and published April 23. Schmidt's paper has a note added in proof with a reference to Curie's paper.

showed that although many elements did not emit this radiation (Marie Curie ultimately tested a very large number of elements), uranium was not unique; thorium emitted it too. In addition to uranium, Schmidt tested thorium oxide, thorium sulfate and thorium nitrate. Marie Curie also tested thorium oxide and thorium sulfate, measuring the leakage current through air exposed to the radiation from the sample. Then in addition to the uranium and thorium salts she tested more samples. She reported results for 10 other ores or minerals including potassium fluoxytantalate with a very low current and pitchblende from three different sources with the sample from Johanngeorgenstadt in Germany showing the highest activity of any sample tested. Commenting on this "very remarkable" fact Marie Curie comments, "these minerals may contain an element much more active that uranium." (Sklodowska Curie M. , 1898, p. 1102) The following paper by Marie Curie, this time co-authored with her husband, amplified this topic.

In that article, published just over three months later, they describe their efforts to chemically separate the highly radioactive component from pitchblende (Curie & S. Curie, Sur une substance nouvelle radio-active, contenue dans la pechblende, 1898a). After a series of steps seeking to retain the radioactive component and chemically separate other elements, they arrived at a mixture of the radioactive component with bismuth. They then separated the sulfate salts of these elements based on different condensation temperatures from their mixed hot gas. They then proposed that they had identified a new, highly radioactive element that they proposed to name polonium. They estimated that its activity was "about 400 times greater than that of uranium." (Curie & S. Curie, Sur une substance nouvelle radio-active, contenue dans la pechblende, 1898a, p. 177).

Five months later the Curies published another article describing the separation of a second new radioactive element, radium (Curie, Curie, & Bémont, Sur une nouvelle substance fortement radio-active, contenue dans la pechblende, 1898b). The Curies worked with 100 kg of mine tailings left after uranium had been extracted from pitchblende mined in Joachimsthal in then Austria-Hungary. With radium, as with polonium, they applied a series of steps to chemically separate a mixture of barium and radium and then they separated the radium and barium based on a different solubility of the hydrochloride salts. Here they estimated that the activity was 900 times greater than uranium. Here, unlike with polonium, a distinctive spectral line of radium was found allowing it to be spectroscopically identified as a new element (Demarçay, 1898).[2] Skeptics did not see the properties of the impure

[2]It is worth noting the treatment of Marie Curie's name in these three articles ((Sklodowska Curie M., 1898), (Curie & S. Curie, Sur une substance nouvelle radio-active, contenue dans la pechblende, 1898a), (Curie, Curie, & Bémont, Sur une nouvelle substance fortement radio-active, contenue dans la pechblende, 1898b)). The third was published only eight months after the first. In the first the author line has her maiden name and her husband's last name in full with no first initial; the second has the initial of her maiden name and her husband's last name in full and no first initial; and the last has her as Mme. P. Curie—her husband's name. My first impression is that she gradually surrendered her identity. Use of her maiden and married names in subsequent articles shows a more complicated pattern.

sample and the spectral line measurement as persuasive evidence that the Curies had detected a new element. The Curies then started working with a much larger quantity of mine tailings so they could extract an adequate quantity of sufficiently pure radium to allow measurement of the atomic mass and satisfy the skeptics.[3]

Within a year, Friedrich Giesel in Braunschweig was also extracting radium from uranium or tailings (Giesel, Einiges über das Verhalten des radioactiven Baryts and über Polonium, 1899) and supplying it for use by his colleagues (Elster & Geitel, 1899) in work discussed later in this chapter. As an aside, it is interesting to note that he had then worked for over 20 years at Chininfabrik Braunschweig as an organic chemist working on synthesis of cocaine and its derivatives and on cholesterol, quinizarin (a dye derived from anthraquinone) and cuscohygrine hydrate (an alkaloid naturally occurring together with cocaine and belladonna). (Stolberg-Wernigerode, 1964).

With the benefit of hindsight, I am picking my way through the primary research articles from the late 1890s. Becquerel and others were looking at other properties and phenomena related to the radioactive elements and the radiation emitted. For example, this included investigation of optical properties (e.g. reflection and refraction of the radiation), fluorescence, and induced or secondary radiation. There was also a suspicion that some gas ("emanation") was released from some samples and caused saturation of the leakage current; air moving between the capacitor plates stopped the saturation.

Investigation of several of these properties and phenomena proved fruitful. Investigation of the emanations led to an independent rediscovery of actinium and to a series of short half-life radioactive decay products. Without looking at those paths, I will look at two things that move my narrative forward in the intended direction.

Separating the Radiation by Stopping Power and Magnetic Deflection

The first involves the stopping power required to block the radiation. Ernest Rutherford, then at McGill University in Montréal, published work completed at Cavendish Laboratories in September 1898 (Rutherford, Uranium Radiation and the Electrical Conduction Produced by It, 1899). The focus of the work is the ionization of the air produced by the radiation from uranium (and thorium); it can be seen both as an extension of earlier work at Cavendish Laboratories on ionization of air by x-rays, and also, as an in-depth analysis of the effect, the change in the conductivity of air because of radiation, discovered by Becquerel and used by Becquerel (with a gold leaf electroscope) and the Curies (with a sensitive electrometer) to measure the

[3]Working with the material left after the Curie's extracted polonium and radium, André-Louis Debierne separated actinium (Debierne, 1900).

intensity of radiation from their samples. There is much more to be found in an article of this length. Rutherford investigates the change in the leakage current as thin sheets of aluminum are placed between the sample and the capacitor plates; at first the leakage current drops off quickly with the addition of thin sheets (about 0.0005 cm thick) but after about six sheets the leakage current drops off very slowly and only falls to one half after about 100 sheets of aluminum (0.05 cm). Here Rutherford concludes that, "These experiments showed that the uranium radiation is complex, and that there are present at least two distinct types of radiation—one that is very readily absorbed, which will be termed for convenience the α radiation, and the other of a more penetrative character, which will be termed the β radiation." (Rutherford, Uranium Radiation and the Electrical Conduction Produced by It, 1899, p. 116). This is the origin of the term α and β radiation. Rutherford also observed that a few millimeters of air or a thin layer of uranium is sufficient to stop α radiation. The stopping power of uranium was such that the intensity of α radiation from a uranium sample depended on the surface area of the sample; using a thicker sample with the same surface area did not increase the intensity of α radiation. In contrast the intensity of β radiation increased with a thicker sample. In addition to uranium, Rutherford worked with thorium[4]. Rutherford reported some difficulties working with thorium, and that compared to uranium it takes three or four times the thickness of aluminum for the intensity of α radiation from thorium to be reduced by one half (Rutherford, Uranium Radiation and the Electrical Conduction Produced by It, 1899, pp. 122–3). With this information we can recognize that in the configuration Becquerel first used to discover radiation from uranium, the black paper surrounding the photographic plate would have stopped the α radiation so the Becquerel was in fact observing the effect of β radiation. The use, shortly later, of an electroscope to show the change in conductivity of the air near the radioactive sample was sensitive to both α and β radiation.

The other thing I wanted to look at is the deflection of the radiation by a magnetic field. In a few months from August 1899 through January 1900 seven people published results of research looking for deflection of radiation by magnetic fields. They are in chronological order: (Elster & Geitel, 1899), (Giesel, Über die Ablenkbarkeit der Becquerelstrahlen in magnetischen Felde, 1899b), (Meyer & Schweilder, Eine weitere Notiz über das Verhalten von Radium in magnetischen Felde, 1899) (Meyer & Schweidler, Über das Verhalten von Rdium und Polonium im magnetischen Felde, 1899), (Meyer & Schweidler, Über das Verhalten von Radium und Polonium in magnetischen Feld (II. Mittleilung.), 1900), (Becquerel, Influence d'un champ magnétique sur le rayonnement des corps radio-actifs, 1899a) and (Curie P. , Action du champ magnétique sur les rayons de Becquerel, 1900). It should be noted that the first authors got a negative result (Elster & Geitel, 1899).

While this burst of activity is impressive, it should not be surprising. Because of recent work with magnetic deflection of cathode rays (Thomson, Cathode Rays,

[4]Note that the announcement of the discovery of radium by the Curies came after the completion of this article by Rutherford and roughly one month before its publication.

1897) and because of general efforts to understand similarities and differences between x-rays, cathode rays and radiation from radioactive substances, it was obvious to investigate whether radiation from polonium and radium was deflected by magnetic fields in that way that cathode rays were. Further, the investigation could not have been done earlier because sufficient quantities of radium and polonium were only then becoming available from Pierre and Marie Curie (Curie, Curie, & Bémont, Sur une nouvelle substance fortement radio-active, contenue dans la pechblende, 1989b) using their method and from Friedrich Giesel (Giesel, Einiges über das Verhalten des radioactiven Baryts and über Polonium, 1899a) using his variant of the Curie's method but working with uranium tailings from a different source. All of the above authors used samples provided by Pierre and Marie Curie or by Friedrich Giesel.

Pierre Curie, in the last paper on this list (Curie P. , Action du champ magnétique sur les rayons de Becquerel, 1900), and Marie Curie (Sklowdowska-Curie, 1900) follow-up on heterogeneity seen in the deflection results from (Meyer & Schweilder, Eine weitere Notiz über das Verhalten von Radium in magnetischen Felde, 1899) and (Becquerel, Influence d'un champ magnétique sur le rayonnement des corps radio-actifs, 1899a). Pierre Curie applied a quantitative technique (again measuring the leakage current produced by undeflected radiation) to measure the intensity of the radiation from a source with and without an applied magnetic field, with and without an absorbing blade of aluminum or black paper, and with different distances traveled through the magnetic field. Pierre Curie found that the most penetrating rays are deflected by the magnetic field and the non-penetrating rays are undeflected[5]. Further the percentage of the intensity deflected drops off with the distance traveled through the magnetic field (suggesting a rough indication of the radius of curvature of the deflection). Finally, Pierre Curie indicates that the radiation from their polonium is only undeflected whereas the polonium from Friedrich Giesel also emits deflectable radiation. (Pierre Curie suggests that the newly prepared polonium from Friedrich Giesel emits deflectable rays that "may be the first to disappear when the activity of the product decreases." We can also suspect that the Curies had a purer sample.) Marie Curie uses their sample of polonium to measure the intensity of undeflectable radiation passing different distances through air and then through a thin piece of aluminum. Next, she uses a magnetic field to first deflect the deflectable radiation from radium and then repeats the measurements for a reduced range of distances; she observes, "The undeflectable rays from radium behave like the rays from polonium."

It can be seen in the previous paragraph that the terms "nonpenetrating and non-deflectable" and "penetrating and deflectable" used by Henri Becquerel, and Pierre and Marie Curie are cumbersome; they also correspond to the α radiation and

[5]With hindsight, we understand that "undeflected" should be replaced with "not measurably deflected", because an experimental configuration suited for producing a conveniently measurable deflection of β rays would produce a deflection of α rays 2.7×10^{-4} as large, without considering differences in velocity, and that would be unmeasurably small.

β radiation described by Rutherford (Uranium Radiation and the Electrical Conduction Produced by It, 1899). It seems likely that Becquerel and the Curies were aware of Ernest Rutherford's terminology. In (Becquerel, Note sur quelques propretés du rayonnement de l'uranium et des corps radio-actifs, 1899b, p. 771), Henri Becquerel refers to an important work published on radiation from uranium by Mr. Rutherford. While the citation provided by Henri Becquerel is incomplete, it must almost certainly refer to (Rutherford, Uranium Radiation and the Electrical Conduction Produced by It, 1899) which was published two months earlier and introduced, for convenience, the terms α radiation and β radiation.

Paul Villard and γ Rays

Continuing the menagerie, while working on optical properties of cathode rays and deflectable rays from radium, Paul Villard discovered a third class of radiation; "the previous facts lead us to allow that the undeflectable portion of the emission from radium contains very penetrating radiation capable of passing through metal plates, radiation that can be detected with the photographic method." (Villard, Sur la réflexion et la réfraction des rayons cathodiques et des rayons déviables de radium, 1900a, p. 1012). In a follow-up paper later that month (Villard, Sur le rayonnement du radium, 1900b), Paul Villard observed that a 0.3 mm thick strip of lead could completely stop the penetrating deflectable radiation but only weakened the penetrating undeflectable radiation. He indicates that the penetrating power of the penetrating undeflectable radiation is comparable to x-rays.

Three years later, Ernest Rutherford provided his definition of γ rays and wrote, "The γ rays which are non-deviable by a magnetic field, and which are of a very penetrating character." (Rutherford, The Magnetic and Electric Deviation of the easily absorbed Rays from Radium, 1903).

Although I have not made a comprehensive search of the works of Paul Villard, Henri Becquerel, and Pierre and Marie Curie for use of the qualifiers α, β and γ, I have carefully searched the subset of works that I consulted for other reasons (mostly between 1896 and 1901) and have not found any instance where they used them, with one exception. In a review article for the 1900 International Congress of Physics in Paris (Becquerel, Sur le rayonnement de l'uranium et sur diverses propriétés physiques du rayonnement des corps radio-actifs, 1900c), Becquerel on page 58 discusses the article by Rutherford and gives the meaning of the terms α and β radiation. Becquerel did publish papers in 1903 and 1906 with α radiation in the title. Rutherford therefore appears to deserve full credit for all three terms: α, β and γ rays.

What Are β Rays?

After considering the terminology introduced by Rutherford, I want to return to consideration of the nature of the penetrating and deflectable rays (β rays in that terminology). The previous discussion established that β rays were deflected when their velocity was perpendicular to an imposed magnetic field. The expected next step would therefore be to establish the sign of the charge and to make quantitative measurements of the charge-to-mass ratio of the β rays produced by radium similar to the measurements with cathode rays by J. J. Thomson (Cathode Rays, 1897) and Walter Kaufmann (Die magnetische Ablenkbarkeit der Kathodenstrahlen und ihre Abhängigkeit vom Entladungspotential, 1897).

Henri Becquerel (Contribution à l'étude du rayonnement du radium, 1900a) described work he had undertaken to clarify the nature and properties of β rays. In a preliminary step (§1), he determined that the effect of air (as compared to vacuum) on the magnetic deflection of β rays in his experimental configuration is negligible. The configuration involved a photographic plate lying flat horizontally with a sample about 1 mm in diameter in a holder on top of the plate, both plate and holder were arranged in a glass tube that could be evacuated and placed between the poles of an electromagnet with a horizontal magnetic field. The tube was evacuated and the magnetic field applied deflecting the β rays in one direction. Then the magnetic field was stopped, air allowed into the tube and the current reapplied to the electromagnet in the opposite sense reversing the direction of the horizontal magnetic field. He found that quantitatively the magnitude of the deflection was the same for both configurations. This would justify the more convenient approach of measuring magnetic deflection in air instead of in vacuum. Continuing (now §2), he measured the magnetic deflection of β rays from two different radium samples; he found that while the intensities of the radiation were different, the magnetic deflection remained the same. In §3, Becquerel described configurations showing the curved trajectories of the β rays from radium samples moving in a magnetic field. Here he placed a horizontal photographic plate in the horizontal magnetic field of his electromagnet with the edge of the plate near the central axis of the electromagnet. He turned the plate with the emulsion side down, placed a lead sheet on the top, glass, side of the plate, and then his small diameter sample on top of that. In that configuration, he showed that the β rays from the sample traveling perpendicular to the magnetic field were brought around full circle to strike the emulsion side of the photographic plate opposite from the position of the sample. In an extension of this configuration, he added a second photographic plate perpendicular to the first and butted against the side of the first plate. In this way he would get images of the β rays partway through their circular trajectory from the sample on top to the emulsion on bottom. There is no figure with the article, and no indication of the pattern formed by the β rays on the vertical photographic plate. It would be interesting to know what this pattern was for reasons discussed below. Let's set aside discussion of §4 for a moment and advance to §5, the last section.

He started the discussion in that section on page 210 by noting that, "the facts which were just presented showed that a portion of the radiation from radium is entirely similar to cathode rays, or to masses of negative electricity transported at high velocity." We need to be careful with this statement to understand what he means by "similar to cathode rays." The quoted sentence refers to negatively charged particles, but not charge-to-mass ratio. This distinction was amplified by referring to a review article by Becquerel from about the same time describing the same series of experiments; in (Becquerel, Sur le rayonnement de l'uranium et sur diverses propriétés physiques du rayonnement des corps radio-actifs, 1900c, p. 69), he writes "To complete the identification of the deflectable radiation from radium and cathode rays, it is sufficient to show, either that the radiation transports negative electrical charges, or that it is deflected by an electrostatic field." (He made a similar statement twice in the preceding pages.) From my perspective, this is not sufficient since I view it as also necessary to show that the charge-to-mass ratio of cathode rays and β rays are the same in order to assure the identification of β rays as electrons.

Continuing the discussion in §5, he used the magnetic deflection of β rays to estimate the mass-to-charge ratio. He supported this statement with the measurement of magnetic field strength and radius of curvature giving a value of the charged particle velocity times the mass-to-charge ratio, $v m/e = 1500$. The remainder of §5, appears to be part plausibility (if the velocity were in a particular range, then the charge-to-mass ratio would agree with previous researchers) and feasibility (the electrostatic deflection could be measured in a vacuum with electric field strength comparable to those used for the cathode rays). This plausibility discussion brings home an important point: he didn't know the velocity of the β rays in his experiment or even whether the velocities were roughly similar or vastly different and he did not know whether he was dealing with a single kind of particle with one charge-to-mass ratio or a mixture of different particles with varied charge-to-mass ratios. Further information about the image on the vertical photographic plate in §3 might give some indication of the size of the difficulty, but it was not provided.

Returning to discuss §4, Becquerel started with a description of images suggesting fuzziness or "dispersion" in the spatial distribution of the β rays. He suggested that this dispersion was because the magnetic field separates a heterogeneous beam of radiation. In hindsight we understand that the heterogeneity is due to a single type of particle having a spectrum of velocities. Here Becquerel did not come close to that understanding and it appears he suggested that the dispersion was due to different types (or subtypes) of radiation and discussed adding thin metal plates to the configuration as filters. While at this point Becquerel had shown that β rays are charged particles, as had been shown with cathode rays, he did not have the value of the charge-to-mass ratio for β rays that could be compared to the charge-to-mass ratio for cathode rays.

While the sign of the charge of β rays could be inferred from the direction of deflection of the β rays in (Becquerel, Contribution à l'étude du rayonnement du radium, 1900a), two months later in March Pierre and Marie Curie (Curie & Curie, Sur la charge électrique des rayons déviables du radium, 1900) presented a method for determining the charge on the β rays that cleverly avoided the need for a vacuum.

This allowed them to show, by a different route, that "The deflectable radiation from radium is negative electrical charges." They had tried to do this by spreading a radioactive substance on one plate of a capacitor with an air gap and using an electrometer to measure the sign and magnitude of the current between the two plates. They were not able to measure the charge with this configuration and they ascribed the difficulty to ionization of the air between the plates of the capacitor leading to changes in its conductivity. One way to overcome this difficulty would be to place the plates of the capacitor in a vacuum eliminating the air. Instead, they filled the space between the plates with a solid dielectric, paraffin or ebonite, to seal out the air. (Their actual configuration was slightly different than this suggests since, among other things, the radium sample was placed against the side of one plate of the capacitor away from the other plate.) The radium sample and the plate were separated by a thin layer of aluminum and ebonite that also stopped the weekly penetrating radiation from reaching the capacitor plate. They were then able to measure a current with a piezoelectric electrometer. Repeating the experiment with x-rays in place of the radium sample, they were not able to measure a current distinguishable from zero. They conclude that, "most likely the radium is the source of a continuous emission of negatively charged particles of matter." We can add to that qualification that they did not yet know the mass, magnitude of the charge, or even the charge-to-mass ratio of the particles.

Later that same month, Henri Becquerel (Becquerel, Déviation du rayonnement du radium dans un champ électrique, 1900b) reported on continuation of his work. His goal was still to be to get "the ratio of the material masses born to the charge that they transport." In this configuration, he directed the beam of penetrating radiation from his radium sample between two conducting plates with a voltage difference applied to them. In his previous paper (Becquerel, Contribution à l'étude du rayonnement du radium, 1900a), Becquerel indicated that it would be necessary to do this in vacuum to achieve the necessary electric field strengths. However, in the fourth paragraph of this paper (Becquerel, Déviation du rayonnement du radium dans un champ électrique, 1900b), Becquerel writes, "After several attempts to get, in vacuum, very large electrostatic fields, I returned to a configuration that I had used three months ago." He had reached a limit in the core competence of his laboratory. He was able by working in air to measure the electrical deflection and get a value proportional to $v^2 m/e$. Again, as with magnetic deflection, he found a spread of electrostatic deflections and tries to filter them with different thicknesses of aluminum; this did not lead to any particular resolution. Using his value for vm/e from the earlier magnetic deflection work with his value for $v^2 m/e$ from this electrostatic deflection work he suggests a value of $v = 1.6 \times 10^{10}$ cm/s. He writes that, "this number is given here only to show the order of magnitude of the velocity."

This in fact shows a limit in what Becquerel could do in his lab. Writing three years later (Becquerel, Recherche sur une propriété nouvelle de la matière, 1903, p. 181), he reviewed some of the difficulties: diffusion of the rays in air, beam collimation, beam intensity requiring long exposures with stable voltages, and the "insufficient resources in my laboratory." He was not the only one to have difficulties. Working at about the same time in Halle, Germany, Ernst Dorn also tried to

Table 8.1 Charge-to-mass ratio reproduced from (Kaufmann, Die magnetische und electrische Ablenkbarkeit der Becquerelstrahlen und die scheinbare Masse der Elektronen, 1901a, p. 152)

Velocity v (10^{10} cm/s)	**Charge to Mass** e/m (10^7 cgs)
2.83	0.63
2.72	0.77
2.59	0.975
2.48	1.17
2.36	1.31

measure the electrostatic deflection of rays from radium (Dorn E. , 1901). He also worked in air and had difficulties with beam intensity. Instead of using photographic plates he used a barium platinum cyanide phosphor screen. The spot on the phosphor could only be seen in a completely dark room with dark-adjusted eyes; it was not possible to make quantitative measurements. Becquerel had needed photographic exposures lasting days. In a letter dated April 15, 1900 Ernst Dorn provided Henri Becquerel the citation for this work and indicated that he had not yet made quantitative measurements (Dorn E. , 1900).

The following year, Walter Kaufmann (Die magnetische und electrische Ablenkbarkeit der Becquerelstrahlen und die scheinbare Masse der Elektronen, 1901a) was able to overcome the challenges that had stopped Becquerel and Dorn. (Kaufmann had provided a preliminary description of the configuration, before adjustment of field values, about five months earlier (Kaufmann, Methode zur exakten Bestimmung von Ladung und Geschwindigkeit der Becquerelstraheln, 1901b).) Notably he put the entire path for the rays being measured in a vacuum; this eliminated any scattering by air and any interference with the imposed electric and magnetic fields due to secondary ions. He also used a fine hole as a diaphragm to produce a narrow beam and exposures lasting four days. He then used crossed electric and magnetic fields to spread the beam by velocity (v) and charge-to-mass ratio (e/m) resulting in a curve on these two axes on a target photographic plate. The result was a clean curve tracing the dependence of the charge-to-mass ratio on velocity of the electron. The spread of the curve along the axis proportional to velocity shows that the particles have a range of velocities when they pass through the crossed electric and magnetic fields, and that the dispersion seen by Becquerel (as a fog in his images) is due to this range of velocities and not to different types of particles within the β rays.

The results are presented in Table I of (Kaufmann, Die magnetische und electrische Ablenkbarkeit der Becquerelstrahlen und die scheinbare Masse der Elektronen 1901a, p. 152) and the last two columns are reproduced above in Table 8.1.

The results again have the same order of magnitude, 10^7 cgs, as found by J. J. Thomson, Lenard, Becquerel and discussed previously. However, it can be immediately seen that the charge-to-mass ratio more than doubles over this small range of velocities, and decreases steadily as the velocity increases. This is a striking feature that needs explanation. Kaufmann presents a theory in which the mass observed during the deflection of the β rays in the crossed electric and magnetic field is equal to the sum of the true mass and a contribution from electromagnetic self-energy. His

Table II presents observed and calculated values according to this theory (they are not reproduced here). The agreement is good, but the number of points is small. However, more than the agreement, the explanation is interesting because it brings electromagnetic self-energy into consideration.

Kaufmann provided further measurements and their analysis in two subsequent papers: (Kaufmann, Die electromagnetishe Masse des Elektrons, 1902) and (Kaufmann, Über die "Elektromagnetische Masse" der Elektronen, 1903). (It should be noted Runge provided an improved analysis of the data presented in Kaufmann's second paper and the methodology and results appear in (Runge, 1903).) Here again a correction was applied for the electromagnetic self-energy: in (Kaufmann, Die electromagnetishe Masse des Elektrons, 1902) it is the factor $\frac{4}{3}\psi(\beta)$ given in equation 2 and in (Kaufmann, Über die "Elektromagnetische Masse" der Elektronen, 1903) the same factor is given in equation 10 and the unnumbered equation immediately below it. In support of this factor, Kaufmann cites Max Abraham (Abraham, Dynamik des Electrons, 1902) and (Abraham, Prinzipien der Dynamik des Elektrons, 1903); we will come back to this theory of electron dynamics by Abraham in Chapter 9.

There is an alternative to this factor, $\frac{4}{3}\psi(\beta)$, and it was presented by Lorentz in a work that will also be discussed in Chapter 9 (Lorentz, Electromagnetic phenomena in a system moving with any velocity smaller than that of, 1904). There Lorentz re-analyzes other data from (Kaufmann, Die electromagnetishe Masse des Elektrons, 1902) removing Kaufmann's correction from Abraham's theory and applying a correction of $\left(1-\left(\frac{v}{c}\right)^2\right)^{-1/2}$, the transformation from the relativistic mass to the rest mass of the electron derived by Lorentz in that paper. The results of the recalculation were presented in Tables III and IV on page 828 of (Lorentz, Electromagnetic phenomena in a system moving with any velocity smaller than that of, 1904) which corresponds to page 277 in Part III of this book. In looking at these tables prepared by Lorentz from Kaufmann's data, the question to ask is whether the value k_2 with the correction from Abraham or the value k_2' with the correction from Lorentz is better at removing the velocity dependence. Lorentz observes, "The constancy of k_2' is seen to come out no less satisfactory than that of k_2." Chapter 9 confirms that Lorentz could have made a much stronger statement.

The Lorentz transformation of the charge-to-mass ratio to the rest frame can likewise be applied to Kaufmann's values from (Kaufmann, Die magnetische und electrische Ablenkbarkeit der Becquerelstrahlen und die scheinbare Masse der Elektronen, 1901a) reproduced in Table 8.1, above. The result is shown in Table 8.2.

Somewhere between Becquerel, Kaufmann and this recalculation by Lorentz, all the pieces are in place for declaring that β rays are electrons like cathode rays; Kaufmann even uses the name electron in his title. The apparent velocity dependence of the β ray charge-to-mass ratio that concerned Becquerel is an effect of special relativity and is accurately explained by the Lorentz transformation between the increased mass of the relativistic electrons traveling through the crossed electric and magnetic fields in the apparatus and the rest frame mass of the electron. The

Table 8.2 Charge-to-mass ratio from (Kaufmann, Die magnetische und electrische Ablenkbarkeit der Becquerelstrahlen und die scheinbare Masse der Elektronen, 1901a, p. 152) after transformation to the rest frame

Charge to Mass e/m_0 (10^7 cgs)
1.432
1.374
1.452
1.562
1.593

While Table 8.2 applies the Lorentz transformation to Kaufmann's stated results, a reanalysis of Kaufmann's data is continued in Chapter 9 (p. 178–83) and a curve is fit to his raw measurements of the electric and magnetic deflections. Those results are in Table 9.1 and the average charge to mass ratio is 1.641×10^7 CGS.

transformation completely removes the velocity dependence of the charge-to-mass ratio once the mass is referred to the rest frame of the laboratory.

The same experiment as (Kaufmann, Die magnetische und electrische Ablenkbarkeit der Becquerelstrahlen und die scheinbare Masse der Elektronen, 1901a) was repeated seven years later by Alfred H. Bucherer with more accurate equipment and using the Lorentz transformation to the rest mass of the electron in the measured charge-to-mass ratio (Bucherer, 1908). He found a value for the rest frame charge-to-mass ratio (e/m_0) of 1.730×10^7 cgs. (This value appears in a supplement that could be a note added in proof, and not in the values in the table.) This is within 2% of the current accepted value, 1.759×10^{11} Ckg^{-1} (National Institute of Standards and Technology, n.d.).[6] Although referencing Kaufmann's work, Bucherer does not mention or discuss the recalculation done by Lorentz. The recalculation by Lorentz of Kaufmann's measurements and this more accurate measurement by Bucherer are perhaps the earliest confirmations of Lorentz transformations and Einstein's theory of special relativity.

Digressions

Zeeman Effect

In a series of papers published in 1897, Pieter Zeeman described the results of experimental work on broadening and splitting of emission spectral lines in strong magnetic fields done while at the University of Leiden. This phenomenon is now known as the Zeeman effect. Pieter Zeeman and Hendrik Lorentz shared the 1902 Nobel Prize in physics for "their researches into the influence of magnetism upon radiation phenomena."

[6]The average result from recalculation in Table 8.1 is within about 7%.

This work deserves to be included in a chapter on discovery of the electron because Zeeman and Lorentz (and also Poincaré, (Poincaré, La théorie de Lorentz et le phénomène de Zeeman, 1899)) interpreted it as related to the movement of electrons within an atom. Unlike other research considered in these chapters, this work at the University of Leiden did not involve free electrons outside an atom. Promptly after learning of Zeeman's results, Lorentz provided a theoretical explanation including a charge-to-mass ratio determined from the experimental results that was close to other values being measured at the time. This was seen then as persuasive evidence that the same electrons observed separately were also present within atoms. However, their interpretation of the results as splitting related to the movement of electrons within an atom is correct only if one doesn't look past the surface. Also, the multiplicity of the splitting of the spectral lines for different atoms (e.g. sodium, iron) and lines, and their optical rotation proved difficult for them to explain.

Looking back, the apparent agreement of the estimate by Lorentz, based on Zeeman's measurements, of the electron charge-to-mass ratio with values obtained by J. J. Thomson can only be seen as fortuitous. There were simply too many factors necessary for an explanation (too much below the surface) that were not accessible.

It would be over 30 years before all the necessary pieces of theory and explanation would be in place. An explanation of the discrete frequencies of the un-split atomic emission lines requires an understanding of the quantized nature of light, and of the quantized electron orbitals in atoms together with their energy levels (the Bohr model of the atom); an explanation of the splitting further requires an understanding of the quantized electron spin of one half (provided by (Dirac, 1928)).

Henri Becquerel, and Pierre and Marie Curie

The work jointly and individually of Henri Becquerel, Pierre Curie and Marie Sklodowska Curie between 1896 and 1903 is a key part of the historical narrative in this chapter. In the record of the published articles, the professional relationship appears to have been collegial, supportive and mutually beneficial. Henri provided clever insights and ideas: looking for x-ray like radiation from natural sources, trying uranium, controlling unused photographic plates, using an electroscope to detect radiation and measuring the charge-to-mass ratio of the deflectable radiation. Pierre provided precision instrumentation for measuring leakage current, mass and penetrating power of radiation. And Marie provided a determination and persistence that allowed her first to test a large number of substances (including thorium and pitchblende ore) for radiation, and then to process a ton of uranium processing tailings (Sklodowska Curie M. , 1899) to extract radium and polonium to determine the atomic weights. The people themselves were as diverse as the talents they contributed. Henri Becquerel was an insider; he was a third-generation member of the Académie des sciences. He was a graduate of a prominent school, the École Polytechnique, as was Henri Poincaré. Pierre Curie was a graduate of the Faculté des sciences de Paris, who saw himself as an outsider, and shy. Marie Curie was a woman, a foreigner, and, until shortly before she won a Nobel Prize, a doctoral student. This mix of talents and people seemed to work well.

Then something changed.

This section has cited numerous papers appearing in the minutes of the weekly sessions of the Académie des Sciences (les *Comptes rendus)*. For a note to appear in the minutes the author had to be a member of the academy or have a member of the academy present it for them. Henri Becquerel was a member throughout this time. Pierre Curie was elected as a member in 1905, the second time he was presented, and died a year later. Marie Curie never became a member, even after her second Nobel Prize. This means that except for one paper in early 1906 Pierre and Marie were dependent on someone else to present their work to the academy. The entries for Pierre and Marie Curie in the author index for volumes 122 to 151 of the minutes (Les Secrétaires Perpétuels, 1927, pp. 161–2) show that from the beginning of 1896 till May 1902 when he was first presented as a candidate for membership in the academy, Pierre Curie had 15 notes published in the minutes where he was the author or co-author. Checking the first page of each article, it can be seen that only two were not presented to the academy by Henri Becquerel. After May 1902 until his death in 1905, nine more notes were published; none of them were presented by Henri Becquerel. For the same two time periods the numbers of articles for Marie Curie (excluding notes where she was a co-author with Pierre Curie to avoid double counting) are less dramatic. Before, of six articles presented, half were presented by Henri Becquerel; after, there were two articles, and neither was presented by Henri Becquerel. There is a clear and clean break in collaboration between Pierre Curie and Henri Becquerel occurring around May 1902. In fact, among the numerous letters of condolence to Marie Curie after the death of Pierre, the index to the Pierre and Marie Curie manuscript archives at the *Bibliothèque nationale de France* does not list one from Henri Becquerel.

What happened?

Something put the relationship between Pierre Curie and Henri Becquerel under strain and it reached a breaking point at the time of the first vote of the *Académie des sciences* on Pierre Curie's membership. Pierre Curie wrote to Georges Gouy in a letter dated March 20, 1902, shortly before this first vote, (Curie P. (n.d.), Pierre et Marie Curie. Papiers. V — DOCUMENTS A CARACTERE PRIVE. CLI Correspondance de Pierre Curie et Georges Gouy.) that a lot of hassle had come with their good fortune and they scarcely had time to breathe. He then writes, "I will certainly please you by telling you that out of everyone Becquerel is the one who bothers us the most and we've had it up to here with him." In a subsequent letter in the same collection, dated June 9, 1902, a week after it was announced that Pierre Curie was not elected member, he again wrote to Georges Gouy and complained about the loss and about the tactics of the winning candidate. He then wrote, "Finally Becquerel, although having declared himself for me, played, I'm sure of it, a double game. In any case I am convinced that he was delighted that I didn't get in and I also think that he must've voted for [the other candidate]." There is nothing in this letter or in subsequent letters in this collection to indicate why Pierre Curie was sure of these assertions against Henri Becquerel.

Health Effects

Henri Becquerel, Pierre and Marie Curie, F. Giesel and others were all exposed to significant quantities of radiation during their work with consequent health effects. Pierre Curie, in confirming an earlier experiment by Friedrich Giesel (Giesel, Ueber radioactive Stoffe, 1900), deliberately exposed his forearm to radiation from radium ("with relatively weak activity") for 10 hours; the resulting lesion had not fully healed 52 days later (Becquerel & Curie, Action physiologique des rayons du radium, 1901). On April 3 to 4, 1901 Henri Becquerel suffered a significant burn from carrying radioactive materials in his vest pocket (Becquerel & Curie, Action physiologique des rayons du radium, 1901). In the same report, it is indicated that Pierre and Marie Curie had both experienced burns to their fingers resulting in blisters or desquamation. Pierre Curie died in April 1906 at 46 years old, but not from radiation; in an apparent incident of distracted walking, he bumped into a horse and fell while crossing a street, and was killed by the traffic. Two years later Becquerel died at an age of only 54 years; although the cause of death was not specified, radiation damage to his heart could be a possibility. Friedrich Giesel died in 1927 of lung cancer that had metastasized from a carcinoma on his right index finger attributed to radiation from radium (Stolberg-Wernigerode, 1964). Marie Curie's situation was likewise unambiguous; she died in 1934 of aplastic anemia, most likely from long-term exposure to radiation. Although Henri Becquerel may have been an early death from radiation poisoning, he was not the first. Clarence Dally, who may have been the first, died in 1904. He was a glassblower working for Thomas Edison to develop and improve tubes to generate X-rays; he was exposed to large quantities of radiation during this work with severe health consequences that led to his death (Brown, 1995).

References

Abraham, M. (1902). Dynamik des Electrons. *Nachrichten von der Gesellschaft der Wissenschaften zu Göttingen*, 20–41.

Abraham, M. (1903). Prinzipien der Dynamik des Elektrons. *Annalen der Physik, Series 4 vol. 10*, 105–179.

Becquerel, H. (1899a). Influence d'un champ magnétique sur le rayonnement des corps radio-actifs. *Comptes rendus de l'Académie des Sciences, 129*, 996–1001.

Becquerel, H. (1899b). Note sur quelques propretés du rayonnement de l'uranium et des corps radio-actifs. *Comptes rendus de l'Académie des Sciences, 130*, 771–7.

Becquerel, H. (1900a). Contribution à l'étude du rayonnement du radium. *Comptes rendus de l'Académie des Sciences, 130*, 206–211.

Becquerel, H. (1900b, March 26). Déviation du rayonnement du radium dans un champ électrique. *Comptes rendus de l'académie des sciences, 130*, 809–815.

Becquerel, H. (1900c). Sur le rayonnement de l'uranium et sur diverses propriétés physiques du rayonnement des corps radio-actifs. In C.-É. Guillaume, & L. Poincaré (Ed.), *Rapports présentés au congrès international de physique réuni à Paris en 1900 sous les auspices de la Société française de physique. III*, pp. 47–137. Paris: Gauthier-Villars (Paris).

Becquerel, H. (1903). Recherche sur une propriété nouvelle de la matière. (A. d. sciences, Ed.) *Mémoires de l'Académie des sciences de l'Institut de France, 46*, 1.

Becquerel, H., & Curie, P. (1901, June 5). Action physiologique des rayons du radium. *Comptes rendus des séances de l'Académie des Sciences, 132*, 1289–91.

Brown, P. (1995). Clarence Madison Dally (1865–1904). *American Journal of Roentgenology, 164*, 237–9.

Bucherer, A. H. (1908). Messungen an Becquerelstrahlen. Die experimentelle Bestätigung der Lorentz-Einsteinschen Theorie. *Physikalishe Zeitschrift, 9*(22), 755–762.

Curie, P. (1900). Action du champ magnétique sur les rayons de Becquerel. *Comptes rendus de l'Académie des sciences, 130*, 73–9.

Curie, P. (n.d.). *Pierre et Marie Curie. Papiers. V — DOCUMENTS A CARACTERE PRIVE. CLI Correspondance de Pierre Curie et Georges Gouy.* Retrieved August 22, 2019, from Source gallica.bnf.fr/Bibliothèque nationale de France: https://gallica.bnf.fr/ark:/12148/btv1b9080328m?rk=21459;2

Curie, P., & Curie, M. P. (1900). Sur la charge électrique des rayons déviables du radium. *Comptes rendus de l'Académie des Sciences, 130*, 647–650.

Curie, P., & S. Curie, M. (1898a). Sur une substance nouvelle radio-active, contenue dans la pechblende. *Comptes Rendus des Séances de l'Académie des Sciences, 127*, 175–8.

Curie, P., Curie, M., & Bémont, G. (1898b). Sur une nouvelle substance fortement radio-active, contenue dans la pechblende. *Comptes Rendus des séances de l'Académie des Sciences, 127*, 1215–17.

Debierne, A.-L. (1900). Sur un nouvel élément radio-actif: l'actinium. *Comptes rendus de l'Academie des Sciences, 130*, 906–8.

Demarçay, E. (1898). Sur le spectre d'une substance radio-active. *Comptes Rendus des séances de l'Académie des Sciences, 127*, 1218.

Dirac, P. A. (1928). The Quantum Theory of the Electron. *Proceedings of the Royal Society of London. Series A, Containing Papers of a Mathematical and Physical Character, 117*, 610–624.

Dorn, E. (1900). Sur les rayons de radium, Lettre de M. E. Dorn à M. H. Becquerel. *Comptes rendus de l'Académie des Sciences, 130*, 1126.

Dorn, E. (1901). Elektrostatische Ablenkung der Radiumstrahlen. *Abhandlungen der Naturforschenden Gesellschaft zu Halle, 22*, 45–50.

Elster, J., & Geitel, H. (1899). Weitre Versuche an Becquerelstraheln. *Annalen der Physik, ser. 3, 69*, 83–90.

Giesel, F. (1899a). Einiges über das Verhalten des radioactiven Baryts and über Polonium. *Annalen der Physik, ser. 3, 69*, 91–94.

Giesel, F. (1899b). Über die Ablenkbarkeit der Becquerelstrahlen in magnetischen Felde. *Annalen der Physik, ser. 3, 69*, 834–836.

Giesel, F. (1900). Ueber radioactive Stoffe. *Berichte der deutschen chemischen Gesellschaft, 33*, 3569–71.

Kaufmann, W. (1897). Die magnetische Ablenkbarkeit der Kathodenstrahlen und ihre Abhängigkeit vom Entladungspotential. *Annalen der Physik, ser. 3, 61*, 544–552.

Kaufmann, W. (1901a). Die magnetische und electrische Ablenkbarkeit der Becquerelstrahlen und die scheinbare Masse der Elektronen. *Nachrichten von der Königl. Gesellschaft der Wissenschaften zu Göttingen*, 2, 143–155.

Kaufmann, W. (1901b). Methode zur exakten Bestimmung von Ladung und Geschwindigkeit der Becquerelstraheln. *Physikalische Zeitschrift, 2*(41), 602–03.

Kaufmann, W. (1902). Die electromagnetishe Masse des Elektrons. *Physikalische Zeitschrift, 4*, 52–57.

Kaufmann, W. (1903). Über die "Elektromagnetische Masse" der Elektronen. *Nachrichten von der Gesellschaft der Wissenschaften zu Göttingen*, 90–103.

Les Secrétaires Perpétuels. (1927). *Table générale des comptes rendus des séances de l'Académie des Sciences; Tomes 122 à 151, 6 janvier 1896 au 27 décembre 1910, Première partie: Auteurs.* Paris: Gauthier-Villars.

Lorentz, H. A. (1904). Electromagnetic phenomena in a system moving with any velocity smaller than that of. *Proceedings of the KNAW [Royal Netherlands Academy of Arts and Sciences], 6 (1903–4)*, 809–831.

Meyer, S., & Schweidler, E. R. (1899). Über das Verhalten von Rdium und Polonium im magnetischen Felde. *Physikalische Zeitschrift, 1*(9), 90–91.

Meyer, S., & Schweidler, E. R. (1900). Über das Verhalten von Radium und Polonium in magnetischen Feld (II. Mittleilung.). *Physikalische Zeitschrift, 1*(10), 113–4.

Meyer, S., & Schweilder, E. R. (1899). Eine weitere Notiz über das Verhalten von Radium in magnetischen Felde. *Anzeiger der Kaiserlichen Akademie der Wissenschaften, 36*, 323–4.

National Institute of Standards and Technology. (n.d.). *CODATA Value: electron charge to mass quotient.* Retrieved November 12, 2019, from https://physics.nist.gov/cgi-bin/cuu/Value?esme

Oudin, & Barthélemy. (1896, January). Communication. *Comptes Rendus des Séances de l'Académie des Sciences*, 150.

Poincaré, H. (1896, January 30). Les rayons cathodiques et les rayons Röntgen. *Revue générale des sciences pures et appliquées, 7*(2), 52–59.

Poincaré, H. (1899). La théorie de Lorentz et le phénomène de Zeeman. *L'Éclaiage électrique, 19* (14), 5–15.

Röntgen, W. (1896, January 30). Une nouvelle espèce de rayons. *Revue Générale des Sciences pures et appliquées, 7*(2), 59–63.

Runge, C. (1903). Über die elektomagnetische Masse der Electronen. *Nachrichten von der Gesellschaft der Wissenschaften zu Göttingen*, 326–330.

Rutherford, E. (1899). Uranium Radiation and the Electrical Conduction Produced by It. *Philosophical Magazine and Journal of Science, Fifth Series, 47*, 109–163.

Rutherford, E. (1903). The Magnetic and Electric Deviation of the easily absorbed Rays from Radium. *Philosophical Magazine and Journal of Science, Sixth Series, 5*, 177–187.

Schmidt, G. C. (1898). Über die von den Thorverbindungen und einigen anderen Substanzen ausgehende Strahlung. *Annalen der Physik ser. 3, 65*, 141–151.

Sklodowska Curie, M. (1898). Rayons émis par les composés de l'uranium et du thorium. *Comptes Rendus des séances de l'Académie des Sciences, 126*, 1101–3.

Sklodowska Curie, M. (1899). Sur le poids atomique du métal dans le chlorure de baryum radifère. *Comptes rendus de l'Académie des Sciences, 129*, 760–2.

Sklowdowska-Curie, M. (1900). Sur la pénétration des rayons de Becquerel non déviables par le champ magnétique. *Comptes rendus de l'Académie des Sciences, 130*, 76–79.

Stolberg-Wernigerode, O. z. (1964). *Neue deutsche Biographie.* Berlin: Historische Kommission bei der Bayerischen Akademie der Wissenschaften. Retrieved August 14, 2019, from https://www.deutsche-biographie.de/pnd116620218.html#ndbcontent

Thomson, J. J. (1897, October). Cathode Rays. *Philosophical Magazine and Journal of Science, Fifth Series, 44*, 293–316.

Villard, P. (1900a). Sur la réflexion et la réfraction des rayons cathodiques et des rayons déviables de radium. *Comptes rendus de l'Académie des Sciences, 130*, 1010–1012.

Villard, P. (1900b). Sur le rayonnement du radium. *Comptes rendus de l'Académie des Sciences, 130*, 1178–9.

Chapter 9
Contributions of Abraham, Lorentz and Poincaré to Classical Theory of Electrons

Introduction

The experimental investigation of cathode rays and of penetrating and electrically deflectable radiation from radioactive decay described in Chapters 7 and 8 of Part II demonstrated that both are moving electrons. This discovery of electrons emphasized a need for a theory of electrons. The article by Poincaré (Poincaré, Sur la dynamique de l'électron 1906), translated in Part I of this book and presenting a theory of electrons, builds on and continues beyond slightly earlier articles by Max Abraham (Abraham, Dynamik des Electrons 1902a) and (Abraham, Prinzipien der Dynamik des Elektrons 1903), and by Hendrik Lorentz (Lorentz, Electromagnetic phenomena in a system moving with any velocity smaller than that of 1904). For the convenience of interested readers, a reformatted copy of Lorentz's article is provided in Part III. These theories are the subject of this chapter.

Lorentz's 1904 paper presents his theory at a mature stage. His earlier work on electrodynamics extended over much of the previous decade with notable contributions in 1892 and 1895 (respectively (Lorentz, La théorie électromagnétique de Maxwell et son application aux corps mouvants 1892) and (Lorentz, Versuch einer Theorie der electrischen und optischen Erscheinungen in bewegten Körpern 1895)). Lorentz's theory and also Larmor's ether vortex theory (Larmor 1893) provide an atomistic basis for electrodynamics (Darrigol, The Electron Theories of Larmor and Lorentz: A Comparative Study 1994). Their theories can be understood as a second phase of understanding and applying James Clerk Maxwell's treatise on electricity and magnetism. In a first phase, Oliver Heaviside, Oliver Lodge and others (together referred to as the Maxwellians) consolidated the theory by reducing the number of equations, clarified it by adopting compact vector notation, and used it to solve practical problems, notably electrical signal propagation in undersea cables (Hunt, 1991). The second phase then involves efforts to understand current and dielectrics in terms of moving bodies (*corps mouvants* or *bewegtern Körpern* in the titles of Lorentz's papers cited above) in place of Maxwell's continuous medium.

B. D. Popp, *Henri Poincaré: Electrons to Special Relativity*,
https://doi.org/10.1007/978-3-030-48039-4_9

Importantly, there were conflicting ideas about whether these moving bodies are a manifestation of the ether or instead have an independent mechanical nature. In connection with Lorentz and this second phase, McCormmach discusses the electromagnetic worldview and states that it "asserted that the only physical realities are the electromagnetic ether and electric particles and that all laws of nature are reducible to properties of the ether, properties which are defined by the electromagnetic field equations." (McCormmach, 1970, p. 459).

This context from the 1890s suggests some issues to look for in reading and comparing the articles by Abraham, Lorentz and Poincaré from 1902, 1904 and 1906 respectively. What is the role of ether, meaning a pervasive, continuous medium? Importantly, how do the theories address the Michelson-Morley experiment that was already 20 years old? What is the origin of mass, is mass exclusively electromagnetic? From a mechanical view, the relation of the theories to established laws of motion are important. Experimental values were available for the electron charge-to-mass ratio and they showed a velocity dependence. Could the theories account for the velocity dependence? What did they say about other properties of the electron such as shape and size?

Max Abraham[1]

In three papers Max Abraham presents his theory of electron dynamics: (Abraham, Dynamik des Electrons, 1902a), (Abraham, Prinzipien des Dynamik des Elektrons, 1902b) and (Abraham, Prinzipien der Dynamik des Elektrons, 1903). These publication dates place him first among the three authors of electron theories presented in this chapter. The three articles present largely the same theory with different levels of discussion and detail, and some differences in secondary matters discussed.

Max Abraham is motivated in this work by measurements in (Kaufmann, Die magnetische und electrische Ablenkbarkeit der Becquerelstrahlen und die scheinbare Masse der Elektronen, 1901) showing that the charge-to-mass ratio of electrons depends on velocity. Working in the context of Lorentz's theory of electrodynamics of moving particles (Lorentz, Versuch einer Theorie der electrischen und optischen Erscheinungen in bewegten Körpern, 1895), Abraham (Dynamik des Electrons, 1902a) asks two questions about this dependence on velocity. "Is it possible to derive that correlation quantitatively from the differential equations of the electromagnetic field? Can the inertia of the electron be completely explained by the dynamic effect of its electromagnetic field, without invoking a mass that is independent of electrical charge?" And indicates that, "Only if these questions can be answered in the affirmative can one see any possibility of an entirely electromagnetic foundation of mechanics." Abraham sees the velocity dependence

[1]Translations from German of quotes used in this section were provided by Nick Hartmann. Personal Communication February 13, 2020.

found by Kaufmann as a means to test, possibly even confirm, a view of mechanics, notably inertial mass, as fundamentally electromagnetic in nature; that is to say, a test of an electromagnetic worldview. Abraham sets for himself the challenge of answering these two questions.

To clarify what is meant by electromagnetic mass, we can refer to equation 3 in (Abraham, Dynamik des Electrons, 1902a, p. 24) which, after adapting the notation, is:

$$(M + m)\frac{\mathrm{d}\boldsymbol{v}}{\mathrm{d}t} = \boldsymbol{F}. \tag{9.1}$$

where, by definition, M is the material mass and m is the electromagnetic mass. Using this notation, the two questions quoted in the previous paragraph ask whether it is possible to find a theoretical expression for $m(\boldsymbol{v})$ that fits the data and whether $M = 0$.

Subsequent discussion in this article and elsewhere refers to transverse and longitudinal mass; this refers to a possible difference in inertia depending upon whether an applied force is perpendicular or parallel to the instantaneous velocity. This results in two equations:

$$(M + m_{\perp})\frac{\mathrm{d}\boldsymbol{v}}{\mathrm{d}t} = \boldsymbol{F}_{\perp}.$$

$$(M + m_{\|})\frac{\mathrm{d}\boldsymbol{v}}{\mathrm{d}t} = \boldsymbol{F}_{\|}.$$

respectively for the case when the force is perpendicular to the velocity and when the force is parallel to the velocity, and where, again by definition, $m_{\perp}$ is the transverse mass and $m_{\|}$ is the longitudinal mass. The case of a force oblique to the velocity becomes more complicated since, as observed by (Abraham, Dynamik des Electrons, 1902a, p. 28), the electromagnetic mass is no longer a scalar.

Almost immediately, Abraham recognizes and acknowledges two significant problems affecting progress on this challenge.

The first significant problem relates to the size of the electron. In (Abraham, Dynamik des Electrons, 1902a, p. 22), he states, “The electron would need to be regarded here not as a point charge but rather as having spatial extent, since a point charge would represent an infinite supply of energy.” The self-energy of the (classical) point charge is infinite. The apparent resolution is to assume that an electron has a shape, which is not a point, and a characteristic dimension, which is not zero. These two assumptions lead to another question: how is the charge distributed over the shape? For a macroscopic, perfectly conducting sphere, the charge is uniformly distributed over the outermost surface. In contrast for perfect insulator the charge is locked in place and might for example be uniformly distributed over the full volume. We're now up to three assumptions involved in calculating the self-energy.[2]

The second significant problem relates to the computational difficulty of calculating the field of the moving electron and the interaction of the electron with its own field. Here, Abraham (Dynamik des Electrons, 1902a, p. 23) observes, "That path appears to be inaccessible, however, given the present-day state of theory: merely calculating the field of a nonuniformly moving electron is extraordinarily complex."[3]

Abraham attempts to get around this problem by considering the total energy in the electric and magnetic fields of an electron moving at a constant velocity. This eliminates acceleration of the electron from consideration and makes the problem independent of time—consequently, no consideration of retarded potentials, for example, is needed, and there is no radiation damping. The core of this approach comes from (Kaufmann, Die magnetische und electrische Ablenkbarkeit der Becquerelstrahlen und die scheinbare Masse der Elektronen, 1901, p. 153), where he presents it in a little more than one page. Kaufmann's approach is built on two parts.

The first is presented in the beginning of equation 13 in (Kaufmann, Die magnetische und electrische Ablenkbarkeit der Becquerelstrahlen und die scheinbare Masse der Elektronen, 1901, p. 153):

$$m_{\|} = \frac{1}{v} \frac{\mathrm{d}W}{\mathrm{d}v}. \tag{9.2}$$

W is the total energy in the electric and magnetic fields of the electron moving with a velocity v. Although (Kaufmann, Die magnetische und electrische Ablenkbarkeit der Becquerelstrahlen und die scheinbare Masse der Elektronen, 1901) does not distinguish longitudinal and transverse mass, it is clear from subsequent discussion that his equation 13 gives the longitudinal mass.

And the other part is the total electric and magnetic field energy of a sphere of radius a and charge e traveling with a constant velocity v, with, $\beta = v/c$. This part is, from equation 12 on the same page:

$$W = \frac{e^2 c^2}{2a} \left[\frac{1}{\beta} \ln \frac{1+\beta}{1-\beta} - 1 \right]. \tag{9.3}$$

Kaufmann cites (Searle, 1897) as the origin of this formula. Adapting for differences in notation, this is the same as equation 25 in (Searle, 1897, p. 340). The context and meaning of the equation discussed by Kaufmann match the

[2]For more recent context for this problem, the reader may consult (Feynman, Leighton, and Sands 1963) Volume II, Chapter 28, Electromagnetic Mass.

[3]And for this problem, the reader may consult (Jackson, Classical Electrodynamics 1999) Chapter 16, Radiation Damping, Classical Models of Charged Particles (roughly sections 16.1–16.4).

development and explanation given by Searle; the formula applies to a charged sphere moving at a constant velocity.

Substitution of equation 9.3 into 9.2 results in:

$$m_{\|} = \frac{e^2}{2a}\frac{1}{\beta^2}\left[\frac{1}{\beta}\ln\frac{1-\beta}{1+\beta} + \frac{2}{1-\beta^2}\right]. \quad (9.4)$$

Equation 9.4, in some form and with the associated assumptions, is, equation 13 in (Kaufmann, Die magnetische und electrische Ablenkbarkeit der Becquerelstrahlen und die scheinbare Masse der Elektronen, 1901, p. 153), equation 25e in (Abraham, Dynamik des Electrons, 1902a, p. 38), equation 10a in (Abraham, Prinzipien des Dynamik des Elektrons, 1902b, p. 61) and equation 16e in (Abraham, Prinzipien der Dynamik des Elektrons, 1903, p. 152).

Abraham, (Dynamik des Electrons, 1902a, p. 21) does mention both papers by Kaufmann and Searle (referenced above) but indicates the need for substantial additional study. His study extends over many pages and I have been selective in looking at his treatment of the velocity dependence of electromagnetic mass and his reasoning contributing to it.

Abraham provides a justification for equation 9.2 by relating the change in the total energy of the electron, including the kinetic energy and the energy in the electric and magnetic fields, to the work done on the electron by a force parallel to the steady velocity of the electron. This leads fairly directly to equation 8 (Abraham, Dynamik des Electrons, 1902a, p. 26), which is the same as equation 9.2 above. This reasoning is filled with conflicting assumptions: the electron is assumed to be traveling at a constant velocity without acceleration, and then assumed to be subject to a force that does work and changes the total energy. The calculation of the differential work is based on the velocity from the ongoing steady displacement and not from a differential velocity produced by the force. Any change in the kinetic energy is neglected; Abraham may even implicitly assume that the kinetic energy is associated only with the mechanical mass. It is hard to see how this mix of assumptions sorts itself out. In (Poincaré, Sur la dynamique de l'électron, 1906, pp. 159–60; p. 82–83) and notably equation 4, Poincaré approaches these assumptions differently. Here he reasons from the electromagnetic action, the speed of the electron and its shape. Recognizing the need, without approximation, to consider the retarded potentials, Poincaré makes the simplifying assumption that the higher order derivatives of the action with respect to the velocity and shape can be neglected compared to the first order derivatives. This eliminates acceleration and is therefore quasi-stationary. He does derive equation 9.2, moving it to more solid dynamical ground. I am skeptical about the usefulness of a quasi-stationary assumption here.[4]

In equation 25d (Abraham, Dynamik des Electrons, 1902a, p. 38) provides a formula for the transverse mass, a companion to Equation 9.4. Here the reasoning

---

[4]For possible comparison with Einstein see §10 of (Einstein, Zur Elektrodynamik bewegter Körper 1905).

seems on more satisfying ground; a force perpendicular to the velocity results in a shift in the direction of motion comparable to a circular motion with a very large radius of curvature. He can relate the force to the centripetal acceleration and the change in angular momentum. Also, here, U is the Lagrangian, so equation 25d could also be based on equating the conjugate momentum to the derivative of the Lagrangian with respect to velocity.[5]

$$m_{\perp} = \frac{e^2}{2a} \frac{1}{\beta^2} \left[\frac{1+\beta^2}{2\beta} \ln \frac{1+\beta}{1-\beta} - 1 \right]. \tag{9.5}$$

Once (Abraham, Dynamik des Electrons, 1902a) makes this distinction between the longitudinal and transverse mass and provides a formula for calculating the transverse mass, Kaufmann only calculates the transverse mass. Equation 9.5 is equation 15 in (Kaufmann, Über die electomagnetische Masse des Elektrons, 1902b, p. 294), equation 2 in (Kaufmann, Die electromagnetishe Masse des Elektrons, 1902a, p. 54), and equation 10 in (Kaufmann, Über die "Elektromagnetische Masse" der Elektronen, 1903, p. 96). Equation 9.5 is also part of Equation 9.7 below.

The point I am emphasizing here is that the only explicit formula for the dependence of (transverse) mass on velocity from Abraham and also the only one used in Kaufmann's analysis of his data is based on the assumption of a spherical electron with a constant radius.

An electron that changed shape as it moved would require associated internal forces and self-energy (now referred to as Poincaré stress and discussed starting on page 189 below) that could not be of electromagnetic origin. Forces that are not of electromagnetic origin are inconsistent with an electromagnetic worldview.

Abraham is more emphatic about this point in (Abraham, Prinzipien der Dynamik des Elektrons, 1903, p. 108), writing, "Also the assumption of a deformable electron seemed to me to be inadmissible for reasons of principle. For it leads to the consequence that in the change of shape work is done by the electromagnetic forces, or against them, that thus besides the electromagnetic energy an inner potential energy of the electron is to be used. If this were really necessary, the electromagnetic justification of the theory of cathode and Becquerel radiation, a purely electrical process, would already be impossible."

An electron that does not change shape avoids that problem, but it is inconsistent with both the observational failure to detect movement relative to the ether and with the principle that absolute motion cannot be detected. While Abraham chooses to be consistent with the electromagnetic worldview, there is clearly no way forward for an assumption of an undeformable electron.

Poincaré's discussion of the forces internal to the electron is covered in the last section of this chapter.

[5] In Equation 9.2 W is the total energy, not the Lagrangian, so this reasoning would not apply there.

Hendrik Lorentz

Lorentz's paper (Lorentz, Electromagnetic phenomena in a system moving with any velocity smaller than that of, 1904) starts with a discussion of experiments to determine the "influence exerted on electric and optical phenomena by a translation, such as all systems have in virtue of the Earth's annual motion." This discussion refers first to the Michelson-Morley experiment—that led Lorentz and Fitzgerald "to the conclusion that the dimensions of solid bodies are slightly altered by their motion through the ether"—and continues with the experiments within the previous few years that also showed no effect. Lorentz is persuaded of the existence of the ether and is seeking to reconcile its existence with negative experimental results.

Continuing, Lorentz presents the equations of electrodynamics with the equation for the electrical and magnetic force per unit charge in his equations 2 and the Lorentz transformations in equations 3 to 5 (in this book, these equations are reproduced in Part III on page 262). He proposes to refer the equations of electrodynamics to a system moving at a constant velocity. Lorentz does not use terms suggesting that the untransformed equations refer to a state of absolute rest, but this is not excluded when reading between the lines. Note that the transformation defined in equations 4 and 5 contain a numerical quantity, l, that will be discussed later. At this point in the article, Lorentz allows that it could be a second or higher order function of the velocity of the moving system.

Lorentz's application of the transformation to the equations of electrodynamics and the electromagnetic force leads to the equations of electrodynamics (equation 9) and the force (equation 10) in the moving system. Upon comparing Lorentz's equations of electrodynamics in the moving system (equation 9) to the equations in the stationary system (equation 2), it is immediately clear, in Lorentz's derivation, that the equations of electrodynamics have been changed by the transformation to the moving system because in the equation for the divergence of the electric displacement in the moving system, the charge density has been multiplied by a factor of $\left(1 - wu'_x/c^2\right)$; the other three equations are however unchanged.

This factor, multiplying the charge density, should not have been there; it results from an error in applying the transformations from the stationary system to the moving system. It is surprising that an article of such historical importance for its presentation of the transformations now named after the author would have an error so evident in hindsight. The error however did not go unobserved; within a year it was found by Poincaré who commented on it in two places.

First Poincaré pointed out this error in a letter to Lorentz from about April 27, 1905 (Kox, 2008, pp. 176–8, letter 126; p. 37–9)[6]. (There is a translation of this letter in Chapter 3 of this book on page 37.) This letter was written just over a month before the weekly session of the *Académie des Sciences* when Poincaré presented (Poincaré, Sur la dynamique de l'électron, 1905). Poincaré quotes the

[6]Here and elsewhere, the first page number refers to the original publication and the second page number (following the semicolon) refers to the page number in Part I of this book.

transformation of the charge density (equation 7) in (Lorentz, Electromagnetic phenomena in a system moving with any velocity smaller than that of, 1904, p. 813; p. 263)[7] and provides his correction. He also provides his corrections to the transformation of the electromagnetic force per unit charge to the moving system. The error and corrections are discussed more below.

Second, in Poincaré's discussion of the transformation and their application to the equations of electrodynamics (Poincaré, 1906, pp. 133–4; p. 49–50), he states that he has a divergence from Lorentz: Lorentz's equations 7 and 8. The first equation (equation 7) relates the charge density in the moving system and the charge density in the stationary system. It is incorrect because it does not reflect a correct calculation of the unit volume of the moving charge seen in the stationary system. Poincaré provides the correct relationship between the two charge densities in his equation 4. Comparing their definitions of the terms, we can see that the factor $\left(1 - wu'_x/c^2\right)$ in Lorentz's notation is written $(1 - \epsilon\xi')$ in Poincaré's notation. Further, using the identity written below Poincaré's equation 4 and the definition of k, we can rewrite this factor as $[k^2(1 + \varepsilon\xi)]^{-1}$, and it becomes clear that Poincaré's correction to the transformation of the charge density eliminates the extraneous factor multiplying the charge density in the equation for the divergence of the displacement. The second equation (equation 8) relates the velocity of the electron in the stationary system to the velocity in the moving system; it reflects an incorrect composition of the two velocities. Poincaré notes that while the two equations are incorrect, the product of the transformations of the charge density and velocity are correct.

This same extra factor in the divergence of the displacement also shows up in equations 10 and 13 (Lorentz, Electromagnetic phenomena in a system moving with any velocity smaller than that of, 1904, pp. 813–814; p. 263–4). In equation 10 for the force on the unit charge, it results in a missing term, $u'_x D'_x$ in the final parentheses of the x component of the force and in extra terms at the end of the y and z components of the force. In equation 13, which gives the displacement in the moving system as a function of the vector and scalar potentials, the term $+\frac{w}{c}\,\mathrm{grad}' A_x$ is extra.

Although the charge density appears again in the formula for the curl of the magnetic field, there it appears as a product with the velocity and is correct because of the compensating errors mentioned at the end of an earlier paragraph.

Taking a step back, perhaps for additional confirmation, Poincaré shows that with his transformation of the charge density, the continuity equation for charge density, satisfied in a stationary system, is also satisfied in the moving system.[8] Stated differently, he confirms the charge is conserved in both the stationary and moving systems.

---

[7]Equation 19a in (Abraham, Dynamik des Electrons, 1902a, p. 33) has the same relation between moving and resting charge density as equation 7 in (Lorentz, Electromagnetic phenomena in a system moving with any velocity smaller than that of, 1904, p. 813; p. 263).

[8]This point is discussed further in a following section on Poincaré's contribution and starts on page 184.

We, also taking a step back, can take a second look at Lorentz's transformations of the spatial coordinates and time given in the equations 4 and 5. First note that there is a typo in the first term on the right-hand side of equation 5; that term, $\frac{l}{k}t$, should be written klt[9]. Substantively, the transformation for the x-coordinate, parallel to the velocity w, does not include this velocity and should be rewritten $x' = kl(x - wt)$. Did this contribute to the errors in his equations 4 and 9?

In any case, the error in Lorentz's equation 9 prevents him from seeing that the equations of electrodynamics are unchanged under his transformation. In contrast, equations 9 and 12 as written even suggest that an experiment might be conceived for determining the direction and magnitude of velocity of the moving system. That would offer a means for measuring the velocity of the laboratory relative to ether at absolute rest. Lorentz does not comment on this possibility.

On the same page, there is a very important concept that Lorentz presents: *local time*. Local time is the time (t') in the moving system and is related to the time in the system at rest by a transformation that mixes the velocity of the moving system with the time and position in the stationary system (Lorentz, Electromagnetic phenomena in a system moving with any velocity smaller than that of, 1904, pp. 812; p. 262, eqn. 5). Local time was an earlier introduction of Lorentz retained in this paper. Poincaré had discussed local time in (Poincaré, La théorie de Lorentz et le principe de réaction, 1900, pp. 262–8; pp. 31–36). That discussion is relevant here and includes setting watches with crossed light signals. Lorentz uses the term, local time, in this article, to indicate the time in the moving system and distinguish it from the time in the stationary system. This is a rudimentary use that does not seem to have benefited from Poincaré's discussion six years earlier. Lorentz does not comment on the resulting mixing of time and position; it is left up to Poincaré and Einstein to pick up the subject.

Next, meaning at the beginning of §6, Lorentz takes up an example of an electrostatic system, which he indicates to mean that the electron in the moving system has no velocity relative to the system. Lorentz indicates quantities measured in this system with a prime and indicates velocity of the electron by the vector $\boldsymbol{u}$. Meaning that he takes $\boldsymbol{u}' = 0$. This effectively removes any influence of the error in the transformation of the charge density (equation 7) from the subsequent discussion, since the error depends on u'_x.

The paragraph between equations 20 and 21 (Lorentz, Electromagnetic phenomena in a system moving with any velocity smaller than that of, 1904, p. 815; p. 265) provides an insight into how Lorentz thinks of his transformation and its use. Over a century later, we are used to thinking, for example, of the transformations as providing a way for two different observers each in an inertial reference frame to reconcile their understanding of a single event they both observe, such as a train arriving at the platform. In this paragraph, Lorentz asks the reader to compare a "moving system" and "another system... which remains at rest." He indicates that

[9] And the typo is worth pointing out because it goes unremarked and repeated by (Miller, 1973) in equation 33d.

the way to do this is by multiplying the dimensions parallel to the direction of motion in the moving system by kl (the inverse of the contraction) and then comparing "corresponding" elements of the two systems. This is recognizably different from an interpretation based on special relativity and our current understanding. One can see how this might be applied to an interferometer with one arm oriented in the direction of the Earth's movement through the ether and the other arm perpendicular to the motion; the two arms then being the corresponding elements. Certainly, there are other situations where identifying the corresponding elements is less clear.

Continuing, Lorentz works towards two conclusions. First, he concludes that the transformation that he proposes for electrons can in fact be applied to systems built up from electrons and even to macroscopic systems such as laboratory experiments, for example for measuring motion relative to the ether. This is a statement of the electromagnetic view. Second, he derives in equation 30 the longitudinal electromagnetic mass, meaning the self-energy, of an electron with charge e and radius R moving with velocity w:

$$m_{\|} = \frac{e^2}{6\pi c^2 R} \frac{\mathrm{d}(klw)}{\mathrm{d}w} \quad m_{\perp} = \frac{e^2}{6\pi c^2 R} kl$$

So far, l has not been determined although it must be one when w is zero and could be a second or higher order function of w/c. Physically l corresponds to a transformation of the dimensions perpendicular to the direction of motion. In the theories of Abraham, Lorentz and Langevin-Bucherer, an electron seen at rest is spherical and an electron moving with a fixed velocity w along the x-axis may have a different shape. Since the transformation is symmetric about the direction of motion, the x-axis, we can consider the two-dimensional shape of the electron in a plane containing the x-axis. At rest, the shape is a circle and in motion it is an ellipse with the ratio of the semi-major axis to semi-minor axis equal to kl/l. Therefore, the moving electron is an ellipsoid, and the aspect ratio is independent of l. For example, Langevin's assumption requiring that the volume of the electron remains unchanged at different velocities is equivalent to requiring that $kl^3 = 1$. Abraham's assumption is that an electron is undeformable, meaning kl and l are each identically one, and therefore remains spherical is incompatible with the Lorentz transformations. Further details are provided in (Poincaré, Sur la dynamique de l'électron, 1906, pp. 154–55; p. 77).

To determine l, Lorentz reasons from transformations of force and acceleration that the factor $\mathrm{d}(klw)/\mathrm{d}w$, appearing in the longitudinal mass above, must be equal to $k^3 l$ and because of the definition of k, this means l must be a constant independent of the velocity and hence $l = 1$.

That brings us to the third divergence[10] raised by Poincaré in the letter of April 27, 1905 (Kox, 2008, p. 176–8, letter 126; p. 37–9): Poincaré does not find Lorentz's reasoning leading to the conclusion $l = 1$ conclusive. In a second letter (undated but presumably between April 27 and June 5, 1905) (Kox, 2008, pp. 178–9, letter 127; p. 39–40), Poincaré tells Lorentz that viewing the transformations as a group requires that $l = 1$. This is a wholly mathematical proof disconnected from discussion of the shape or volume of the moving electron.

My discussion so far of Lorentz's model has focused on some serious problems in the details of (Lorentz, Electromagnetic phenomena in a system moving with any velocity smaller than that of, 1904). I should not allow that to be the last word. These are in fact details. The achievement of the paper and the reason why it is now, justifiably well-known is because Lorentz presents the transformations, which Poincaré named after him, and shows that they explain why motion through the ether cannot be detected. This is a big achievement which Poincaré and history noted.

Section 11 starts with the statement that "It is easily seen that the proposed theory can account for a large number of facts." The examples in the section all relate to accounting for why laboratory experiments have not detected motion relative to the ether. This is exactly the point that I emphasized in the previous paragraph as a success of Lorentz's theory.

In the next to last section, Lorentz reprocesses the data for the β ray charge-to-mass ratio data presented in (Kaufmann, Die electromagnetishe Masse des Elektrons, 1902a). This is a very interesting step, and seems to get overlooked. Kaufmann's published results used the relation from Abraham's theory (i.e. $\psi(\beta)$) to attempt to remove the velocity dependence from the charge-to-mass ratio. (Lorentz, Electromagnetic phenomena in a system moving with any velocity smaller than that of, 1904, pp. 828–30; p. 277–9) reproduces the results from Kaufmann with the symbols β, and k_2 where the first is proportional to the velocity and the second to the rest-frame charge-to-mass ratio based on Abraham's formula. They are shown in the first and third columns of the tables. It needs to be understood that in these tables $\psi(\beta)$ is a value calculated from β and not a separate measured quantity. Equation (35) presents the relationship between the charge-to-mass ratio and $\psi(\beta)$, but note that there are other factors included in the proportions related to fitting the data including k_1 and also s which is given in the table titles. The last two columns in the tables β' and k_2' are the results of reprocessing by Lorentz. Here $\beta' = s \cdot \beta$ and k_2' is calculated with Lorentz's formula 38. In Abraham's theory k_2 should be constant, and in Lorentz's theory k_2' should be constant. (Lorentz, Electromagnetic phenomena in a system moving with any velocity smaller than that of, 1904) states, "The constancy of k_2' is seen to come out no less satisfactory than that of k_2." Looking

[10]If you're keeping count and wondering what the second divergence is, it is that charged electrons are not stable if there are only electrical forces. This is discussed starting page 189 below. The additional forces required for stability are now called the Poincaré stresses. Poincaré brings the need for this additional force to Lorentz's attention in a third letter in the same timeframe (Kox 2008, p. 179, letter 128; p. 40). There is no extant reply from Lorentz to Poincaré to any of these three letters.

at the columns of numbers it is easy to agree with this observation. It is possible to make a better comparison, and my comparison is shown in the next section of this chapter. That comparison shows that Lorentz could have made a much stronger statement. His theory does in fact explain the results of Kaufmann's experiment much better than analysis with Abraham's theory.

In the final section of (Lorentz, Electromagnetic phenomena in a system moving with any velocity smaller than that of, 1904), Lorentz returns to experiments for detection of the Earth's motion through the ether and suggest that the expected results of Trouton's experiment are too small to measure.

In concluding this section, I think back to Poincaré's letter (Kox, 2008, p. 176–8, letter 126; p. 37–39) where he states, concerning the work Lorentz presented in this paper, "I agree with you on all the essential points; however there are a few divergences in the details." Some of the essential points in this article by Lorentz are: the concept of local time, the recognition that the transformations do make it impossible to detect motion relative to the ether, and that the same transformations demand an increase of mass with velocity. The step of re-analyzing Kaufmann's data with his theory instead of Abraham's is also clever and, as discussed in the following section, provides stronger support for his theory then Lorentz realized.

Reanalyzing Kaufmann's Data

In Chapter 8, pages 158–160 I discussed experimental data from Kaufmann and also Lorentz's reprocessing of that data in connection with confirming that β rays have a single rest frame charge-to-mass ratio (i.e. there aren't multiple types of β rays) and also that the rest frame charge-to-mass ratio for β rays matched the ratio for electrons from other contexts. Here I take another look at Kaufmann's data from (Kaufmann, Die electromagnetishe Masse des Elektrons, 1902a) and (Kaufmann, Über die "Elektromagnetische Masse" der Elektronen, 1903); my intent is to compare how well Abraham's and Lorentz's theories (discussed in the previous two sections) explain the velocity dependence of the charge-to-mass ratio of the β ray moving through the crossed electric and magnetic fields of the experiment, and also to determine a value for the rest-frame ratio.

Kaufmann's experimental apparatus exposed a photographic plate to β rays from a radium source that had passed through crossed electric and magnetic fields. After developing the plate, a curve was visible and the data analysis started with measuring the coordinates of selected points on the curve in centimeters relative to an arbitrary origin. This raw data is reproduced, for example, in the first two columns of Tables I to III in (Kaufmann, Die electromagnetishe Masse des Elektrons, 1902a). The z and y values from these two columns are related to the deflections of the β ray by the electric field and by the magnetic field. In (Kaufmann, Die electromagnetishe Masse des Elektrons, 1902a) the deflections are represented by ζ and η where the relation of these quantities to the velocity and to the moving charge-to-mass ratio are given by equations 5 and 6. In (Kaufmann, Über die "Elektromagnetische Masse" der Elektronen, 1903), the deflections are represented by z' and y' where the corresponding equations are 8 and 9. (In these equations, F and H are the magnitudes of the electric and magnetic fields in the apparatus.)

From Abraham's theory discussed in the first section of this chapter we expect that:

$$\frac{e}{m} = \frac{4}{3} \cdot \frac{e}{m_0} \cdot \frac{1}{\psi(\beta)} \tag{9.6}$$

where

$$\psi(\beta) = \frac{1}{\beta^2}\left[\frac{1+\beta^2}{2\beta} \cdot \ln\left(\frac{1+\beta}{1-\beta}\right) - 1\right] \tag{9.7}$$

and where β is the ratio of the speed of the electron seen in the laboratory frame to the speed of light, m is the mass of the moving electron, m_0 is the mass of the electron referred to the rest frame and e is the charge of the electron. Equations 9.6 and 9.7 are equivalent to Equation 9.5 above (noting that $m_0 = 2e^2/3a$); as discussed there, the equation is based on the assumption of an undeformed sphere of radius a moving at a constant speed.

Using the equations referenced above from Kaufmann, we know that e/m is proportional to ζ^2/η. Since the magnitude of the laboratory electric and magnetic fields enters the proportion, data from different photographic plates cannot be combined. I am not otherwise concerned at this point with other constant factors occurring outside $1/\psi(\beta)$. To determine $1/\psi(\beta)$, I use the value of β recalculated by Lorentz, or in (Runge, 1903) or calculated by Kaufmann, in order of preference, and recalculate the function myself using Equation 9.7.

Equation 9.6 can then be written in terms of the measured quantities as:

$$\frac{\zeta^2}{\eta} = a\frac{e}{m_0} \cdot \frac{1}{\psi(\beta)} \tag{9.6$'$}$$

where a is a constant of proportionality.

Here I can plot ζ^2/η against $1/\psi(\beta)$. Based on Equation 9.6′, I expect the plot to show a linear relationship and the quality of the linear fit to the data is a test of the agreement with the data of Abraham's theory relating the moving electron mass to the rest mass.

From Lorentz's theory, the corresponding relationship is:

$$\frac{e}{m} = \frac{e}{m_0} \cdot \sqrt{1 - \beta^2} \tag{9.8}$$

where the same symbols have the same meaning as above.

The same observations as above apply to *e*/*m* and also apply to β which is proportional to ζ/η. Equation 9.8 can then be rewritten in terms of the measured quantities and yields:

$$\eta^2 = a\zeta^2 + b\left(\frac{e}{m_0}\right)^{-2}\zeta^4 \tag{9.8$'$}$$

where again *a* and *b* are constants.

Lorentz (Electromagnetic phenomena in a system moving with any velocity smaller than that of, 1904, p. 829; p. 278) has the same equation as Equation 9.8′ here, except it doesn't explicitly show the rest frame charge-to-mass ratio.

In this case I can plot η^2 against ζ^2. Based on Equation 9.8′, I expect this plot to show a quadratic relationship and the quality of the fit to the data is a test of the agreement with the data of Lorentz's theory relating the moving electron mass to the rest mass.

Fitting the data to Lorentz's relationship involves two free parameters. Fitting the data to Abraham's relationship involve only one free parameter. However, the calculation of β by Lorentz or Runge used a least-squares fit that had one free parameter. Comparing the fits should be fair.

Figures 9.1 and 9.2 show results for a dataset from 1902 (Kaufmann, Die electromagnetishe Masse des Elektrons, 1902a, p. 55, Table IV).

Figures 9.3 and 9.4 show results for a dataset from 1903 (Kaufmann, Über die "Elektromagnetische Masse" der Elektronen, 1903, p. 99, plate 18).

I have only presented two pairs of comparisons and both show a substantially better fit to the Lorentz transformation than to Abraham's theory. I have however done this comparison for all datasets from (Kaufmann, Die electromagnetishe Masse des Elektrons, 1902a) and (Kaufmann, Über die "Elektromagnetische Masse" der Elektronen, 1903). In all cases, the Lorentz transformation is a better fit to Kaufmann's data from 1902 and 1903 relating electron charge-to-mass ratio and velocity. The statement by Lorentz quoted above, "to come out no less satisfactory than that" of Abraham, is supported by this analysis; in fact, a much stronger statement by Lorentz is justified.

Equation 9.8′ and the fits to the quadratic curves in 9.4 provide a means to determine the electron rest frame charge-to-mass ratio from Kaufmann's data. Returning to equations 5 and 6 in (Kaufmann, Über die "Elektromagnetische Masse" der Elektronen, 1903) and Equation 9.8 above, referring to the coefficient

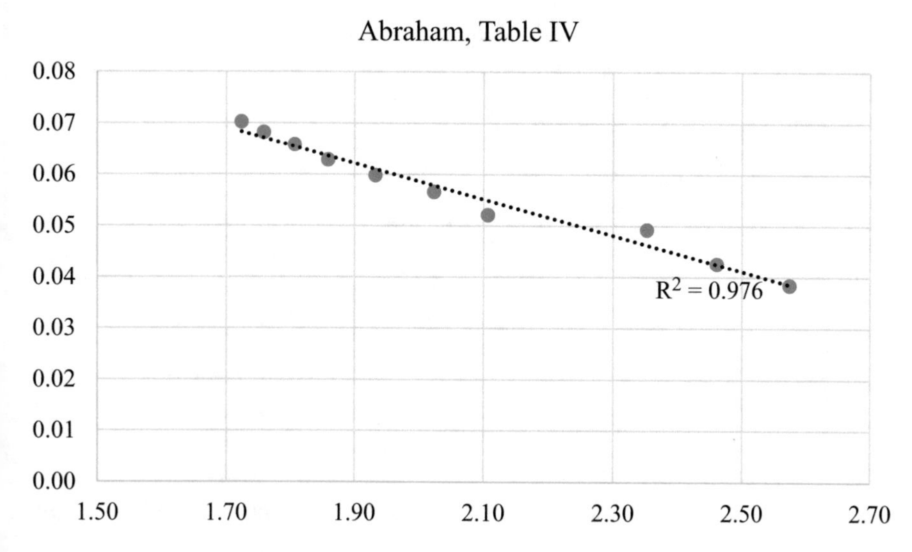

Figure 9.1 Data from (Kaufmann, Die electromagnetishe Masse des Elektrons, 1902a, p. 55, Table IV) and fitting of β from (Lorentz, Electromagnetic phenomena in a system moving with any velocity smaller than that of, 1904, p. 828; p. 278) with calculation to show relationship expected in Equation 9.6′. All units are arbitrary.

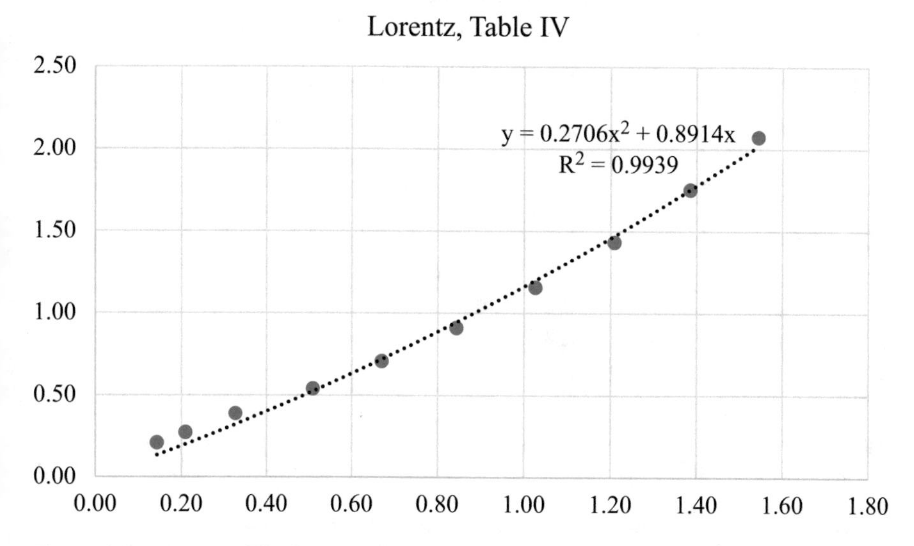

Figure 9.2 Data from (Kaufmann, Die electromagnetishe Masse des Elektrons, 1902a, p. 55, Table IV) with calculation to show relationship expected in Equation 9.8′. All units are arbitrary.

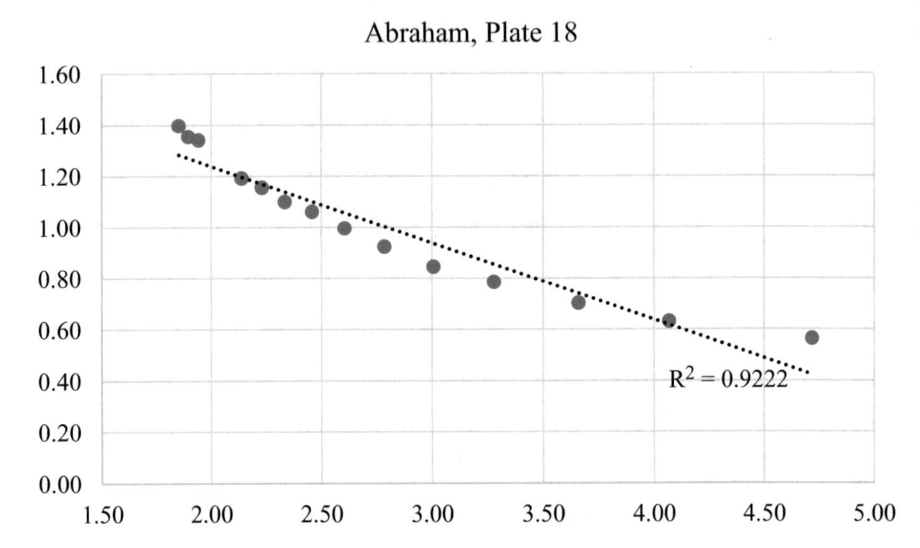

Figure 9.3 Data from (Kaufmann, Über die "Elektromagnetische Masse" der Elektronen, 1903, p. 99, plate 18) and fitting of β from (Runge, 1903, p. 329) with calculation to show relationship expected in Equation 9.6′. All units are arbitrary.

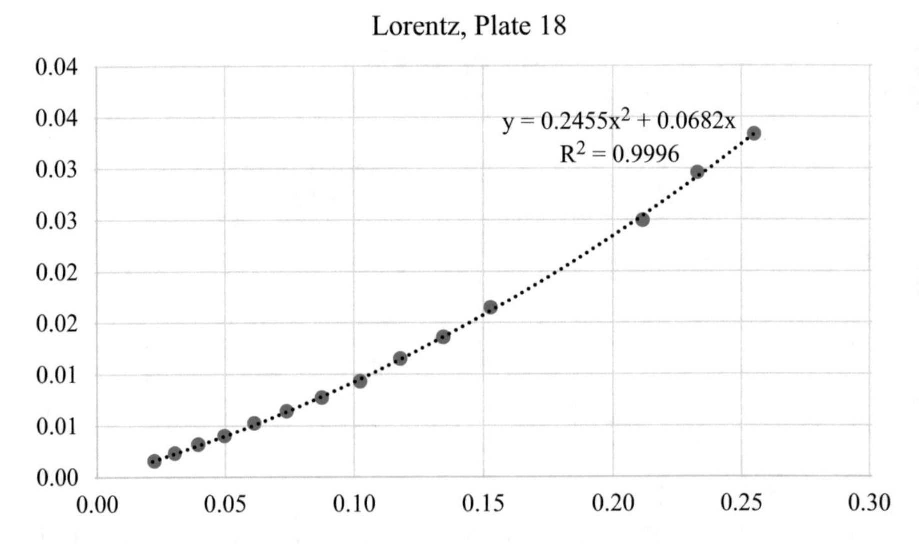

Figure 9.4 Data from (Kaufmann, Über die "Elektromagnetische Masse" der Elektronen, 1903, p. 99, plate 18) with calculation to show relationship expected in Equation 9.8′. All units are arbitrary.

Table 9.1 Charge-to-mass ratio reproduced from (Kaufmann, Die magnetische und electrische Ablenkbarkeit der Becquerelstrahlen und die scheinbare Masse der Elektronen, 1901, p. 152)

				e/m_0	(Kaufmann, 1903)	(Runge, 1903)
Plate	H	C_1	C_2	in 10^7 cgs units		
15	200	0.0572	0.3356	1.483	1.74	1.669
18	200	0.0682	0.2455	1.894	1.97	1.917
19	200	0.0542	0.2787	1.584	1.80	1.755
24	199	0.0475	0.2414	1.602	1.73	1.653
		Average:		1.641	1.810	1.749
		Standard Deviation:		0.177	0.111	0.121

of the linear term in the fit to a quadratic equation in Figure 9.4 as C_1 and the coefficient of the quadratic term as C_2, it can be shown that:

$$\frac{e}{m_0} = \left(\frac{2c}{x_2^2 + x_1 x_2}\right) H^{-1} \sqrt{\frac{C_1}{C_2}} \tag{9.9}$$

The constants x_1 and x_2 depend on the configuration of the equipment and are given in (Kaufmann, Über die "Elektromagnetische Masse" der Elektronen, 1903) between equations 5 and 6. The speed of light is c. H is the magnetic field strengthen; it is given in Kaufmann's Table VI. The numerical constant in parenthesis is therefore 7.186×10^{10} cgs units.

The results of the recalculation for the four plates from (Kaufmann, Über die "Elektromagnetische Masse" der Elektronen, 1903) are gathered in the Table 9.1, below.[11] The results from Kaufmann and the recalculation by (Runge, 1903) are included for comparison with the reminder that they are calculated using Abraham's formula.

For comparison with the average value 1.641×10^7 cgs units from Table 9.1, note that Alfred Bucherer (Messungen an Becquerelstrahlen. Die experimentelle Bestätigung der Lorentz-Einsteinschen Theorie., 1908) got a value of 1.730×10^7 cgs units, and that the current accepted value is 1.758×10^7 cgs units (National Institute of Standards and Technology, n.d.).

This work by Kaufmann and Lorentz, followed by more accurate measurements by (Bucherer, 1908), are an early experimental confirmation of special relativity.

[11] I have done the corresponding analysis with the data from (Kaufmann, Die electromagnetishe Masse des Elektrons, 1902a), but the standard deviation is much larger so it didn't seem worth presenting the results.

Henri Poincaré

In (Poincaré, Sur la dynamique de l'électron, 1906), Poincaré took up and moved forward topics raised by Abraham and Lorentz discussed above, including his divergences from Lorentz discussed in the context of the three letters from Poincaré to Lorentz. That is the subject of this section. Poincaré also made a significant departure from Abraham and Lorentz concerning electromagnetic mass and ether, and brings gravitation into consideration. That is one part of Chapter 10.

Poincaré started his §1 by presenting the equations of electrodynamics, the Lorentz force and the Lorentz transformations respectively in equations 1, 2 and 3 (Poincaré, Sur la dynamique de l'électron, 1906, p. 132; p. 48). The inverse of the Lorentz transformation is presented in equation $3'$. These equations correct the two points noted above on page 175: in the transformation of the time, the factor k is moved from the denominator to numerator and in the transformation of the x-coordinate the term proportional to the time, t, is added reflecting the motion of the reference frame. The first correction is a simple typo. The second correction assures that, in the limit of small velocities, the Lorentz transformation of the coordinate along the direction of movement reduces to a Galilean coordinate transformation; this correction is also simple.

Transformation of Charge Density—First Divergence

In Equation 7, Lorentz (Electromagnetic phenomena in a system moving with any velocity smaller than that of, 1904, p. 813; p. 263) indicated, without further explanation, that the transformation of the charge density between the moving and stationary systems was proportional to kl^3. Presumably, Lorentz assumed that the charge on the electron would be the same in the moving and stationary systems and that the volume in the stationary system would be proportional to xyz and the volume in the corresponding ellipsoid in the moving system would be proportional to $x'y'z'$. Then, applying his transformations from his equation 4, gives $x'y'z' = kl^3xyz$ and hence the factor in Lorentz's transformation of the charge density.

This is the first divergence that Poincaré wrote about in a letter to Lorentz and which was mentioned in the section on Lorentz above. The letter is also translated in Part I, Chapter 3.

Poincaré was explicit in his calculation of the volume. He started with the equation of a sphere moving with a velocity u in one coordinate system and substituted in the transformation into a second coordinate system (primed) moving at a constant velocity w relative to the first. That gave him the equation for an ellipsoid[12] moving at a constant velocity in the second coordinate system and he then

[12]For possible comparison with Einstein concerning shape see §4 of (Einstein, Zur Elektrodynamik bewegter Körper 1905).

calculated the volume of the ellipsoid from its equation. This resulted in a factor for the density (inverse of the volume relationship) of $k(1 + uw/c^2)/l^3$. This means that:

$$\rho' = \frac{k}{l^3}\left(1 + \frac{uw}{c^2}\right)\rho$$

which is Poincaré's equation 4 in §1.[13]

At this point, Poincaré also calculated separately the transformation of the velocity of the moving sphere. In a Galilean transformation the velocity in the second system would be $u + w$, a direct sum of the velocities. In equation 8, Lorentz (Electromagnetic phenomena in a system moving with any velocity smaller than that of, 1904, p. 813; p. 263) directly stated (again without explanation) that the transformation is:

$$u' = k^2(u + w) = \frac{u + w}{1 + w^2/c^2}.$$

In contrast (Poincaré, Sur la dynamique de l'électron, 1906, p. 133; p. 50) shows the derivation of the composition of velocities:

$$u' = \frac{dx'}{dt'} = \frac{\frac{k}{l}d(x + wt)}{\frac{k}{l}d(t + xw/c^2)} = \frac{u + w}{1 + uw/c^2}. \tag{9.10}$$

This formula has the result that a sphere moving with velocity u transformed to coordinates moving with velocity w has the same composed velocity as a sphere moving with velocity w transformed to coordinates moving with velocity u; this is not true of Lorentz's formula for the transformation of the velocity.

I think this may be a good point to consider what the origin of this first divergence might be. In choosing words to describe Lorentz's steps above, I have tried to reflect his view of corresponding systems. Seen that way, Lorentz was trying to compare the volume of corresponding spheres—actually one sphere and one ellipsoid. Contraction of the semi-minor axes by the factors k and l result in a new volume proportional to the product of the semi-minor axes and the proportion kl^3, above, found by Lorentz. Thus, the perspective of comparison of corresponding objects leads to Lorentz's comparison of the two volumes and a divergence from Poincaré. Poincaré used the Lorentz transformations to change from one coordinate system to another. (We can see that they are inertial since his derivation of the transformation of the velocity u to u' depends on the assumption that $dw/dt = 0$.) Here the coordinate system transformation can be seen since Poincaré wrote the equation for a moving sphere and applied the transformation to the equation which results in an equation for a moving ellipsoid; from each equation he could determine the volume of the shape.

[13]For possible comparison with Einstein see §9 of (Einstein, Zur Elektrodynamik bewegter Körper 1905).

The perspective of transforming coordinate systems can also be seen in the discussion at the end of the previous paragraph comparing the transformation of velocity. Poincaré's version of the velocity transformation has a symmetry (in the term uw/c^2) that is absent in Lorentz's version (which has the term w^2/c^2).

This aspect of the divergence comes into clearer focus in Poincaré's discussion at the beginning of §4 (Poincaré, Sur la dynamique de l'électron, 1906, p. 144; p. 64). There he wrote that the Lorentz transformations form a group and argued that consequently a transformation from a first coordinate system (without primes) to a second coordinate system (with primes) with a relative velocity $\varepsilon c = u$ followed by a transformation to a third coordinate system (with double primes) with a relative velocity $\varepsilon' c = w$ must have the same result as a transformation directly from the first coordinate system to the third coordinate system with a velocity $\varepsilon'' c = u'$ that is a composition of u and w. An immediate consequence is that the composition of velocities is correctly given by equation 9.10 above, which Poincaré wrote as:[14]

$$\varepsilon'' = \frac{\varepsilon + \varepsilon'}{1 + \varepsilon\varepsilon'}. \tag{9.10'}$$

Poincaré had a fundamentally different understanding of the meaning and use of the transformations than Lorentz did. That is the divergence. It led Lorentz into a misunderstanding that Poincaré avoided.

Further, this transformation involves a four-vector: one temporal and three spatial coordinates. Additionally, the transformation mixes spatial and temporal coordinates from one system to another. To the extent that he noticed it, Poincaré did not comment on the mixing in the articles translated in Part I. This stands in contrast Einstein who started his 1905 paper (Einstein, Zur Elektrodynamik bewegter Körper, 1905) with a discussion of time and simultaneity, and the loss of simultaneity when observing from different reference frames.

It is a recognized consequence of equation 9.10 that adding two velocities, even close to the speed of light, cannot result in a velocity greater than the speed of light. A decay product shot forward from a decaying particle traveling at nearly the speed of light will not reach a speed greater than the speed of light, even if the product is itself an energetic photon.

Continuing in §1, Poincaré's values for ρ' and u' (above) can be substituted into the left-hand side of the continuity equation for charge in the moving coordinates:

$$\frac{\mathrm{d}\rho'}{\mathrm{d}t'} + \sum \frac{\mathrm{d}\rho' u'}{\mathrm{d}x'} = 0$$

which simplifies to

[14]For possible comparison with Einstein see §5 of (Einstein, Zur Elektrodynamik bewegter Körper 1905).

$$\frac{1}{l^4}\left[\frac{\mathrm{d}\rho}{\mathrm{d}t}+\sum\frac{\mathrm{d}\rho u}{\mathrm{d}x}\right]=0$$

showing that if the continuity equation for charge is satisfied in one system then it is satisfied in the other. That is to say, the continuity equation is unchanged under Lorentz transformation[15]. As Poincaré observed, "With the hypothesis from Lorentz, this condition would not be satisfied."

Unchanged under Lorentz Transformation (Covariant)

With the correct transformation for the charge density, Poincaré was able to apply the Lorentz transformations to the equations of electrodynamics and the Lorentz force and confirmed that they are unchanged by the transformations.[16] Lorentz had not been able to confirm that these equations were unchanged under Lorentz transformation because of the first divergence.

As noted in the previous subsection, Poincaré also showed that the continuity equation for charge density was invariant under Lorentz transformation and how to compose velocities.

As discussed in the following subsection, Poincaré's group-based arguments showed that the Minkowski space-time interval (see Equation 9.11, below) is unchanged.

It should also be noted that Poincaré showed that the conditions for electron stability are unchanged under Lorentz transformation, showing that an electron with only electromagnetic forces is unstable at rest and in motion.

In this article, Poincaré used the action and electromagnetic Lagrangian and they became important tools in several sections. He used them to re-derive the equation for the Lorentz force, the form of the Poincaré stress and the longitudinal and transverse mass in quasi-stationary motion. It is therefore worth taking a look at his demonstration in §3 that the action and Lagrangian are unchanged under Lorentz transformation. He did this by presenting the electromagnetic action in the first equation in the section and applying the Lorentz transformations to the fields and to the differentials for the integration. This leads directly to a proof that the Lagrangian is unchanged up to a factor of l^4 (and he subsequently shows that $l = 1$) and that the action is unchanged, without qualification. Lastly, he needed to show that the variant of the action (δJ) given in equation 2 of §3 is unchanged under Lorentz transformation; this is similar to the demonstration for equation 10 in §2. He

[15]Because the continuity equation is unchanged by Lorentz transformation, it indicates that $(c\rho, \boldsymbol{J})$ is a four-vector. For more, see (Jackson, Classical Electrodynamics, 1999, pp. 554–555, eqn. 11.129).

[16]For possible comparison with Einstein see §6 and §9 of (Einstein, Zur Elektrodynamik bewegter Körper, 1905).

showed this with an application of the Lorentz transformations correctly including composition of velocities.

Transformations as Group with Invariants—Third Divergence

As noted above on page 176, the third divergence of Poincaré from Lorentz arises because Poincaré did not find Lorentz's argument that $l = 1$ persuasive. l is the longitudinal component of the transformation; if it is 1, then the transverse dimensions of the moving electron are unchanged. To address this, Poincaré provided an entirely theoretical proof that $l = 1$ based on the recognition that the coordinate transformations must form a group, as discussed in §4. The heart of the argument is discussed above in the paragraph introducing Equation 9.10′. He reasoned that the transformation from system A to B and then system B to C must have the same result as a transformation directly from system A to C, if the velocities are combined correctly. And Equation 9.10′ shows the correct way to combine the velocities. This result was confirmed for the calculation of k, since it indicated that after the successive transformations from A to B then B to C the resulting value of k is $1/\sqrt{1+\varepsilon''^2}$ as expected for a direct transformation from A to C using the composed velocity from equation 9.10′. However, for l, the value after applying the two transformations is ll' which must equal l''. This is only possible for $l = l' = l'' = 1$.[17]

Poincaré continued discussion of the group implications in this section with the infinitesimal generator of the group. He recognized that these generators are equivalent to infinitesimal translations along the axes and infinitesimal rotations around the axes. In this context he notes that a reversal of the direction of motion (by a 180° rotation) must be equivalent to the inverse transformation providing a second group-based argument that $l = 1$. This equality indicates that there is no amplification by the translation. He also noted that these transformations must leave the quadratic form

$$x^2 + y^2 + z^2 - c^2t^2 \tag{9.11}$$

unchanged.

In §9 where Poincaré presented work on a theory of gravitation that is unaffected by Lorentz transformation in the same way as the electromagnetic forces[18], Poincaré (Sur la dynamique de l'électron, 1906, p. 168; p. 93) returned to look for "invariants of the Lorentz group." He repeated that the transformations in this group are linear and do not change the quadratic form given in Equation 9.11 above. He asked us to consider

[17]For possible comparison with Einstein see last pages of §3 of (Einstein, Zur Elektrodynamik bewegter Körper, 1905) for an argument that $\varphi(v) = 1$ based on a translation and its inverse.

[18]There is limited further discussion of this work in the following chapter, Chapter 10.

$$x, y, z, ict$$

as the coordinates of a point in four-dimensional space. In the context of the theory of gravitation, he was looking for invariants that involved the position and velocity at both the attracting and attracted body and invariants that involved the force and time rate of change of energy ($T = \boldsymbol{F} \cdot \boldsymbol{v}/c$). This led to the four invariants respectively in equation 5 and equation 7 in (Poincaré, Sur la dynamique de l'électron, 1906, pp. 169–70; p. 94–5) in §9. Each invariant has an associated four-vector.

Electron Stability: Poincaré Stress—Second Divergence

This brings us to the first discussion by Poincaré of the second divergence from Lorentz. Poincaré (Sur la dynamique de l'électron, 1906, p. 136; p. 53) presented the conditions for the stability of an electron that "is exclusively of electromagnetic origin" and "only subject to forces of electromagnetic origin." The conditions are unchanged under Lorentz transformation. The conditions are only satisfied if there is no charge. This is worse than suggested by Abraham in the quote on page 172 above; there Abraham required electrons be undeformable to avoid the implications of work done to change the shape of electrons. Instead, in the electromagnetic world view, electrons are unstable at any speed.

Poincaré concluded §1 with the statement: "One therefore has to accept that in addition to electromagnetic forces, there are either other forces or bonds. One must then look at what conditions these forces or bonds must satisfy for the equilibrium of the electrons to be undisturbed by the transformation." These "other forces or bonds" required for stability of a classical electron are now known as Poincaré stress. Poincaré also took up the discussion of the necessary "forces or bonds" in §6 as part of the discussion of the shape of stationary and moving electrons.

Although the title of §6 is "Contraction of Electrons", the subject of the section is now actually known as Poincaré stress; it is the subject of the second divergence mentioned by Poincaré in the letter to Lorentz from about April 27, 1905 (Kox, 2008, pp. 176–8, letter 126; p. 37–39) discussed above.

For purposes of discussion and generality, Poincaré brought an arbitrary l back into consideration as part of a discussion of electron shape changing with velocity. In this section, he also referred to an ideal immobile electron and a real moving electron.[19] Poincaré reviews the shape of the electron in the theories from Abraham, Lorentz and Langevin. As noted in the first section of this chapter, Abraham's electron is an undeformable sphere and that assumption is unacceptable on nearly

[19] In fact, the discussion of real and ideal electrons reads as if it were written in the context of Lorentz's corresponding states; it is a notable contrast to the discussion of the volume of a moving sphere and the derivation of the transformation of charge density and the equation of continuity for charge.

all counts. Lorentz's electron is spherical at rest and deformed, according to the Lorentz transformation, in motion ("real electron"). Langevin (Sur l'origine des radiations et l'inertie électromagnétique, 1905), in a theory of electromagnetic radiation, proposed an electron with constant volume so $kl^3 = const.$

Poincaré wrote the electric field for an ideal electron, meaning at rest, with no magnetic field and an electric field defined in terms of an unspecified potential. He then determined the corresponding electric and magnetic fields for a real electron, meaning moving, using the Lorentz transformation of the fields. (This is equation 1 (Poincaré, Sur la dynamique de l'électron, 1906, p. 152; p. 73).) He gave the longitudinal and transverse electrical energy and the magnetic energy (respectively A, B and C) in the moving system in terms of integrals of the fields over all space and time, and used the previously determined transformations to relate these quantities to the corresponding energies in the immobile system. The total energy is then $E = A + B + C$ and the Lagrangian (Poincaré calls it the "action per unit time") is $H = A + B - C$.

Next, Poincaré takes the electromagnetic momentum (in his notation D) as the component of the Poynting vector in the direction of motion integrated over all space.[20] then, in the first part of equation 2 Poincaré (Sur la dynamique de l'électron, 1906, p. 153; p. 76) required that $D = \mathrm{d}H/\mathrm{d}\varepsilon$. In an exercise left to the reader, Poincaré states that this is only satisfied if $l = k^{-1/3}$. This condition means that the volume of the electron must be constant; this is Langevin's assumption or, in the limiting case of $l = 1$, it is Abraham's undeformable electron.

Either way this result is a problem, since it is inconsistent with the Lorentz transformation. Looking for solutions to this problem, Poincaré first switched to a different Lagrangian and with it explored different options. For the Lagrangian itself, he used the Lagrangian for a spherical electron moving at a constant velocity from (Abraham, Prinzipien des Dynamik des Elektrons, 1902b, p. 37), equation 25.[21] This is the same Lagrangian as discussed in the first section of this chapter on page 172 in connection with equation 9.5.

[20]This is not correct and introduces an incorrect factor of order unity. The core of the problem is an incorrect application of the Lorentz transformation; an approach similar to that taken by Poincaré with the transformation of the charge density in §1 and discussed above is needed. That is to say a manifestly covariant four-vector approach is needed. Poincaré's equation for the momentum is the same as, for example, equation 2–13 in (Rohrlich 2007, p. 15). But as Rohrlich indicates, it is not correct. Rohrlich on page 17 provides a brief history of the multiple times this problem and the correction were found and provides a detailed discussion of the resolution in section 6–3, pages 129–134.

[21]Interestingly, Poincaré, while noting that this formula applies to a spherical electron, does not follow the reference to Searle mentioned by Abraham at the top of the following page and provided a few pages earlier which does have the corresponding potential for an ellipsoid (specifically equation 23 (Searle 1897, p. 340)); he manages to make do.

In a first exploration with the second Lagrangian, Poincaré (Sur la dynamique de l'électron, 1906, p. 155; p. 78) asked us to "imagine that the electron is subject to a binding force, such that there is a relation between r and θ." Here, θ is the eccentricity of the ellipsoidal moving electron. For Abraham, the electron is undeformable, the moving electron remains a sphere, and therefore $\theta = 1$. For Lorentz, the eccentricity of the moving electron is due to the Lorentz contraction and therefore $\theta = k$. This binding force would maintain the electrostatic stability of the stationary electron. Poincaré found that this does not change the earlier conclusion that the volume of the electron must be constant. The static binding force does not resolve the electron stability problem.

Now, Poincaré (Sur la dynamique de l'électron, 1906, p. 157; p. 80) asked, "what additional forces, other than the binding forces, would need to be involved to incorporate Lorentz's law"? Poincaré answered this by adding an additional potential, $F(\theta, r)$, to the Lagrangian based on Abraham's equation 25, and first retains the assumption that there is a binding force relating r and θ, such that $r = b\theta^m$. Additionally requiring that $k = \theta$, consistent with the Lorentz transformations, Poincaré finds for the potential:

$$F = \frac{a}{3bk}$$

where a is a numerical constant entering through Abraham's Lagrangian.

Finally, Poincaré assumed that there is no binding relation between r and θ and assumes a general potential of the form:

$$F = Ar^{\alpha}\theta^{\beta}.$$

Conditions on the Lagrangian were then used to determine the parameters A, α, and β, again in the assumption of the Lorentz transformations. The result is:

$$F = \frac{a}{3b^4}k^2r^3.$$

With the Lorentz transformations, stability of the electron requires "*adding an additional potential proportional to the volume of the electron*" (Poincaré, Sur la dynamique de l'électron, 1906, p. 158; p. 81), his emphasis. Later (page 165; page 90), Poincaré considers the Lagrangian in the limit of small velocities and determines that b is inversely proportional to the rest mass, which he refers to as the "experimental mass".

In summary, Poincaré determines that the stability of classical, relativistic electrons depends on a negative internal pressure[22] proportional to:

$$m_0{}^4 k^2 r^3.$$

This is the Poincaré stress. Stability of electrons, in pre-quantum theory, requires a force that is not of electromagnetic origin. Poincaré has shown that it is impossible to have both Lorentz transformations (whether to make the ether undetectable or absolute motion undetectable) and an electromagnetic world view.

Synopsis

The three theories of electrons considered in this chapter, from Abraham, Lorentz and Poincaré, lead to different perspectives based on different constraints each person brought to the theory.

Abraham favored an explanation of mass entirely based on electromagnetic forces and saw an opportunity to explain the recently discovered dependence of electron mass on velocity. He recognized that any change in shape of a moving electron would result in a change in self-energy that was not due to electromagnetic forces so he assumed an undeformable, spherical electron. He was able to develop a formula for the dependence of electromagnetic mass on velocity and the formula was used by Kaufmann in several papers for analyzing his data on electric and magnetic deflection of high-speed electrons.

Lorentz, continuing his earlier work on electrodynamics of moving bodies, presented the transformations that Poincaré named after him. Lorentz used these transformations to explain why the ether could not be detected. His electron was spherical at rest, became ellipsoidal in motion, and its change in mass with velocity was a consequence of the transformations. Reanalyzing some of Kaufman's data, Lorentz showed that his theory explained the experimental results at least as well is Abraham's theory. In fact, a fuller reanalysis shows that Lorentz's theory provides a much better explanation.

Poincaré indicated his general agreement with Lorentz's theory but indicated three divergences from Lorentz. With a more insightful application of the transformations, Poincaré obtained the correct transformation of the charge density. He was therefore able to show that the equations of electrodynamics, the continuity equation for charge density, the Lorentz force and electromagnetic Lagrangian are all unchanged under Lorentz transformation. Poincaré recognized that the

[22]In the Introduction, (Poincaré, Sur la dynamique de l'électron 1906, p. 130; p. 46) referred to it as an external pressure. Positive external pressure and negative internal pressure largely amounts to the same thing. Since his argumentation is based on the Lagrangian, he does not have a specific mechanism to distinguish between an external and internal origin of the pressure.

transformations form a group; this offered him a better argument that there is no change in the dimensions perpendicular to the direction of motion and Poincaré explicitly looked for invariants of the group finding that the space-time interval is an invariant. Poincaré (Sur la dynamique de l'électron, 1906, p. 163; p. 88) wrote that this means that "Lorentz's hypothesis is the only one which is compatible with the impossibility of showing absolute motion." Poincaré also showed that a completely electromagnetic explanation of the electron mass is impossible and determines the additional non-electromagnetic potential which is needed to assure the stability of electrons, the Poincaré stress.

References

Abraham, M. (1902a). Dynamik des Electrons. *Nachrichten von der Gesellschaft der Wissenschaften zu Göttingen*, 20-41.

Abraham, M. (1902b, October). Prinzipien des Dynamik des Elektrons. *Physikalische Zeitschrift, 4* (1b), 57-63.

Abraham, M. (1903). Prinzipien der Dynamik des Elektrons. *Annalen der Physik, Series 4 vol. 10*, 105-179.

Bucherer, A. H. (1908). Messungen an Becquerelstrahlen. Die experimentelle Bestätigung der Lorentz-Einsteinschen Theorie. *Physikalishe Zeitschrift, 9*(22), 755-762.

Darrigol, O. (1994). The Electron Theories of Larmor and Lorentz: A Comparative Study. *Historical Studies in the Physical and Biological Sciences, 24*(2), 265-336.

Einstein, A. (1905). Zur Elektrodynamik bewegter Körper. *Annalen der Physik, 17*, 891-921.

Feynman, R. P., Leighton, R. B., & Sands, M. (1963). *The Feynman Lectures on Physics*. Reading, MA: Addison-Wesley Publishing Company.

Hunt, B. J. (1991). *The Maxwellians*. Ithaca, NY: Cornell University Press.

Jackson, J. D. (1999). *Classical Electrodynamics* (Third ed.). New York: John Wiley & Sons.

Kaufmann, W. (1901). Die magnetische und electrische Ablenkbarkeit der Becquerelstrahlen und die scheinbare Masse der Elektronen. *Nachrichten von der Königl. Gesellschaft der Wissenschaften zu Göttingen, 2*, 143-155.

Kaufmann, W. (1902a). Die electromagnetishe Masse des Elektrons. *Physikalische Zeitschrift, 4*, 52-57.

Kaufmann, W. (1902b). Über die electomagnetische Masse des Elektrons. *Nachrichten von der Gesellschaft der Wissenschaften zu Göttingen*, 291-6.

Kaufmann, W. (1903). Über die "Elektromagnetische Masse" der Elektronen. *Nachrichten von der Gesellschaft der Wissenschaften zu Göttingen*, 90-103.

Kox, A. J. (2008). *The Scientific Correspondence of H. A. Lorentz* (Vol. 1). New York: Springer Science+Business Media.

Langevin, P. (1905). Sur l'origine des radiations et l'inertie électromagnétique. *J. Phys. Theor. Appl., 4*, 165-183. doi:https://doi.org/10.1051/jphystap:019050040016500

Larmor, J. (1893). A Dynamical Theory of the Electric and Luminiferous Medium. *Proceedings of the Royal Society of London, 54*, 438-461.

Lorentz, H. A. (1892). La théorie électromagnétique de Maxwell et son application aux corps mouvants. *Archives néerlandaises des Sciences exactes et naturelles, 25*.

Lorentz, H. A. (1895). *Versuch einer Theorie der electrischen und optischen Erscheinungen in bewegten Körpern*. Leiden: E. J. Brill.

Lorentz, H. A. (1904). Electromagnetic phenomena in a system moving with any velocity smaller than that of. *Proceedings of the KNAW [Royal Netherlands Academy of Arts and Sciences], 6 (1903-4)*, 809-831.

McCormmach, R. (1970). H. A. Lorentz and the Electromagnetic View of Nature. *Isis, 61*(4), 459-497.

Miller, A. I. (1973). A study of Henri Poincaré's "Sur la Dynamique de l'Électron". *Arch. Hist. Exact Sci., 10*, 207-328.

National Institute of Standards and Technology. (n.d.). *CODATA Value: electron charge to mass quotient*. Retrieved November 12, 2019, from https://physics.nist.gov/cgi-bin/cuu/Value?esme

Poincaré, H. (1900). La théorie de Lorentz et le principe de réaction. *Archives néerlandaises des sciences exactes et naturelles, 5*, 252-78.

Poincaré, H. (1905). Sur la dynamique de l'électron. *Comptes rendus de l'Académie des Sciences, 140*, 1504-1508.

Poincaré, H. (1906). Sur la dynamique de l'électron. *Rendiconti del circolo matematico di Palermo, 21*, 129-176.

Rohrlich, F. (2007). *Classical Charged Particles* (Third Edition ed.). Hackensack: World Scientific Publishing Co.

Runge, C. (1903). Über die elektomagnetische Masse der Electronen. *Nachrichten von der Gesellschaft der Wissenschaften zu Göttingen*, 326-330.

Searle, G. F. (1897). On the Steady Motion of an Electrified Ellipsoid. *Philosophical Magazine and Journal of Science, Fifth Series, 44*, 329-341.

Chapter 10
Poincaré as a Physicist

Introduction

In the previous chapter, Chapter 9, I gave some attention to distinguishing the contributions of Abraham, Lorentz and Poincaré to a pre-quantum theory of electrons. That work involved a close, objective consideration of what each had done. This chapter, in contrast, steps back and takes a broader, more subjective view of how Poincaré approached and carried out this work. For me, two things are notable. On the positive side, there is his work to expound on and extend the work of Lorentz. His choice to name the transformations after Lorentz is warranted; Poincaré's choice to refer to additions and corrections he provided as Lorentz's theory seems very charitable. This is especially so when Poincaré states that absolute motion cannot be detected when Lorentz was arguing only that the ether cannot be detected. In contrast, Poincaré seems to have missed opportunities to make a number of strong predictive statements: for example, "there is no need for an ether." There are other things that are notable. Why didn't he make these predictions?

A Koan

> Let us imagine an astronomer before Copernicus who was thinking about the Ptolemaic system; he would notice that for all the planets one of the two circles, epicycle or deferent, is traversed in the same time. That cannot be by chance; there is therefore some unknown mysterious link between all the planets.

B. D. Popp, *Henri Poincaré: Electrons to Special Relativity*,
https://doi.org/10.1007/978-3-030-48039-4_10

> Copernicus, by simply changing the coordinate system regarded as fixed, made this appearance disappear; each planet now describes only one circle and the periods of revolution become independent (until Kepler reestablished the link between them that was thought to have been destroyed).
>
> (Poincaré, Sur la dynamique de l'électron, 1906, p. 131; p. 47)[1]

The above quote from the introduction is frequently discussed with varied interpretations. These interpretations may say more about the person interpreting the quote than about the person who wrote it.

Let me start with a fairly obvious remark: the time to traverse the epicycle or deferent Poincaré refers to is 365 days, one year. Since the Earth is stationary in the Ptolemaic system, the orbital period of the Earth shows up in other places instead. When Copernicus re-centered the model of the solar system on the Sun, the orbital period of the Earth became explicitly associated with the Earth and no longer with the other planets.

Poincaré was advocating looking for things that cannot be by chance and finding the unknown link.

As a general statement, that is sound advice for physicists. In the specific context of the following paragraphs what does Poincaré mean? Well, that's a little bit more complicated.

In my reading, Poincaré is first referring to the last section of the paper where he extends "Lorentz's theory" to all forces and therefore develops a theory of gravitation that is unchanged under Lorentz transformation. This brings the speed of light into the law of gravitation and Poincaré is wondering why the speed of light appears in both electrodynamics and gravitation. Poincaré argues against a purely electromagnetic explanation of all forces and mechanics, so he might reasonably ask why the velocity of electromagnetic waves appears in a theory of gravitation. Perhaps it points to an underlying connection between fundamental forces.

My readers certainly may still apply their own interpretation to the koan.

Advocating for Others

A key positive aspect of Poincaré's work is the steps he takes to expound on, clarify and extend the work of others. In the documents translated in Part I, those others are notably Lorentz and Langevin. In Chapter 9, in the summary of the work by Poincaré on a theory of electrons, I structured my exposition by focusing on the three divergences from Lorentz that Poincaré discussed in three letters written to Lorentz in the spring of 1905. (I translated those letters in Part I, Chapter 3.) That suited my purpose of distinguishing what Lorentz and Poincaré had done in treating the subject matter, since it served to show that Poincaré had gone far beyond what Lorentz did or

[1]Here and elsewhere, the first page number refers to the original publication and the second page number (following the semicolon) refers to the page number in Part I of this book.

even aspired to do. It seems plausible to think that Poincaré could have similarly focused on his contributions and emphasized how what he did was new and different. My point here is that Poincaré did not.

As an example, consider Poincaré's derivation of the Lorentz force on a charged particle subject to electric and magnetic fields in (Poincaré, Sur la dynamique de l'électron, 1906, §2). In the second sentence Poincaré writes, "I will however go back over the question because I prefer to present it in a slightly different form which will be useful for my purpose." Poincaré thus acknowledges the work (Lorentz, Contributions to the theory of electrons. I, 1902), which might not get much attention otherwise, and credits Lorentz with the idea of using a least action, Lagrangian approach that Poincaré uses extensively in the remainder of this paper. This promotes Lorentz and does nothing to diminish Poincaré.

For other examples involving Poincaré's discussion of Lorentz's work, the situation is substantially different. Consider discussion of the application of the Lorentz transformations to the equations of electrodynamics. Lorentz does do this, but because of an error ("the first divergence" between Lorentz and Poincaré) he gets equations that are changed by the transformation. He does not discuss the meaning of these changes. Therefore, when Poincaré identifies the correction (transforming the charge density so that the equation of continuity for charge is unchanged under the Lorentz transformation, which is related to treating charge density and current as a four-vector) and shows that the equations of electrodynamics are unchanged, he is on new ground. Additionally, when he concludes that this means that the equations of electrodynamics, and therefore electromagnetic and optical phenomena also, are compatible with the impossibility of determining absolute motion, he moves further into new territory. It seems hard to imagine that he would have been justifiably criticized for putting his name on this new territory. He does not do that. Instead, he refers to it as part of Lorentz's theory.

Poincaré first brings the key issues to Lorentz's attention in private letters that are notably polite and collegial. The journal article that follows downplays the issues (they are divergences, not errors) and does not mark out how far Poincaré had gone beyond what Lorentz did. In fact, I think this is very charitable, more than just collegial. The tenor of the discussion of Poincaré's contribution to the understanding of special relativity (discussed in Chapter 11) would be very different if the same charity he showed to Lorentz were applied to him.

Poincaré has a full section (Poincaré, Sur la dynamique de l'électron, 1906, pp. 146–151; p. 67–70) discussing the theory of electromagnetic radiation by Langevin (Sur l'origine des radiations et l'inertie électromagnétique, 1905). As with Lorentz this is an example of Poincaré presenting and amplifying the work of a colleague. A close comparison of the two papers would take me too far afield. A more casual reading of the section still shows that Poincaré put his mark on the presentation of Langevin's work. In one case, Poincaré shows how to use the Lorentz transformations to make the calculation of the field due to a uniformly moving electron easier. He starts with a well-known solution for the electric field of a stationary electron he next applies Lorentz transformations to the electric and magnetic field for transformation to coordinates where the electron has the required uniform velocity. The result is the magnetic and electric fields for a uniformly moving electron (Poincaré, Sur la dynamique de l'électron, 1906, p. 149; p. 70, equation 4). Since the equations are unchanged under Lorentz transformation, a valid solution of the equations must still be a valid solution after transformation. Poincaré used this to find a solution to the equations in coordinates that were easy to work in and then transform this solution to the coordinates needed for Langevin's work. This makes for a very interesting alternate approach to a part of what Langevin did.

Latter sections "Predictions" and "Attitude", looks at this from a different side. When Poincaré goes beyond explaining and extends to new material, why doesn't he put his name on it and make a prediction with it?

Underlying Principles

Olivier Darrigol (Henri Poincaré's Criticism of Fin de Siècle Electrodynamics, 1995) states that, "Poincaré favoured a 'physics of principles' in which the compatibility of theories with general principles was more important than the completeness of their physical picture." This preference for principles can be seen in the work translated in Part I. A key example is Poincaré's concern with ad hoc assumptions in Lorentz's work to explain why the ether cannot be detected; Poincaré's evident response is to propose the relativity postulate which declares the impossibility of detecting absolute motion. The ad hoc assumption is no longer needed. Similarly arguments for and use of the principle of conservation of momentum are essential to the reasoning in (Poincaré, La théorie de Lorentz et le principe de réaction, 1900).

Lorentz and Fitzgerald had introduced the contraction of the length of bodies moving through the ether as an ad hoc assumption for explaining why experiments for detecting the absolute motion of the Earth through a pervasive ether, such as the Michelson-Morley experiment, had failed. In caricature, it was a means conceived to explain an experiment that contradicted what they were persuaded was true. Understandably, for two reasons, Poincaré did not like this approach. First, the introduction of this ad hoc assumption contradicted his preference for basing theories in physics on principles. Also, he did not share their opinion about the existence of the ether (discussed below).

In the first two paragraphs of (Poincaré, Sur la dynamique de l'électron, 1906, p. 129; p. 45), we find that Poincaré did not see the experiments that did not "provide us a means for determining the absolute motion of the Earth" as a difficulty to be addressed with an explanation. Instead, Poincaré saw the experiments as pointing to a general law of nature that he is "naturally lead to accept," the Relativity Postulate.

In that paragraph of the introduction, Poincaré defines the relativity postulate as the "impossibility of showing the absolute motion of the Earth," although elsewhere he does state it in other forms. For example, Poincaré (Sur la dynamique de l'électron, 1905, p. 1505; p. 41) states the relativity postulate as, "it is *completely* impossible to determine absolute motion" (his emphasis). Poincaré indicates that it is a postulate because it can still be confirmed or rejected by experiment.[2]

Let us look at another statement of the relativity postulate "Lorentz's hypothesis is the only one which is compatible with the impossibility of showing absolute motion" (Poincaré, Sur la dynamique de l'électron, 1906, p. 163; p. 88). Miller (A study of Henri Poincaré's "Sur la Dynamique de l'Électron", 1973, p. 296) paraphrases it as, "Therefore, only if $l = 1$ can the equations of motion transform... in such a way as to exclude the possibility of detecting the ether." This paraphrase misrepresents Poincaré's statement in three ways. It is useful to look at this misrepresentation more closely because it serves to expose the force of Poincaré's statement of the postulate of relativity. First is the issue of what is being evaluated: "Lorentz's hypothesis is the only one" (broad) or "$l = 1$" (very narrow). The next is the strength of the criterion being applied. With Poincaré, "compatibility with the impossibility" is called for. Miller looks for a "way to exclude." Last, in the criterion, Poincaré did not use the word ether (a point discussed below); he wrote "absolute motion" and that is less ambiguous (e.g. which concept of ether, how might it be detected) and much broader. Poincaré is in fact using the relativity postulate as a strong criterion to reject Abraham and Langevin's theories, and any others that might allow a way to detect absolute motion.[1]

Continuing to look at Poincaré's insistence on underlying principles, let us look at his defense of conservation of momentum. Poincaré (La théorie de Lorentz et le principe de réaction, 1900, p. 270; p. 29) asks, "Why is conservation of momentum obvious to our thinking?" His explanation is that a violation of the conservation of momentum would lead fairly directly to perpetual motion; perpetual motion is even less palatable. Further, conservation of momentum appears to be a consequence of conservation of energy. For theories of electrodynamics Poincaré here again applies a principle as a criterion for rejecting theories, stating "all theories which do not respect this principle [of conservation of momentum] would be condemned as a group" (Poincaré, La théorie de Lorentz et le principe de réaction, 1900).

Methods

It is expected that Poincaré, as a mathematician, would bring mathematical tools and techniques to his work in physics.

[2]For possible comparison with Einstein see the definition on the first page of (Einstein, Zur Elektrodynamik bewegter Körper, 1905), "conjecture that not only in mechanics, but in electrodynamics as well, the phenomena do not have any properties corresponding to the concept of absolute rest."

A notable feature of (Poincaré, Sur la dynamique de l'électron, 1906) is the use of the Lagrangian and the related principle of least action "Principle of Least Action" is even the title of §2. This method is another, more subtle and very distinctive way that Poincaré argues by principles. It provides a means to derive fundamental relations using the energy of a system and a variational principle minimizing a function of this energy. This function of the energy is the Lagrangian and the integral of the Lagrangian over all space and between two defined times is the action. As an example, in §2 just mentioned, Poincaré uses the energy in the electric and magnetic field and the work done on a charged particle to re-derive the formula for the electric and magnetic force per unit mass acting on the moving particle, called the Lorentz force.[3] The usefulness of the method can be seen, since relatively simple assumptions were used to arrive at a useful, general formula governing the motion of the particle.

In the following section Poincaré shows that the Lagrangian for $l = 1$ (meaning that the Lorentz transformations do not change the dimensions perpendicular to the direction of motion) and the action in general are unchanged under Lorentz transformation. This is necessary to establish the usefulness of the Lagrangian method in contexts where v^2/c^2 approaches 1.

Continuing to use the Lagrangian, Poincaré (Sur la dynamique de l'électron, 1906 §6) derives the additional potential (Poincaré stress) needed to stabilize an electron against repulsive Coulomb forces. While this proof does not require any additional assumptions, it does not provide any information about the origin of the force or the mechanism for stabilizing the electron either.

Poincaré (Sur la dynamique de l'électron, 1906 §4) recognizes that the Lorentz transformations form a group. He is then able to apply methods relevant to groups and use them in several different contexts. First, as discussed in Chapter 9, it is the basis of his argument that $l = 1$, since he had not been persuaded by other considerations proposed by Lorentz. Further, the identification of this group feeds two other considerations: looking for invariants and a geometric interpretation.

Looking for and using invariants is a method that Poincaré keeps returning to.[4] the first invariant is $x^2 + y^2 + z^2 - c^2t^2$; Poincaré calls it a quadratic form. It was later named the space-time invariant by Hermann Minkowski. No matter what un-accelerated reference frame you measure it in, the wavefront of a flash of light in space time is a sphere. Einstein (Zur Elektrodynamik bewegter Körper, 1905 §3) used this invariant to argue for the plausibility of these transformations. It also means that the distance between any two points in space-time does not change with the observer. As discussed in Chapter 9, Poincaré also determines other invariants from the Lorentz transformation group.

Poincaré's recognition that the Lorentz transformations form a group lead him to a geometric interpretation of spacetime in which he considered x, y, z, and ict "as the

[3]The earlier derivation of the Lorentz force using least action is in (Lorentz, Contributions to the theory of electrons. I, 1902).

[4]See for example (Poincaré, On the Three-Body Problem and the Equations of Dynamics, 2017).

coordinates of [a point] in four-dimensional space. We have seen that the Lorentz transformation is solely a rotation of this space around the origin, which is regarded as fixed." (Poincaré, Sur la dynamique de l'électron, 1906, p. 168; p. 94). Poincaré does not expand on this idea. Famously Hermann Minkowski (Minkowski, 1908) did, independently. This seems to be a missed opportunity, to be considered with others in a following section.

Related to this geometric interpretation of space-time is a change in how Poincaré used the Lorentz transformations. This is seen in how Poincaré used the Lorentz transformations to discuss charge and current seen by different observers; he does this in a way that departs from Lorentz's corresponding states. In Lorentz's view, the transformation allows comparison of a moving electron or arm of an interferometer with a corresponding one at rest with the ether. While Poincaré does at times use the transformations this way, it is clear that, as in this example, he does use the transformations in a geometric sense as a transformation between observations of one event by two different observers. Related, but secondary, there are some hints that Poincaré is close to recognizing some covariant four-vectors; due emphasis must be placed on my choice of the word "hints." Some examples, notably charge and current density, are discussed in Chapter 9.

Predictions

Poincaré (La théorie de Lorentz et le principe de réaction, 1900, p. 242; p. 13) says, "*Good theories are flexible*." (His emphasis.) They also make predictions that are verifiable or falsifiable. For example, Abraham (Dynamik des Electrons, 1902) proposed an explanation and formula for explaining the dependence of β ray mass on velocity seen in the data from (Kaufmann, Die magnetische und electrische Ablenkbarkeit der Becquerelstrahlen und die scheinbare Masse der Elektronen, 1901), and the agreement between the two appeared satisfactory. Lorentz (Electromagnetic phenomena in a system moving with any velocity smaller than that of, 1904) provided a different explanation and a formula in better agreement with Kaufmann's data.

What are Poincaré's predictions?

A massive body can't be accelerated to the speed of light. Poincaré (La dynamique de l'électron, 1908, p. 396; p. 120) writes, "Thus the mass, the momentum and the energy become infinite when the velocity is equal to that of light. The result of this is that *no body can reach a velocity greater than that of light by any means*." (His emphasis.)

As noted above, Poincaré viewed the principle of conservation of momentum as an important underlying principle. In (Poincaré, La théorie de Lorentz et le principe de réaction, 1900), he uses this principle to show that electromagnetic radiation must transfer momentum from the emitter to the receiver producing a recoil in the emitter. Further, he concludes that this transfer of momentum also conserves angular momentum. Next, Poincaré estimates the magnitude of the effect and shows that a

1 kg machine emitting 3 kW in a single direction produces a recoil with the force of 10^{-5} N, which he concludes is unmeasurably small. He recognized the connection of this transfer with the known phenomenon of radiation pressure. This phenomenon was used, for example, to explain the formation of comet tails: momentum transferred from solar radiation to gas and particles released from the comet push them away from the Sun. As discussed further below, Einstein (Das Prinzip von der Erhaltung der Schwerpunktsbewegung und die Traegheit der Energie, 1906) takes the further step of connecting this transfer of momentum found by Poincaré with the transfer of mass and its energy content between the emitter and receiver.

In the previous chapter, Chapter 9, Max Abraham had recognized that a requirement that mass be solely of electromagnetic origin (which was key to the electromagnetic worldview) was a constraint on theories of electrons and consequently adopted a model electron that was an undeformable sphere. Poincaré showed that Abraham's model electron was not consistent with the relativity postulate, so that the mass of an electron could not be exclusively of electromagnetic origin. This means that Poincaré had shown there was a serious defect with the electromagnetic worldview. It wasn't the only problem he observed. Poincaré (La dynamique de l'électron, 1908, p. 390; p. 109) observes that "the *total* mass of a [proton] is much larger than that of a negative electron." Poincaré allows that there could be multiple explanations. For example, they could have the same electromagnetic mass, but significantly different "real mass" accounting for the difference in total mass. Or, the proton could be much smaller, resulting in a much larger electromagnetic self-energy. (And he emphasizes, "I definitely mean much smaller.") Feynman adds an interesting twist to this argument by comparing electrons and muons (Feynman, Leighton, & Sands, 1963, pp. 28–12). Their masses are very different, but otherwise the properties are nearly identical. As Poincaré concluded, there is much more going on with mass than just electromagnetic self-energy.

In fact, Poincaré goes farther than that summary. At the beginning of his discussion of gravitation, Poincaré (Sur la dynamique de l'électron, 1906, p. 166; p. 91, §9) writes, "But there are other forces to which an electromagnetic origin cannot be attributed, such as gravitation, for example. It can in fact happen that two systems of bodies produce equivalent electromagnetic fields, meaning exerting the same action on charged bodies and on currents and that however these two systems do not exert the same gravitational action on Newtonian masses. The gravitational field is therefore distinct from the electromagnetic field." This statement is in direct contradiction with the electromagnetic worldview and positions gravitation as an equal, fundamental force with the electromagnetic force. In itself, I don't see this dramatic change in perspective as leading to a prediction. However, someone who thinks that all forces and masses are of electromagnetic origin does not need to look at a separate theory of gravitation. Poincaré does not agree with that view and therefore immediately turns his attention to gravitation.

In the same paragraph as cited above, Poincaré also writes, "*Forces of any origin, and in particular gravitation, are affected by a translation* (or, if you prefer, by the Lorentz transformation) *in the same way as the electromagnetic forces*." (His emphasis.) This is related to one of the frequently discussed contrasts between

Einstein and Poincaré. Einstein started from principles (e.g. the relativity postulate: the speed of light is always constant) and arrived at the Lorentz transformations as a kinematic consequence. Poincaré started with electromagnetic and optical phenomena and showed that the Lorentz transformations were required so that absolute motion could not be detected. From the above quote, it is clear that Poincaré, although starting from a different point and moving in a different direction, has arrived at the same conclusion: whatever the origin of the force, it must not be possible to detect absolute motion.

Arriving at things in this direction, Poincaré recognizes that there is an important consequence to requiring that the Lorentz transformations apply to gravitational forces: the gravitational force will additionally have to depend on the relative motion of the attracting bodies via a retarded potential. Poincaré moves forward with developing a theory of gravitation that is unchanged under Lorentz transformation. Among other things, this leads to predictions that gravitation propagates with the speed of light and that there are gravitational waves. Poincaré ends the article expressing the hope that the divergence between his covariant theory of gravitation and Newton's law of gravitation will be large enough to be observable. In a following paper, Poincaré (La dynamique de l'électron, 1908, pp. 399–400; p. 124–125, §14) §14 discusses the 38″ anomaly in the precession of Mercury's perihelion as a possible test. He indicates that his theory can explain 7″ precession. It should however be noted that Mercury's motion is accelerated so a theory based on Lorentz transformations and therefore constant velocities is not applicable. Poincaré has pointed in a new direction and taken a first step.

Note that I'm not counting Poincaré stress as a prediction; they are consequences of the theory, but it's hard to see how, even in principle, the stabilizing pressure could be measured or falsified.

I do want to look for possible missed opportunities, but I need to be cautious about the enormous benefit of hindsight. The following items are all ones that Einstein did say, and that Poincaré would have had a sound basis for saying independently because of his own work.

There is no need for a luminiferous ether. Certainly Lorentz contractions brought ether theories closer to being unfalsifiable, which would seem to make the theories as a class a candidate for elimination. As noted at the beginning of Chapter 7, work reported in (Thomson, Cathode Rays, 1897) sought to determine whether cathode rays are wholly processes in the ether or wholly material. In less than a decade the answer is fully settled. Poincaré's electron theory is entirely particle based and has no place for processes in the ether. Does this leave any need for the ether in theory or principle? Darrigol (Henri Poincaré's Criticism of Fin de Siècle Electrodynamics, 1995) states that, "Poincaré early adopted a skeptical attitude toward ether, and identified progress with a gradual elimination of this concept." For example, Poincaré uses the term "ether" three times in (Poincaré, Sur la dynamique de l'électron, 1906): the first two times are in the introduction and the second time is in §6 as part of the discussion of the shape of the electron and the theories of Abraham, Lorentz and Langevin-Bucherer. It might seem that he was eliminating the word from his vocabulary, but then we turn to (Poincaré, La dynamique de l'électron, 1908) and he uses the word almost 50 times. Some of those times he uses the word to explain or argue against a theory of the ether, but not always. Poincaré had sound reasons and opportunity in one or both of these papers to make a strong statement against the ether; he did not.

Einstein (Zur Elektrodynamik bewegter Körper, 1905) states as a postulate that "in empty space light is always propagated with a definite velocity V which is independent of the state of motion of the emitting body." Poincaré's approach to the argumentation does not need this postulate. It is however a consequence of the law of addition of velocities that Poincaré derived (e.g. Equation 9.10′) and Poincaré does not comment on it.

While it is going too far to consider this a missed prediction, it is still worth noting that Einstein (Das Prinzip von der Erhaltung der Schwerpunktsbewegung und die Traegheit der Energie, 1906) showed that (Poincaré, La théorie de Lorentz et le principe de réaction, 1900), discussed above, was one step from showing that since electromagnetic radiation transferred momentum between bodies it also had to transport mass between them with an energy content mc^2. Edmund Whittaker (Whittaker, 1953, p. 51) even allows it to be understood that Poincaré had taken this last step six years before Einstein. Instead, it must not be overlooked that Einstein wrote (Das Prinzip von der Erhaltung der Schwerpunktsbewegung und die Traegheit der Energie, 1906) months after the last of his five famous papers from 1905. Einstein had written in 1905 about the energy content of matter. The understanding gained from that paper had to have helped Einstein understand what Poincaré had come close to, and to write about it himself in the 1906 paper.

While not a prediction, Poincaré, as discussed in the previous section, did recognize that the Lorentz transformations correspond to a rotation of coordinates in four-dimensional spacetime. Poincaré would seem to have been in an excellent position to explore and develop a mathematical theory of spacetime. He did not and this opportunity glanced at by Poincaré, was left to Hermann Minkowski two years later.

The point of this section is to present some subjects where Poincaré came close to making some important predictions. And that serves to prepare the ground for the discussion in the following section that asks, "why didn't he?" The positive side of this section is that Henri Poincaré was dealing with the right subjects in an important field at the right time.

Attitude

The premise of the previous section is that, after having developed and collected postulates, then having applied his methods to them, Poincaré should have been in a position to say something akin to, "If I've understood correctly, then when you look here, you will see this." I referred to the "look here—see this" as predictions and asked that they be verifiable or falsifiable. In my analogy that is to say, if you measure the mass of the electron as a function of the velocity with this equipment then the experimental data will be represented by this function. This is exactly the point of Lorentz's reprocessing of Kaufman's data; it allowed Lorentz to write in response to Abraham: the experimental data is represented at least as well by my function.

So when I chose attitude as the title for this section, I was thinking, asking myself, what was it in Poincaré's thought process, in his personality, in his view of his role in a community of scientists that kept him from making one or more of the statements that in the previous section I suggested he could have made?

Before getting to that question, let me look at other factors that I think are relevant.

There are some formerly useful ideas that Poincaré could throw overboard, but holds on to. I've included the declaration that there is no ether among the missed predictions above. Chapter 12 includes a discussion of vector notation. Poincaré was writing out vectors (e.g. for electric and magnetic fields) and derivatives in Cartesian coordinates in work published in 1890 and he was still doing that in 1906, even though, in between, he had clearly been exposed to and understood more compact and less cumbersome formalism familiar to us. Using Cartesian coordinates is clearly a case of Poincaré being conservative and sticking with what was working well for him. Poincaré's attachment to the principle of conservation of momentum is a positive side of this trait.

Poincaré, in his letters to Lorentz translated in Chapter 3, referred to "differences in detail" and an argument that "does not seem conclusive." The differences Poincaré indicated are discussed in Chapter 9. The first one goes to the core of understanding how to apply the Lorentz transformations and prevents Lorentz from recognizing that the equations of electrodynamics are unchanged under Lorentz transformation. It is hard to see that as a matter of detail; Poincaré seems to have some reason for downplaying the differences between him and Lorentz.

Consider also the last line of the introduction to (Poincaré, La théorie de Lorentz et le principe de réaction, 1900). There, Poincaré wrote, I'm asking the reader "to forgive me for having presented at such length ideas with so little novelty." This is very self-abasing. The article recognizably has significant novelty in the discussion of momentum transport by radiation and the concept of local time.

Why is Poincaré taking such a deferential approach to Lorentz and to his readers (who are presumably physicists)?

It suggests several possibilities to me. Poincaré may not want to step out of place as an outsider, may want to avoid offending anyone and especially not the insiders like Lorentz. There is also modesty, in the sense of "Sorry to bother you, I just need a few minutes of your time." He may be reluctant to overstate the case or even claim that he has something important to contribute to physics.

This diffidence and reluctance seen in his writing may be a symptom. His failure to make the predications discussed above—predictions that seem well justified in his writing and in hindsight—may also be symptoms. What is causing the symptoms?

Really, this is something of a mystery.

Howard Stein (Physics and Philosophy Meet: the Strange Case of Poincaré, 2014) refers to the discussion of (Poincaré, Sur la dynamique de l'électron, 1906) (including this situation that I just described where he did not make justified predictions) as strange and adds, "But I have not seen it pointed out just how strange; I know of nothing like it in the entire history of physics."

Howard Stein sees an explanation in the role that Poincaré takes for himself in his philosophy of science. He states this as, "The basic mistake that I ascribe to Poincaré is that of seeing the significance of theoretical work as residing *essentially and exclusively* in its function in *organizing* knowledge." He continues, "Indeed, I am inclined to venture the psychological hypothesis that Poincaré, whose confidence in his own *mathematical* powers was very great indeed, had some diffidence about trespassing on the domain of *physical prediction*." (Stein, 2014, p. 23, footnote 55) (His emphasis.) I started the previous section with the statement that good theories made predictions that could be verified or falsified. In the same footnote, Howard Stein indicates that in Poincaré's philosophy of science, "Poincaré did not regard such new consequences as a desideratum for a theory, indeed that in a certain sense he viewed it as unreasonable and illegitimate for a theory to be expected to have such consequences."

Olivier Darrigol (The Mystery of the Einstein-Poincaré Connection, 2004), referring to a different but related mystery, states, "Poincaré and Einstein belonged to the fringes of the physics community. One was a foremost mathematician with a special interest in physics, the other a young patent clerk trained at the Zurich Polytechnikum. They both lacked some of the prejudices of established physicists; they both took the stance of an impartial judge; they both preferred to reformulate existing theories in their own way rather than digging out the original motivations of their authors."

Through the selection of similar words like organizing, judge, trespassing and fringes by Darrigol and Stein, we can see commonality in their perception of characteristics that might lead Poincaré to not make predictions in physics theories.

I think there is another related symptom that is commonly stated in the form of one of several questions: why didn't Poincaré respond to Einstein's paper? Why didn't Poincaré make a further elaboration of his theory? Why didn't he do something, anything?

First, it's possible that Poincaré thought of his 1906 work as an elaboration or clarification of Lorentz's theory and that he had already done his part with the exposition he provided. (This understates the difference between Lorentz's ad hoc approach—this is what is needed to explain why the ether hasn't been detected—Poincaré's postulate approach—it is what is required by the relativity postulate.) If Poincaré does not think of it as *his* theory then it is understandable that he would not see a need to defend or for further explain. He can leave that to Lorentz or someone else. This is consistent with a perception that he is an outsider who shouldn't overstep his position, or that his role is organizing knowledge and not advancing it.

Let me offer an alternative, affirmative explanation.

Poincaré demands that his readers work. Close and careful reading of Poincaré is rewarded; it also requires readers to provide many of their own road signs: "Slow, sharp curve ahead," "Turnout for scenic vista, next right," "Major conclusion on next page." For example, readers of (Poincaré, On the Three-Body Problem and the

Equations of Dynamics, 2017, p. 103, Equation 18) who don't recognize the Duffing equation may easily miss that Poincaré is discussing fully chaotic behavior of a well-known non-linear oscillator, and then answer incorrectly the question, "did Poincaré describe a fully chaotic system?" For a more relevant example, look at (Gray, 2013, p. 374). There he quotes Poincaré (Poincaré, Sur la dynamique de l'électron, 1905, p. 1506; p. 43), "*the deformable and compressible electron is subject to a kind of constant external pressure.*" (He also has a similar quote from (Poincaré, Sur la dynamique de l'électron, 1906, p. 130; p. 46).) In these quotes, Poincaré is referring to the Poincaré stress discussed here in Chapter 9 in the subsection "Electron Stability." Classical electrons are not stable with only electromagnetic forces; the electromagnetic worldview is not possible. The constant external pressure, the Poincaré stress, is what would be needed to stabilize the electron. Also, because the observed shape of a moving electron changes with the relative velocity, the observed self-energy and Lagrangian of the electron also change. Jeremy Gray appears to miss this point and concludes his discussion (on the same page) with the sentence, "It is as if the poor moving electrons are doing work battling through the ether, which appears as some sort of head wind." This is far removed from what Poincaré wrote.

Does Poincaré understand how hard he is making his readers work? Maybe not. He may not realize how much more facile his command of the subject matter is. This could be a factor in Poincaré's continued use of the convention of writing out vector partial-differential equations in Cartesian coordinates without a separate symbol for partial derivatives. (This point is discussed in Chapter 12.) Perhaps he is fully comfortable with this notation when others, even Oliver Heaviside, felt a pressing need for a more compact and expressive notation to follow the reasoning.

Is my list of missed predictions in the previous section because I and other readers have not provided for ourselves the signs that Poincaré thought would be clearly understood? I don't know, but I suggest that it may be a partial explanation.[5] Clearly reading, understanding and explaining Poincaré's work is a matter of what we each bring to that effort. As suggested above, it would be interesting if we read his work with the same attitude that Poincaré brought to (Lorentz, Electromagnetic phenomena in a system moving with any velocity smaller than that of, 1904).

For balance, I need to point out a competing negative explanation. This explanation sees Poincaré as not adequately understanding what he had come across, and indeed written. The missed predictions that I listed in the previous section are then seen as physics that was beyond Poincaré's ability to understand, even with his great gifts as a mathematician. This view could be summarized by a statement that Poincaré was a much better mathematician that a physicist.

There is still plenty of room for discussion and no resolution of the mystery in sight.

---

[5]Yes, I know that misunderstood genius is a cliché.

Conclusion

It is hard to come up with a concise description of Poincaré as a physicist. In part this is because Poincaré is a complex subject for a biography is his own right.

Getting a clear view of Poincaré as a physicist is made harder because we are near (or partially in) a polarizing subject, priority for the theory of special relativity. That priority dispute is something that I cover in the following chapter, Chapter 11.

I started this chapter with a quote from Poincaré that I described as a koan. I think I need to leave you at the end of the chapter with an even bigger mystery and very little to guide your assessment.

Notes

1. As mentioned in the preface, (Miller, 1973) and (Whittaker, 1953) should be approached with caution as sources of information about what Henri Poincaré wrote and meant. Each appears to have their own significant biases. The following chapter, Chapter 11, provides some indication of the nature of their biases. Please note that (Miller, 1973) was "Communicated by G Holton." Gerald Holton and Edmund Whitaker are both principles in the polemic concerning priority for the theory of special relativity. When reading their works and this chapter (or indeed this book), I suggest you carefully consider explicit, implicit and unsuspected biases. For me, please consider both that I have a deep interest in Henri Poincaré's work and what I have written in the preface.

References

Abraham, M. (1902). Dynamik des Electrons. *Nachrichten von der Gesellschaft der Wissenschaften zu Göttingen*, 20–41.

Darrigol, O. (1995). Henri Poincaré's Criticism of Fin de Siècle Electrodynamics. *Stud. Hist. Phil. Mod. Phys., 26*(1), 1–44.

Darrigol, O. (2004). The Mystery of the Einstein-Poincaré Connection. *Isis, 95*, 614–626.

Einstein, A. (1905). Zur Elektrodynamik bewegter Körper. *Annalen der Physik, 17*, 891–921.

Einstein, A. (1906). Das Prinzip von der Erhaltung der Schwerpunktsbewegung und die Traegheit der Energie. *Annalen der Physik, 20*, 627–633.

Feynman, R. P., Leighton, R. B., & Sands, M. (1963). *The Feynman Lectures on Physics*. Reading, MA: Addison-Wesley Publishing Company.

Gray, J. (2013). *Henri Poincare: a scientific biography.* Princeton, NJ: Princeton University Press.

Kaufmann, W. (1901). Die magnetische und electrische Ablenkbarkeit der Becquerelstrahlen und die scheinbare Masse der Elektronen. *Nachrichten von der Königl. Gesellschaft der Wissenschaften zu Göttingen, 2*, 143–155.

Langevin, P. (1905). Sur l'origine des radiations et l'inertie électromagnétique. *J. Phys. Theor. Appl., 4*, 165–183. doi:https://doi.org/10.1051/jphystap:019050040016500

Lorentz, H. A. (1902). Contributions to the theory of electrons. I. *Proceedings of the KNAW [Royal Netherlands Academy of Arts and Sciences], 5*, 608–28.

Lorentz, H. A. (1904). Electromagnetic phenomena in a system moving with any velocity smaller than that of. *Proceedings of the KNAW [Royal Netherlands Academy of Arts and Sciences], 6 (1903–4)*, 809–831.

Miller, A. I. (1973). A study of Henri Poincaré's "Sur la Dynamique de l'Électron". *Arch. Hist. Exact Sci., 10*, 207–328.
Minkowski, H. (1908). Die Grundgleichungen für die elektromagnetischen Borgänge in bewegten Körpern. *Nachrichten von der Gesellschaft der Wissenschaften zu Göttingen*, 53–111.
Poincaré, H. (1900). La théorie de Lorentz et le principe de réaction. *Archives néerlandaises des sciences exactes et naturelles, 5*, 252–78.
Poincaré, H. (1905). Sur la dynamique de l'électron. *Comptes rendus de l'Académie des Sciences, 140*, 1504–1508.
Poincaré, H. (1906). Sur la dynamique de l'électron. *Rendiconti del circolo matematico di Palermo, 21*, 129–176.
Poincaré, H. (1908). La dynamique de l'électron. *Revue générale des sciences pures et appliquées, 19*, 386–402.
Poincaré, H. (2017). *On the Three-Body Problem and the Equations of Dynamics.* (B. D. Popp, Trans.) Springer.
Stein, H. (2014). *Physics and Philosophy Meet: the Strange Case of Poincaré.* Retrieved January 5, 2020, from PhilSci-Archive: http://philsci-archive.pitt.edu/id/eprint/10634
Thomson, J. J. (1897, October). Cathode Rays. *Philosophical Magazine and Journal of Science, Fifth Series, 44*, 293–316.
Whittaker, E. (1953). *A History of the Theories of the Aether and Electricity, The Modern Theories.* London: Thomas Nelson and Sons Ltd.

Chapter 11
Einstein, Poincaré and the Origins of Special Relativity

Introduction

Two papers I translated in this book have had considerable notoriety thrust upon them because of the historical discussion of priority for the discovery of the theory of special relativity. The publication date of the first paper serves to establish a first-to-publish date for Henri Poincaré just a few months before Albert Einstein's publication and the second paper by Poincaré provides details for this discussion. The second paper (Poincaré, Sur la dynamique de l'électron, 1906a) is commonly called the Palermo paper, a reference to the journal in which it was published. The corresponding paper by Einstein (Einstein, Zur Elektrodynamik bewegter Körper, 1905)[1] is one of his famous series of papers from 1905.

Many gallons of ink have been used to argue about the position of Poincaré's and Einstein's 1905 papers in the discovery of special relativity and precedence for that discovery. A good deal of the ink in fact seems to be quite toxic which makes the discussion unattractive.

If, unlike me, you are interested in the discussion of precedence, then I hope that the translations of relevant papers by Poincaré provided in Part I contributes to your understanding of what Poincaré wrote.

I also find that discussion to be a distraction from several important topics that Poincaré does treat and which I do find interesting. In fact, the other chapters in Part II treat topics that I find both interesting and overlooked. What is an electron? What holds it together? What did Hendrik Antoon Lorentz and Henri Poincaré contribute to reworking electrodynamics without an ether and with discrete charges? Further, there are additional topics here that I haven't considered: Henri Poincaré in these papers is also trying to understand the origin of mass, and the unification of gravitation and electromagnetism. Any of these are meritorious in their own right.

[1]Quotations from Einstein's 1905 paper on special relativity used in this chapter are taken from (Einstein, The Collected Papers of Albert Einstein, 1989).

B. D. Popp, *Henri Poincaré: Electrons to Special Relativity*,
https://doi.org/10.1007/978-3-030-48039-4_11

Therefore, my reason for preparing this book is not to get involved in that discussion of precedence for special relativity; it is instead to point out some of the interesting things done by Poincaré (and Lorentz) that have been drowned out by the discussion. However, if I were to write this book ignoring the noise from the discussion, it would leave a rather large elephant in the room. The purpose of this chapter is therefore to acknowledge the presence of the elephant, look it over, and then get it out of sight.

A prominent feature of some contributions to this discussion is a checklist of items seen as necessary for having a full theory of special relativity. (The checklist may in some situations only be metaphorical. Alternatively, Jeremy Gray (Henri Poincare: a scientific biography, 2013, p. 368) writes, "the pro-Poincaré faction can... cut and paste their man's words into a fairly impressive list of insights.") Some items on the checklist are Lorentz transformations, demonstration that Maxwell's equations of electrodynamics are invariant under Lorentz transformation, and a statement that the speed of light is the same in all inertial reference frames. Checkmarks are then placed next to these items in a Poincaré column and in an Einstein column. In this approach to the comparison, there are many checks in the Poincaré column. Some contributors do see enough checks in the Poincaré column to assert that Poincaré did have a full theory of special relativity and maybe even some glimmers on the horizon in the direction of the general theory of relativity.

This checklist approach seems to have three main shortcomings. First, it passes over a sizeable portion of what Poincaré did write about that isn't needed for the special relativity checklist. Other chapters in this book help draw attention to some of this material that has been passed over. The next shortcoming is that the reductionist nature of a checklist leaves out consideration of holistic issues: what did Poincaré and Einstein each write in their respective papers about what they were trying to do and how they saw the items from the checklist fitting together. I agree with the statement in (Darrigol, The Mystery of the Einstein-Poincaré Connection, 2004, p. 618), "in order to compare Poincaré's and Einstein's theories properly one must read every one of their statements in context, taking into account both the inner logic of their investigations and the contemporary problematics to which they were responding." I hope the translations in Part I, and Chapters 9 and 10 are an aid to readers wishing to follow Darrigol's prescription. The third shortcoming lies in understanding how other people saw and worked with what Poincaré and Einstein had each written.

In this last case, the answer is fairly clear. Poincaré's paper from 1906 ("the Palermo paper") and his follow-up from 1908 seems to have largely languished until Edmund Whittaker (Whittaker, 1953) brought attention to them (discussed later in this chapter). In contrast, Einstein's readers, after a delay, tried to assimilate, write about and use Einstein's paper. First Hermann Minkowski in 1908 and then Arnold Sommerfeld, Paul Langevin and others went down this path. Their efforts and publications added weight behind Einstein's work. This path then became the history of the adoption and acceptance of the theory of special relativity. Books have been written about that history and I'm not going to look at it further here.

Instead, returning to the second shortcoming, I now want to look at what Poincaré and Einstein wrote in their papers about what they were trying to do. This is necessarily a discussion of what they wrote in a particular place at a particular time. It does not consider their notes and ideas before they were organized and exposed for publication; it does not consider how they presented their work at a different time in a later publication. Then, to frame the discussion, I will look at some of the earliest contributions to the back-and-forth discussion about priority and also recent discussion about the discussion, and bring the chapter to a close.

Poincaré on What He Was Trying to Do

To get a more holistic view of what Poincaré thought he was trying to accomplish it is useful to look at what he wrote in the introduction (Poincaré, Sur la dynamique de l'électron, 1906a). There he lays out his plan for the paper, and several key considerations or approaches to his work are evident.

Poincaré starts his introduction with a summary of the evidence—ending with Michelson's experiment—that it is impossible to detect experimentally the absolute motion of the Earth. He accepts this as an experimental law and calls it the Relativity Postulate. The contraction proposed by Lorentz and Fitz Gerald accounts for the results of Michelson's experiment; this must be generalized to the Lorentz transformation to bring it into agreement with the full generality of the Relativity Postulate. Since electromagnetic phenomena are not altered by the Lorentz transformations (Poincaré will prove in this paper that Maxwell's equations are covariant), Poincaré concludes that electromagnetic phenomena in stationary and moving systems are indistinguishable ("the exact image of each other").

In brief, Poincaré has forged a chain from experimental evidence, with the relativity postulate and the Lorentz transformation, to the indistinguishability of inertial reference frames.

Poincaré continues the introduction with an application of the Lorentz transformation to electrons and consideration of the experimental results of Walter Kaufmann. This leads Poincaré to consider the shape and electromagnetic mass of moving electrons. He arrives at the conclusion that subatomic charged particles subject only to electromagnetic forces are not stable. For them to be stable, some additional, non-electromagnetic force is required that is "[comparable] to a constant external pressure"—this force is now referred to as "Poincaré stress."

The evidence with which Poincaré started the introduction dealt with electromagnetic and optical phenomena. Next in the introduction, Poincaré considers whether the relativity postulate and Lorentz transformation apply to phenomena with a different origin (such as gravitational phenomena), whether inertial mass is solely of electromagnetic origin, and how Newton's law of gravitation might need to be modified to be consistent with the relativity postulate and Lorentz transformation.

Poincaré concludes the introduction by asking whether there is some underlying explanation for the appearance of the speed of light as the speed of propagation of gravitational phenomena. Since the speed of light appears in descriptions of electromagnetic phenomena and gravitational phenomena, is there a connection between the two?

This leads to an analogy to Copernicus's work at the end of the introduction that some people have found obscure. Before Copernicus, in the Ptolemaic system, the Earth was taken as the center of the solar system and the apparent position of the planets calculated based on their motion around the Earth. Copernicus recast the calculations of the apparent positions in terms of the Sun at the center of the solar system and the Earth and other planets revolving around the Sun. In the Copernican system the motion of the Earth in a circle around the Sun in 365 days was explicitly present and appeared in the calculation of the Earth's position. In the calculation of positions in the Ptolemaic system, the circular orbit of the Earth with a 365-day period was implicitly present in the calculation of the position of each planet. A person looking at the Ptolemaic calculations might have reasonably asked why certain patterns repeated for each of the planets. Therefore, in his analogy, Poincaré looks back to Copernicus who by placing the Sun at the center of the solar system used the annual circular motion of the Earth around the Sun to explain the identical, repeated circles and travel times in the Ptolemaic system. We can look forward and, anachronistically, rephrase Poincaré's question to ask whether the place of the speed of light in an underlying theory explains its seemingly separate appearance in both electromagnetic and gravitational phenomena.

In this synopsis of the introduction to (Poincaré, Sur la dynamique de l'électron, 1906a), we can recognize several considerations important to Poincaré.

The first consideration is experimental evidence.

Immediately in the first paragraph of the introduction, we see the importance of reasoning from experimental results. The names of two experimentalists are cited in this paragraph; two more are named on the following page. Based on Michelson's experiment, Poincaré states the Relativity Postulate; Poincaré calls it a postulate because it could still "be confirmed or rejected by more precise experiments."[2] The Lorentz contraction is presented as a way to take into account the result of Michelson's experiment. We then observe Poincaré reasoning from this experimental basis to see what can be deduced and developed.

Next in the introduction, Poincaré refers to experiments by Kaufmann (measuring the charge-to-mass ratio and the dependence of electron mass on velocity). Poincaré mentions Kaufmann twice in the introduction but does not mention him again in the body of this paper. In the paper, (Poincaré, La dynamique de l'électron, 1908), Poincaré mentions Kaufmann by name nine times including in the title of §X. This is a point of departure for discussing the properties (radius and mass in particular) under Lorentz transformation.

The second consideration is the work of others in the field.

[2]In my paraphrase, his postulate is a statement about what nature has shown us (by experiment or observation) and not a statement about how nature is, fundamentally.

Lorentz's earlier paper (Lorentz, Electromagnetic phenomena in a system moving with any velocity smaller than that of, 1904) is critical to this paper by Poincaré (Poincaré, Sur la dynamique de l'électron, 1905a). Poincaré builds on it and, throughout he corrects, amplifies and clarifies. Poincaré also summarizes the work by Max Abraham on electron shape and Paul Langevin on electromagnetic waves.

The third consideration is an attitude of modesty seen as respect or deference to his colleagues.

Consider the statement at the end of the introduction to (Poincaré, La théorie de Lorentz et le principe de réaction, 1900, pp. 252, p. 8). Poincaré indicates that his objections (concerning the need for Lorentz's theory to conserve momentum) allow Lorentz's theory to show its hidden virtues and asks the reader, "to forgive me for having presented at such length ideas with so little novelty." Poincaré's handling of his divergences from Lorentz discussed at length in Chapter 9 in the letters translated in Chapter 3 reinforce this point.

The fourth consideration is identifying and using invariants.

Although not evident from the introduction, it is clear from Poincaré's other work and again here that he chooses to look for invariants. Here, (Poincaré, Sur la dynamique de l'électron, 1906a, p. 168; p. 93), below equation 4′, Poincaré looks for invariants of the Lorentz group. He finds that $x^2 + y^2 + z^2 - (ct)^2$ is an invariant. (Recall that in the units chosen by Poincaré, $c = 1$.) This is the spacetime interval that is invariant in Minkowski space. He also finds several other invariants and they are given in equations (5) and (7).

The fifth consideration is retaining what is thought to be known and established and working effectively.

As specific examples this means conservation of momentum (*principe de réaction*) and conservation of energy—stated differently these are both invariants of motion—and Newton's law of gravitation. A key consideration in (Poincaré, La théorie de Lorentz et le principe de réaction, 1900) is that Lorentz's theory of electricity and magnetism of moving bodies be adapted to conserve momentum. In adapting to satisfy this consideration, Poincaré concludes that electromagnetic radiation must transport momentum.

The negative side of this consideration is that Poincaré continues to retain and use the term ether in (Poincaré, La dynamique de l'électron, 1908). In adopting the relativity postulate, Poincaré denies the possibility of detecting absolute motion and it would seem that the ether should be abandoned too. Einstein writes that the ether is "superfluous" and makes a clean break. Poincaré could have made a clean break too; he continues to use the term. A physicist seeking to understand a phenomenon can certainly have a preference for one reference frame over another even if nature and the phenomenon do not impose that preference. Still his continued use of the term ether, which may previously have been a useful consideration, is problematic.

Einstein on What He Was Trying to Do

Turning now to Einstein's introduction to his work (Einstein, Zur Elektrodynamik bewegter Körper, 1905), he starts with the statement of a problem. To illustrate the problem, he describes a scenario with interaction between a magnet and a conductor. In one version of the scenario, the conductor moves at a constant velocity past a stationary magnet. In the other version, the conductor is stationary while a magnet moves past at a constant velocity. In both versions a current is produced in the conductor and the resulting magnitude and direction of the current produced is the same. The current produced cannot be used to determine whether the magnet or the conductor is moving; only the velocity of whichever one is moving can be determined. In contrast to the phenomenon just described the description of the two versions in "electrodynamics—as usually understood at present—" is different. Einstein characterizes this difference as an asymmetry. And so in his problem statement in the first sentence he writes, "electrodynamics... when applied to moving bodies, leads to asymmetries that do not seem to attach to the phenomena."

Following the discussion of symmetry in relative motion in electrodynamics in the first paragraph, Einstein observes in the second paragraph the failure to detect absolute motion of the Earth relative to the ether. Einstein says that this is a similar kind of example. Here the experimental results show that the ether (or equivalently absolute motion) cannot be detected. The phenomenon (aberration) is symmetric; it cannot show whether the distant star that is the source of the light is stationary and the Earth is moving or vice versa.

Einstein uses these two examples of symmetry as a basis for a conjecture that the phenomena do not have any properties corresponding to absolute rest. Accepting this conjecture, he raises it to the status of a postulate that he calls "the principle of relativity." He adds a second postulate to this: in empty space light always travels with a definite velocity independent of the motion of the emitting body. Armed with these two postulates, Einstein would seem to be ready to apply them to electrodynamics to investigate the consequences. Instead, he turns his attention to clocks and coordinate systems in the following section and says that insufficient attention to them is at the root of the difficulties. And that is the conclusion of the introduction.

At the time of writing this introduction and for the next several years, Einstein was employed as a patent examiner in the Swiss Patent Office in Bern. He was therefore familiar with the organization used in drafting patents. In general, that organization has an abstract, a statement of the field of the invention, a discussion of the prior art (meaning relevant, published work in the field of the invention) along with a discussion of an unresolved problem or opportunity in the prior art, a brief description of the invention addressing this problem or opportunity, a detailed description of the invention and finally the claims setting out the boundaries that characterize the invention. Looking at the beginning of Einstein's article in Annalen der Physik and applying this analogy with patent drafting, we see that Einstein has indicated the field of his work (electrodynamics, Maxwell's theory as currently understood), provided a brief statement of the problem present in the prior art (asymmetries in the treatment of phenomena involving relative motion) and then a brief description of the "invention" (the postulate of relativity and the postulate that the speed of light is a universal constant). He accomplishes this in a few more than 400 words; for comparison, a patent abstract (which Einstein does not provide) must preferably not exceed 150 words.

On the other hand, he has not identified or discussed the prior art, that is the scientific literature in the field. Famously, we do not know which (if any) works by Lorentz or Poincaré, or anyone else, he was familiar with. Seen one way, this allows some mythologizing to exist. As an undergraduate physics major, I heard that Einstein was in some way more philosopher, working on a different plane than ordinary physicists, and this was reflected in his considerations of symmetry, simultaneity and the speed of light. A stated consequence is that without Einstein's work on the subject special and general relativity would not have been developed until decades later and would have a substantially different form. Seen another way, the absence of a discussion by Einstein of what was in the prior art complicates our assessment and understanding of his reasoning and his understanding of the context of his theory. Both perspectives make assessment of priority more difficult.

A patent examiner, including Einstein, examining a patent application would look at the prior art (e.g. references and citations) disclosed in an application and would on their own search the published literature for additional prior art to identify the knowledge available to a person working in the field. The examiner organizes the result of this search in a search report. Comparison of the invention against the knowledge available to a person working in the field allows the examiner to determine whether the claimed invention is novel and nonobvious. It seems very curious that Einstein has not done this in his own work.

In the section following the introduction, Einstein starts his detailed description of the invention. There, he famously instructs his readers on how to tell time (and the importance of simultaneity). "We have to bear in mind that all our propositions involving time are always propositions about simultaneous events. If, for example, I say that 'the train arrives here at 7 o'clock,' that means, more or less,' the pointing of the small hand of my clock to 7 and the arrival of the train are simultaneous events."[3] This lesson in telling time applies to a person with a watch in hand standing on the platform next to the train pulling into the station. The situation becomes more complicated if the observer, clock and event are at separate locations. Here, Einstein uses surveyor's rods and light signals to set out a reference frame with synchronized clocks.

[3]An alternate translation, "We must take into account that all of our judgments in which time plays a role are always judgments about simultaneous events. If, for example, I say, 'That train arrives here at 7 o'clock,' it essentially means, 'The train arriving and the small hand of my watch pointing to 7 are simultaneous events.'" provided by Ken Kronenberg. Personal communication, December 2016.

With the synchronized clocks and surveyed framework in place, he can then use the two postulates (the relativity postulate and that the speed of light in empty space is a universal constant, independent of movement of the emitter) to determine the Lorentz transformations as a consequence.

As with Poincaré above, in this synopsis of the introduction to (Einstein, Zur Elektrodynamik bewegter Körper, 1905), we can recognize several considerations important to Einstein.

Directly with the first sentence, Einstein brings up the importance of alignment between natural phenomenon and theoretical explanation. He rejects a theoretical explanation that is asymmetric for a natural phenomenon that is symmetric. When a conducting coil moves past a stationary magnet at a constant velocity, a certain current is produced in the coil and the same current is produced when a magnet moves past a conducting coil at a constant velocity. An essential minimum constraint on the theory is that the theory correctly explains the magnitude of the current in both configurations. Einstein is further demanding that the explanation intrinsic to the theory be symmetric for both configurations. His consideration of this symmetry in the phenomenon motivates him to incorporate the principle of relativity into his theory.

The second consideration is that the speed of light is constant. Where Poincaré anticipates some underlying connection or significance for the speed of light, Einstein promotes it to a fundamental, universal constant.

The third consideration relates to reference frames. Here Einstein starts by explaining how to tell time at one point: it involves the simultaneity of the hands on his watch and the position of the train next to him and then sets out a reference frame with synchronized clocks. The importance of simultaneity becomes clear now because two events that are simultaneous in one reference frame might not be simultaneous in a different reference frame moving with a constant velocity. This leads to aptly emphasizing the importance of identifying the relevant reference frames and correctly using them. This is a key step in understanding many paradoxes presented in special relativity. Is the half-life of the meson formed high in the Earth's atmosphere measured in the reference frame of the observer on the Earth's surface or in the reference frame of the meson traveling at a high velocity relative to the observer?

The Start of the Dispute: Edmund Whittaker

The title of E. T. Whittaker's book, "A History of the Theories of Aether and Electricity, the Modern Theories, 1900–1926" (Whittaker, 1953), is curious. In fact, it is a second volume; about 50 years earlier Whittaker wrote a first volume covering a long swath of the history of physics with understanding of the ether as an organizing theme. The preface to this second volume starts, "The purpose of this volume is to describe the revolution in physics which took place in the first quarter of the 20th century." For a book covering this period, the use of ether in the title seems anachronistic. Whether you look to Poincaré (Poincaré, Sur la dynamique de l'électron, 1906a) or Einstein (Einstein, Zur Elektrodynamik bewegter Körper, 1905), ether appears to receive a death blow at the beginning of the quarter-century he proposes to cover. For Poincaré, the failure to detect absolute motion relative to the ether leads to the Relativity Postulate. This makes the ether unnecessary for Poincaré, although it does continue to appear in his writing, in some cases just to argue for the impossibility of the properties it would be required to have. For his part, Einstein writes, "introduction of a 'light ether' will prove superfluous" because there is no space at absolute rest. How can a book about physics in the 20th century have ether in the title?

The title of the second chapter of Whittaker's book is likewise curious, and also provocative; it is "The Relativity Theory of Poincaré and Lorentz." Why isn't it, Einstein's Theory? A plausible first reaction might be that Whittaker is trying to completely write Einstein out of the history of physics. Referring to the table of contents, we can see that Einstein's name appears there nine times in subchapter titles. So, we see that Einstein has not been written out, but his position in the history of both special and general relativity has been greatly marginalized. Within the second chapter, Einstein's name appears five times. When it first appears, Whittaker writes, "In the autumn of [1905]... Einstein published a paper which set forth the relativity theory of Poincaré and Lorentz with some amplifications, and which attracted much attention." The same paragraph ends with the second mention of Einstein's name and credits him with modifications made to the formulas for aberration and Doppler effect. (Einstein's name appears once more in this context and twice in connection with the formula "$E = mc^2$.") It is a stunning demotion and calling it provocative seems like understatement.

At this point a digression seems well justified. Whittaker (A History of the Theories of the Aether and Electricity, The Modern Theories, 1953, p. 51), writes, "In 1900 Poincaré,[3] referring to the fact that in free ether the electromagnetic momentum is $(1/c^2)$ times the Poynting flux of energy, suggested that electromagnetic energy might possess mass density equal to $(1/c^2)$ times the energy density: that is to say, $E = mc^2$ where E is energy and m is mass." His footnote 3 is a reference to (Poincaré, La théorie de Lorentz et le principe de réaction, 1900) translated in Part I. In that paper, Poincaré is concerned with the transport of momentum by electromagnetic radiation, which he shows is necessary for conservation of momentum. He does relate this to the Poynting vector; he does not take the additional step of connecting the momentum of the electromagnetic radiation to an inertial mass. As Whittaker continues his discussion on the following page, he references in footnote 4 (Einstein, Das Prinzip von der Erhaltung der Schwerpunktsbewegung und die Traegheit der Energie, 1906). In the introduction to that paper, Einstein states that "the conclusion that the mass of a body changes with the change in its energy content... is the necessary and sufficient condition for the law of the conservation of motion of the center of gravity to be valid" in systems with both mechanical and electromagnetic processes. This is closely related to the statement that conservation of momentum requires that electromagnetic radiation carry momentum. Einstein in the next sentence states, "the simple formal considerations that have to be carried out to prove this statement are in the main already contained in a work by H. Poincaré" and references (Poincaré, La théorie de Lorentz et le principe de réaction, 1900).[4] Up to a point Einstein and Poincaré follow similar reasoning. Einstein takes the additional step of relating the inertial mass and the energy. Had Poincaré thought of it, it would have been a small step beyond what he had done to relate the momentum of the electromagnetic radiation to an inertial mass and show that the mass is equal to E/c^2. He didn't take the step. We need to be careful not to credit Poincaré with a discovery that is implicit in the physics but not explicit in his writing because our knowledge of subsequent work allows us to see what is close at hand but not firmly grasped. As for Whittaker's statement quoted above, I have not found support, which is clear and does not need hindsight, in the paper by Poincaré he referenced, or in other papers I looked at.

Returning to the analogy of patent examination used above, it can be seen that there is one thing of value that Whittaker has provided in this chapter: a search report. In the footnotes, among other citations, Whittaker cites the following from before July 1905: (Poincaré, La théorie de Lorentz et le principe de réaction, 1900), (Poincaré, Électricité et optique, 1901), (Poincaré, L'état actuel et l'avenir de la physique mathématique, 1904), (Poincaré, Sur la dynamique de l'électron, 1905a), (Lorentz, Electromagnetic phenomena in a system moving with any velocity smaller than that of, 1904), two earlier works each by Poincaré and Lorentz and one work by Joseph Larmor. In preparing a search report, the patent examiner looks for the most relevant published sources from before the priority date and avoids accumulating redundant sources. Unlike a patent examiner, Whittaker was not explicitly preparing a search report, instead he was citing references that support his historical narrative of the development of the theory. This does not answer questions about what Einstein knew about this context, and when he knew it; it does show that other people were publishing papers on questions that in hindsight we can recognize as related.

Inclusion of a source in a search report indicates that the source needs to be considered in evaluating the patent but does not demand a particular conclusion by the examiner. Also, the examiner may well find things that were not unknown to the inventor; this reflects a difference in their roles and perspectives.

There is a place for the patent examiner to provide a written opinion on the patentability of the patent application and to reject, object or grant the application. In this chapter of his book, Whittaker has written his view of the historical development of the theory of special relativity. It is that view and the perceived provocation, that led to the voluminous discussion of priority.

[4]This appears to be the only citation by Einstein up to at least 1920 of a publication by Poincaré.

Contemporary Reactions from Physicists

Some of the earliest reactions to Whittaker's demotion of Einstein came from physicists whose remarks seem to get slipped into other writings as short paragraphs.

Chronologically the first reaction appears to be from Louis de Broglie. It came roughly within the year after publication of Whittaker's book and appeared in a preface that de Broglie wrote for volumes IX and X of Poincaré's collected papers (de Broglie, 1954). In a paragraph in the preface, de Broglie wrote:

> *"Poincaré's celebrated paper on the Dynamics of the Electron published in 1906 in the minutes of the mathematics circle of Palermo, after having been summarized in a Note to the Minutes [of the Académie des sciences], is still very interesting to reread today. Commenting on the Lorentz transformation and the ideas of the illustrious Dutch physicist on the contraction of moving bodies and on local time, Henri Poincaré completely developed the new dynamics of the electron which followed from it: he made a connection to the theory of electromagnetic radiation that Paul Langevin had just laid out in a beautiful work and he compared the various assumptions that one could make about the structure of the electron and its deformation resulting from its movement. Poincaré thus established the new relativistic dynamics of the electron on a solid base which even now has many applications: he thus accomplished a work of major importance, but at the same time, perhaps because he was more mathematician than physicist, he did not grasp the general viewpoint supported by a minute critique of the measurement of distances and times that the young Albert Einstein discovered with an inspired intuition and which led to a complete transformation of ideas about space and time. Poincaré did not take this decisive step, but he is, with Lorentz, the one who most contributed to making it possible. Let us remark on an important point from Poincaré's paper: the discovery of the fact that an electron as Lorentz had conceived it is not stable under the action of electromagnetic forces alone, that its stability requires the involvement of another force, of unknown nature, deriving from a potential proportional to the volume of the electron. This 'Poincaré pressure', which can be interpreted as indicating the incomplete nature of our usual understanding of the electromagnetic field, still has even now all of its importance and it is often a question in the most recent works on the structure of the electron. This was a major discovery in physics by the leading mathematician."*

In the first edition in 1962 (Jackson, Classical Electrodynamics, 1962), Jackson wrote on page 353, "By supposing that all matter was essentially electromagnetic in origin and so transformed in the same way as Maxwell's equations, Lorentz was able to deduce the contraction law (11.10). Then Poincaré showed that the transformation of charges and current densities could be made in such a way that all the equations of electrodynamics are invariant in form under Lorentz transformations. In 1905, almost at the same time as Poincaré and without knowledge of Lorentz's paper, Einstein formulated special relativity in a general and complete way, obtaining the results of Lorentz and Poincaré, but showing the ideas were of much wider applicability. Instead of basing his discussion on electrodynamics, Einstein showed that just two postulates were necessary one of them involving a very general property of light." This statement does not appear in the third edition (Jackson, Classical Electrodynamics, 1999).

Chapters 15 and 16 of The Feynman Lectures on Physics, vol. 1 (Feynman, Leighton, & Sands, 1963) contain scattered references to contributions by Poincaré to special relativity. Most notably, Chapter 16 starts, "In this chapter we shall continue to discuss the principle of relativity of Einstein and Poincaré." The following paragraph in that chapter quotes Poincaré's statement of the principle of relativity from (Poincaré, L'état actuel et l'avenir de la physique mathématique, 1904).[5]

Gerald Holton's Response to Edmund Whittaker

The first specific response to the provocation by Whittaker comes from Gerald Holton (Holton, 1960), which falls chronologically between the quotes from de Broglie and Jackson above. In his paper, Holton takes up the challenge in the last titled section "Whittaker's Account of the Origins of Einstein's Work." Holton states the topic of the dispute as, "to what extent Einstein's work was original rather than anticipated by, or specifically based on other published work." In (Whittaker, 1953), there is not a comparable statement of intent; maybe Whittaker saw this as a case of near simultaneous, independent discovery and told the history from the perspective of the first to submit a manuscript to the publisher or again maybe Whittaker was biased against Einstein, due to his attachment to the ether that extended to the title of his book, because Einstein called the ether superfluous (see above) whereas Poincaré had not fully abandoned it. It seems very hard to read into (Whittaker, 1953) a suggestion of misconduct by Einstein.

Holton does not indicate what use "of other published work" was suggested by Whittaker—although he does repeatedly insist that Einstein had not seen or read (Lorentz, Electromagnetic phenomena in a system moving with any velocity smaller than that of, 1904)—and refers instead to bias and the need to analyze Whittaker's work by dealing "with the prior commitments and prejudices of the scholar himself." (Holton, 1960, p. 634) Holton then provides seven main findings from considering Whittaker's work in light of potential bias.

Instead of considering that direction further, I want to turn to what Holton both does and does not say about Poincaré's work. After Holton notes that Whittaker has not recanted his earlier provocation, Holton states that Whittaker "repeats that Poincaré in a speech in St. Louis, USA, in September 1904(27) had coined the phrase 'principle of relativity.'" Footnote 27 cites (Poincaré, L'état actuel et l'avenir de la physique mathématique, 1904) and the English translation by G. B. Halsted in (Poincaré, The Principles of Mathematical Physics, 1905b). Holton in this quote is denying that Poincaré coined (or at least used) the phrase "principle of relativity" in 1904. About a page later in his second finding, he again refers to (Poincaré, L'état actuel et l'avenir de la physique mathématique, 1904). He first states that it "turns out

[5]The translation quoted by Feynman is accurate but does not exactly follow the translation by G. B. Halsted published in (Poincaré, The Principles of Mathematical Physics, 1905b) and in (Poincaré, The Value of Science, 1907), or the translation by J. W. Young in (Poincaré, The Present and the Future of Mathematical Physics, 1906b).

not to enunciate the new relativity principle" and then provides a direct quote from Halsted's translation starting, "'the principal of relativity, according to which the laws of physical phenomena should be the same whether for an observer fixed or for an observer carried along in a uniform motion [of] translation…'." Astonishingly, this is the same statement used in full by Feynman (Feynman, Leighton, & Sands, 1963) at the beginning of chapter 16, as noted above. On the other hand, Holton does not cite (Poincaré, Sur la dynamique de l'électron, 1906a); this work would be highly relevant to any discussion of independent discovery, but would not be relevant to discussion of the bases of Einstein's work since it had been submitted for publication but not yet published at the time Einstein submitted his paper. In an article communicated by Gerald Holton, Arthur Miller (A study of Henri Poincaré's "Sur la Dynamique de l'Électron", 1973) does discuss this article by Poincaré, but the general tone seems less than kind.

Conclusion

Provocation of the Einstein camp by Whittaker was met by provocation of the Poincaré camp by Holton; polemic was met with polemic; and the ink started to flow. The result is on the whole unappealing.

The translations in Part I accompanied by the discussion in the two previous chapters, Chapters 9 and 10, can help the reader do their own careful alignment of what Poincaré and Einstein each wrote in 1905. This is a necessary step for any meaningful comparison and the result will certainly show areas of substantial agreement, allowing for differences in modes of expressing their ideas.

Poincaré's writing is clear but concise. It still leaves plenty of room for his readers to fill in steps and look between the lines, a point that was brough up late in Chapter 10. This contributes an interesting challenge in reading Poincaré as there is also room for misinterpretation or misunderstanding. It also provides insight into how Poincaré reasoned and mustered his arguments.

Chapter 10 looks at points where Poincaré appeared to be well positioned for making clear predictions or breaking from formerly useful ideas that had lost their place. This may lead you to ask, along with me, "Why doesn't he just say that there is no need for an ether that cannot be detected?" The section "Attitude" in that chapter mentions some reasons (although it might be better to call it speculation) why he didn't. There is no satisfactory answer. Perhaps the reason is that Poincaré saw his role in physics as collecting, cataloging and explaining, and as excluding predicting and extrapolating. That could also point the direction to an answer to the question, "Why didn't Poincaré make a response in some form to Einstein's 1905 and 1906 papers?"

My discussion of (Poincaré, Sur la dynamique de l'électron, 1906a) in Chapter 9 indicates in footnotes sections from (Einstein, Zur Elektrodynamik bewegter Körper, 1905) that may align with Poincaré's work being discussed. It may show areas where Poincaré is ahead of Einstein; it does show areas, notably Einstein's sections 1 and

2, that don't seem to have anything with which they could be aligned in (Poincaré, Sur la dynamique de l'électron, 1906a).

When Poincaré wrote about the space-time invariant and invariance of the continuity equation for charge, he may have been close to anticipating four-vectors introduced by Minkowski two years later. Poincaré certainly wrote about a variety of subjects—notably electron shape and stability—that were outside the more focused scope covered by Einstein. The discussion in Chapter 9 of the "first divergence" of Poincaré from Lorentz and his proof of the space-time invariant indicates that Poincaré understood the Lorentz transformations as transformations between coordinate systems including time and three spatial coordinates and even understood that this resulted in mixing space and time coordinates. Poincaré's discussion of local time in (Poincaré, La théorie de Lorentz et le principe de réaction, 1900) also needs to be mentioned in this context. It is still clear that Poincaré did not give the consideration to time and the impact on simultaneity of this mixing that Einstein did.

It therefore has to be noted that Poincaré has no dramatic statements about trains arriving in stations, no discussion of simultaneity and no suggestion that conclusions about simultaneity depend on the inertial reference frames selected. There does not seem to be anything in Poincaré's writing translated in Part I of this book that is comparable to (Einstein, Zur Elektrodynamik bewegter Körper, 1905) sections 1 and 2.

References

Darrigol, O. (2004). The Mystery of the Einstein-Poincaré Connection. *Isis, 95*, 614–626.

de Broglie, L. (1954). Préface. In G. Petiau, editor, *Oeuvres de Henri Poincaré* (Vol. IX, pp. VII-XIII). Paris: Gauthier-Villars.

Einstein, A. (1905). Zur Elektrodynamik bewegter Körper. *Annalen der Physik, 17*, 891–921.

Einstein, A. (1906). Das Prinzip von der Erhaltung der Schwerpunktsbewegung und die Traegheit der Energie. *Annalen der Physik, 20*, 627–633.

Einstein, A. (1989). *The Collected Papers of Albert Einstein* (Vols. Volume 2: The Swiss Years: Writings, 1900–1909 (English translation supplement)). (A. Beck, Trans.) Princeton, NJ: Princeton University Press. Retrieved from https://einsteinpapers.press.princeton.edu/vol2-trans/154

Feynman, R. P., Leighton, R. B., & Sands, M. (1963). *The Feynman Lectures on Physics*. Reading, MA: Addison-Wesley Publishing Company.

Gray, J. (2013). *Henri Poincare: a scientific biography*. Princeton, NJ: Princeton University Press.

Holton, G. (1960). On the Origins of the Special Theory of Relativity. *American Journal of Physics, 28*, 627–636.

Jackson, J. D. (1962). *Classical Electrodynamics* (First ed.). New York: John Wiley & Sons.

Jackson, J. D. (1999). *Classical Electrodynamics* (Third ed.). New York: John Wiley & Sons.

Lorentz, H. A. (1904). Electromagnetic phenomena in a system moving with any velocity smaller than that of. *Proceedings of the KNAW [Royal Netherlands Academy of Arts and Sciences], 6 (1903–4)*, 809–831.

Miller, A. I. (1973). A study of Henri Poincaré's "Sur la Dynamique de l'Électron". *Arch. Hist. Exact Sci., 10*, 207–328.

Poincaré, H. (1900). La théorie de Lorentz et le principe de réaction. *Archives néerlandaises des sciences exactes et naturelles, 5*, 252–78.

Poincaré, H. (1901). *Électricité et optique* (éd. Deuxiène Édition, revue et complétée). (J. Blondin, & E. Néculéa, Éds.) Paris: Gauthier-Villars.

Poincaré, H. (1904). L'état actuel et l'avenir de la physique mathématique. *Bulletin des sciences mathématiques, 28*, 302–324.

Poincaré, H. (1905a). Sur la dynamique de l'électron. *Comptes rendus de l'Académie des Sciences, 140*, 1504–1508.

Poincaré, H. (1905b, January). The Principles of Mathematical Physics. *The Monist, XV*(1), 1–24.

Poincaré, H. (1906a). Sur la dynamique de l'électron. *Rendiconti del circolo matematico di Palermo, 21*, 129–176.

Poincaré, H. (1906b). The Present and the Future of Mathematical Physics. *Bulletin of the American Mathematical Society, 12*, 240–260.

Poincaré, H. (1907). *The Value of Science.* (G. B. Halstead, Trans.) New York: Science Press.

Poincaré, H. (1908). La dynamique de l'électron. *Revue générale des sciences pures et appliquées, 19*, 386–402.

Whittaker, E. (1953). *A History of the Theories of the Aether and Electricity, The Modern Theories.* London: Thomas Nelson and Sons Ltd.

Chapter 12
Adoption of Vector Notation for Classical Electrodynamics

Considerations

Soon after starting to prepare these translations, I was faced with a choice about how to deal with Poincaré's vector and differential notation. Poincaré followed conventions in this matter which have not stood the test of time. The choice was therefore whether to retain the notational conventions used by Poincaré or to rewrite his equations (and affected text) to use notation more familiar and comfortable for me and my readers. Having reached this point, you're certainly aware that I retained the original notational conventions.

When I made that choice, I recognized that I would need to provide readers of these translations with some guidance; that is one purpose of this chapter.

Beyond providing guidance on understanding the notation that Poincaré used, this chapter looks at the notation used in (Lorentz, Electromagnetic phenomena in a system moving with any velocity smaller than that of, 1904), since Poincaré was certainly familiar with that notation in 1905, and then steps back over 15 years earlier to look at the notational choices made and developed by Joshua Willard Gibbs, Oliver Heaviside and Henri Poincaré in their efforts to understand James Clerk Maxwell's work on electricity and magnetism. All three resist the use of quaternions; only Poincaré continued Maxwell's use of Cartesian coordinates and explicit derivatives. In 1890 the right question is, why hadn't Poincaré seen and adopted Heaviside's notation? The answer may be related to a larger question of how readily scientific and technical knowledge passed across the English Channel at that time. By 1905 that question is no longer relevant and is replaced by the question, why did Poincaré continue to use the same notation after having seen a more compact and effective alternative? Now the answer lies somewhere in the range of preference, comfort and personal choice.

B. D. Popp, *Henri Poincaré: Electrons to Special Relativity*,
https://doi.org/10.1007/978-3-030-48039-4_12

Poincaré's Notation

While Poincaré is familiar with the difference between full and partial differentials, this is not reflected in the notation where d is used for both. In general (and there may be exceptions) it is safe to assume that when Poincaré writes $\frac{\mathrm{d}}{\mathrm{d}t}$ he in fact means $\frac{\partial}{\partial t}$ (in our notation). Please be aware that there are occasions when he does use the symbol ∂ with a different meaning which he defines on its first use.

If that were the only, or even the main, difference in notation a footnote would suffice and this chapter would be unnecessary. Poincaré writes out vectors and differentials by their individual, Cartesian components. In order to have a specific example, refer to (Poincaré, Sur la dynamique de l'électron, 1906, p. 132; p. 48–49)[1] written with Poincaré's notation.

Lorentz adopted a specific system of units so as to make the factors of 4π disappear in the formulas. I will do the same and additionally I will choose the units of length and time such that the speed of light is equal to one. Under these conditions, by calling: f, g, h the electric displacement; α, β, γ the magnetic force; F, G, H the vector potential; ψ the scalar potential; ρ the electric charge density; ξ, η, ζ the electron velocity; and u, v, w the current, then the fundamental formulas become:

$$
\begin{gathered}
u = \frac{\mathrm{d}f}{\mathrm{d}t} + \rho\xi \frac{\mathrm{d}\gamma}{\mathrm{d}y} - \frac{\mathrm{d}\beta}{\mathrm{d}z}, \alpha = \frac{\mathrm{d}H}{\mathrm{d}y} - \frac{\mathrm{d}G}{\mathrm{d}z}, f = -\frac{\mathrm{d}F}{\mathrm{d}t} - \frac{\mathrm{d}\psi}{\mathrm{d}x} \\
\frac{\mathrm{d}\alpha}{\mathrm{d}t} = \frac{\mathrm{d}g}{\mathrm{d}z} - \frac{\mathrm{d}h}{\mathrm{d}y}, \frac{\mathrm{d}\rho}{\mathrm{d}t} + \sum \frac{\mathrm{d}\rho\xi}{\mathrm{d}x} = 0, \sum \frac{\mathrm{d}f}{\mathrm{d}x} = \rho, \frac{\mathrm{d}\psi}{\mathrm{d}t} + \sum \frac{\mathrm{d}F}{\mathrm{d}x} = 0, \qquad (1) \\
\square = \Delta - \frac{\mathrm{d}^2}{\mathrm{d}t^2} = \sum \frac{\mathrm{d}^2}{\mathrm{d}x^2} - \frac{\mathrm{d}^2}{\mathrm{d}t^2}, \square\psi = -\rho, \square F = -\rho\xi.
\end{gathered}
$$

An element of matter of volume $\mathrm{d}x\mathrm{d}y\mathrm{d}z$ experiences a mechanical force whose components $X\mathrm{d}x\mathrm{d}y\mathrm{d}z$, $Z\mathrm{d}x\mathrm{d}y\mathrm{d}z$, $Y\mathrm{d}x\mathrm{d}y\mathrm{d}z$ are determined from the formula:

$$X = \rho f + \rho(\eta\gamma - \zeta\beta). \qquad (2)$$

These equations are subject to a remarkable transformation discovered by Lorentz and which are of interest because they explain why no experiment is able to let us know the absolute motion of the universe. Let us set:

(continued)

[1]Here and elsewhere, the first page number refers to the original publication and the second page number (following the semicolon) refers to the page number in Part I of this book.

$$x' = kl(x + \varepsilon t), t' = kl(t + \varepsilon x), y' = ly, z' = lz, \tag{3}$$

where l and ε are arbitrary constants, and where

$$\mathrm{k} = \frac{1}{\sqrt{1 - \epsilon^2}}.$$

As we read through the first paragraph, we first see that Poincaré adopts electrostatic units for charge and takes $c = 1$. Next, he tells us that the components of the electric field and magnetic field are respectively: f, g, h and α, β, γ. We would normally write these as $\boldsymbol{E}$ and $\boldsymbol{B}$, respectively. He provides the components of the vector potential and also the scalar potential, although these go unused. There are the components of the electron velocity ξ, η, ζ or $\boldsymbol{v}$. And finally, below equation (1) are the components of the electromagnetic force on the electron X, Y, Z or $\boldsymbol{F}$.

This isn't quite enough to allow us to make sense of equations (1) as two choices for notational compactness need to be pointed out. First, note that each of the first four equations refers to only the first component of three vector equations. This means that the first equation (dropping the current density and using the partial derivative symbol) needs to be expanded from one component

$$\frac{\partial f}{\partial t} + \rho\xi = \frac{\partial \gamma}{\partial y} - \frac{\partial \beta}{\partial z}$$

to include the other two components

$$\frac{\partial g}{\partial t} + \rho\eta = -\frac{\partial \gamma}{\partial x} + \frac{\partial \alpha}{\partial z},$$
$$\frac{\partial h}{\partial t} + \rho\zeta = \frac{\partial \beta}{\partial x} - \frac{\partial \alpha}{\partial y}.$$

There is a vector cross product hidden in here.

Second, the next three equations are all scalar equations, but in the summations there is no indication that the sum should be done over the three components of the vectors. This means that, correctly understood, the second of these vector equations should be written:

$$\frac{\partial f}{\partial x} + \frac{\partial g}{\partial y} + \frac{\partial h}{\partial z} = \rho.$$

Here there is a vector dot product. Applying this understanding, the Laplacian (Δ) in the first equation in the last row will be understood as:

$$\Delta = \frac{\partial^2}{\partial x^2} + \frac{\partial^2}{\partial y^2} + \frac{\partial^2}{\partial z^2}$$

The last step is then to introduce the vector differential operator del (or nabla):

$$\nabla = \left(\frac{\partial}{\partial x}, \frac{\partial}{\partial y}, \frac{\partial}{\partial z} \right)$$

Using del and replacing the components with vectors, we can write these three examples as:

$$\begin{aligned} &\frac{\partial \boldsymbol{E}}{\partial t} + \rho \boldsymbol{v} = \nabla \times \boldsymbol{B}, \\ &\nabla \cdot \boldsymbol{E} = \rho, \\ &\nabla^2 \end{aligned}$$

Pulling all this together, we can rewrite this example in familiar notation:

Lorentz adopted a specific system of units so as to make the factors of 4π disappear in the formulas. I will do the same and additionally I will choose the units of length and time such that the speed of light is equal to one. Under these conditions, by calling: $\boldsymbol{E}$, the electric field; $\boldsymbol{B}$, the magnetic field; $\boldsymbol{A}$, the vector potential; ψ, the scalar potential; ρ, the electric charge density; $\boldsymbol{v} = (v_x, v_y, v_z)$, the electron velocity; and $\boldsymbol{J}$, the current density, then the fundamental formulas become:

$$\begin{aligned} &\boldsymbol{J} = \frac{\partial \boldsymbol{E}}{\partial t} + \rho \boldsymbol{v} = \nabla \times \boldsymbol{B}, \;\; \boldsymbol{B} = -\nabla \times \boldsymbol{A}, \;\; \boldsymbol{E} = -\frac{\partial \boldsymbol{A}}{\partial t} - \nabla \psi, \\ &\frac{\partial \boldsymbol{B}}{\partial t} = -\nabla \times \boldsymbol{E}, \;\; \frac{\partial \rho}{\partial t} + \nabla \cdot (\rho \boldsymbol{v}) = 0, \;\; \nabla \cdot \boldsymbol{E} = \rho, \;\; \frac{\partial \psi}{\partial t} + \nabla \cdot \boldsymbol{A} = 0 \\ &\Box = \nabla^2 - \frac{\partial^2}{\partial t^2} = \sum \frac{\partial^2}{\partial x_i^2} - \frac{\partial^2}{\partial t^2}, \qquad \Box \psi = -\rho, \qquad \Box \boldsymbol{A} = -\rho \boldsymbol{v}. \end{aligned} \tag{1}$$

(continued)

An element of matter of volume $dxdydz$ experiences a mechanical force $\boldsymbol{F}dxdydz$ determined from the formula:

$$\boldsymbol{F} = \rho \boldsymbol{E} + \rho(\boldsymbol{v} \times \boldsymbol{B}). \tag{2}$$

These equations are subject to a remarkable transformation discovered by Lorentz and which are of interest because they explain why no experiment is able to let us know the absolute motion of the universe. Let us set:

$$x' = kl(x + \varepsilon t), \quad t' = kl(t + \varepsilon x), \quad y' = ly, \quad z' = lz, \tag{3}$$

where l and ε are arbitrary constants, and where

$$k = \frac{1}{\sqrt{1 - \epsilon^2}}.$$

Note that the vector and scalar potentials ($\boldsymbol{A}$ and ψ) can be eliminated resulting in the familiar four Maxwell's equations for the electric and magnetic fields and a continuity equation for the charge density.

Lorentz's Notation

The previous subsection describes Poincaré's choice of notation that involves writing the components of vectors and derivatives individually, only writing one component of vector equations and using an ambiguous summation for scalar product of two vectors.

In writing (Poincaré, Sur la dynamique de l'électron, 1905) and (Poincaré, Sur la dynamique de l'électron, 1906), Poincaré heavily references Lorentz's paper from the previous year (Lorentz, Electromagnetic phenomena in a system moving with any velocity smaller than that of, 1904). We can therefore be confident that Poincaré fully understood the notation that Lorentz had used even though he did not adopt it. It is therefore worth looking at the notation in Lorentz's paper.

There, Maxwell's equations appear on page 811 (in this book it is page 261; the equation numbers are unchanged) as equations (2) together with the formula for electromagnetic force per unit charge ($\mathfrak{f}$). In Lorentz's notation the equations are:

$$\begin{aligned}
\operatorname{div}\mathfrak{d} &= \varrho, \quad \operatorname{div}\mathfrak{h} = 0,\\
\operatorname{rot}\mathfrak{h} &= \frac{1}{c}\left(\dot{\mathfrak{d}} + \varrho\mathfrak{v}\right),\\
\operatorname{rot}\mathfrak{d} &= -\frac{1}{c}\dot{\mathfrak{h}},\\
\mathfrak{f} &= \mathfrak{d} + \frac{1}{c}[\mathfrak{v}\cdot\mathfrak{h}].
\end{aligned}$$

Two notational clues are provided by Lorentz on the previous page. He states, "a vector will be denoted by a German letter" and that the notation $[\dot{\mathfrak{v}}\cdot\mathfrak{h}]$ is the "vector [cross] product." rot (for rotation) is clearly curl. Replacing fraktur with bold uppercase Roman letters, a dot over a quantity with $\frac{\partial}{\partial t}$ and using curl and $\times$, these equations become:

$$\begin{aligned}
\text{div } \boldsymbol{D} &= \rho, \quad \text{div } \boldsymbol{H} = 0,\\
\text{curl } \boldsymbol{H} &= \frac{1}{c}\left(\frac{\partial \boldsymbol{D}}{\partial t} + \rho\boldsymbol{v}\right),\\
\text{curl } \boldsymbol{D} &= -\frac{1}{c}\frac{\partial \boldsymbol{H}}{\partial t},\\
\boldsymbol{F} &= \boldsymbol{D} + \frac{1}{c}\boldsymbol{v}\times\boldsymbol{H}.
\end{aligned}$$

Because of a difference in their definitions, $\mathfrak{f} = X/\rho$. Since both Poincaré and Lorentz are working in units where the vacuum permittivity and permeability are by definition 1, $\boldsymbol{D} = \boldsymbol{E}$ and $\boldsymbol{H} = \boldsymbol{B}$ in vacuum.

Lorentz introduces the vector and scalar potentials on page 813 (locally, page 264) in equations (11) and (12); the Laplacian and grad are defined on page 814 (also, page 264) below equation (14).

Unlike Poincaré, Lorentz's notation here shows Heaviside's influence although it has acquired a German accent. The accent comes through most clearly in the use of fraktur for vectors. As will be seen below, Augustus Föppl (Föppl, 1894) could be a plausible source. There are a few differences: Heaviside and Föppl used a large V for the vector product (instead of the square bracket notation) and used curl, not rot; Heaviside disliked the use of fraktur because it was difficult to read and write.

This would seem to be the right point to step back to about 1890.

Understanding J. C. Maxwell

Situating our attention in 1890 places us amidst several endeavors to understand *A Treatise on Electricity and Magnetism* (Maxwell, 1873) then over 15 years old. The effort appears to have had three main components: finding the focus by reducing the number of equations for the electric and magnetic fields to four; organizing and clarifying the exposition of the ideas; and providing a notational machinery driven by the demands of the physics. The result of this effort was several major works on electricity and magnetism including: (Poincaré, Électricité et optique, I Les théories de Maxwell, 1890), (Heaviside, 1893) and (Föppl, 1894). This list is not comprehensive but does include major works from three different countries in as many languages. This was not exclusively the province of English scholars.

The effort to understand Maxwell was seen as challenging. Poincaré introduces his discussion of Maxwell's theory (Poincaré, Électricité et optique, I Les théories de Maxwell, 1890, p. v) in 1890 with the statement "The first time a French reader opens Maxwell's book, a rising unease and often even distrust initially mixes with their admiration. It is only after a prolonged exchange and at the cost of great effort that this feeling passes. Some eminent minds still have it even now." Although they are not French, it is easy to imagine Gibbs, American, and Heaviside, English, agreeing with this sentiment. In the case of Heinrich Hertz in Germany, there is no need to speculate since he wrote (Hertz, Electric Waves, 1893, p. 27), "If we read Maxwell's equations and always interpret the meaning of the word 'electricity' in a suitable way, nearly all the contradictions which at first are so surprising can be made to disappear. Nevertheless, I must admit that I have not succeeded in doing this completely, or to my entire satisfaction; otherwise instead of hesitating, I would speak more definitely." This is followed by a footnote where Hertz, apparently in reference to the above quotation from Poincaré or to its spirit, writes, "Poincaré . . . expresses a similar opinion."

Returning to the focus of this chapter, the following sections look at the different notations used by J. W. Gibbs, O. Heaviside, A. Föppl and H. Poincaré, and secondarily at notation used by others whose work Poincaré likely read. The point here is to survey the notational choices at that time in order to compare them with the notation subsequently used by Poincaré in 1905.[2]

[2]For a full and comprehensive history of vector analysis, (Crowe, 1985) is an essential source.

J. W. Gibbs

Of the authors in these sections, J. W. Gibbs alone did not write a book inspired by his efforts to understand and explain Maxwell's work, even though that was the context of his work on vector notation. However, he and Heaviside were the first using vectors and vector analysis publicly in 1879 and 1882[3] respectively. In 1884 Gibbs had a book (Gibbs, 1884) on vector analysis printed privately, but copies were somewhat widely circulated. There is no positive indication that Poincaré had a copy of the book. For many years running Gibbs taught a course on this subject at Yale University; his course material forms the basis for a book (Wilson, 1901) by a former student that was published and released through normal channels.

In (Gibbs, 1884) on pages 16 and 17, Gibbs defines the *derivative* (we would say *gradient*) of a scalar u (∇u), and the *divergence* and *curl* of a vector ω (respectively $\nabla \cdot \omega$ and $\nabla \times \omega$) in terms of Cartesian unit vectors and derivatives with respect to Cartesian variables. When writing equations, he consistently uses dell (∇) and not abbreviations like grad, div, curl or rot, and uses $\cdot$ and $\times$ for the vector dot product and cross product respectively. Also notice that Gibbs (as we have also seen with Poincaré) does not notationally distinguish full and partial derivatives.

This choice of notation matches what an upper-level undergraduate physics major or graduate student during the last 50 years would have seen in (Jackson, Classical Electrodynamics, 1999). In that book Jackson's definitions corresponding to the ones from Gibbs presented in the previous paragraph appear inside the back cover.

For a second comparison, consider the older classic (Morse & Feshbach, 1953). On page 31 (equation 1.4.1) they define the gradient of the scalar ψ and use both gradψ and $\nabla\psi$. Similarly, the divergence of a vector $\boldsymbol{F}$ is defined on page 35, equation 1.4.5 with both div$\boldsymbol{F}$ and $\nabla \cdot \boldsymbol{F}$ and on page 41 with both curl$\boldsymbol{F}$ and $\nabla \times \boldsymbol{F}$. The notation using ∇ appears to have been provided for the information of readers who might encounter it somewhere else. For example, when it comes to writing Maxwell's equations, Morse and Feshbach on page 205 (equation 2.5.11) use the notation curl$\boldsymbol{E}$ and div$\boldsymbol{B}$. As will be seen next curl and div were introduced by Heaviside.

O. Heaviside

In 1882, Oliver Heaviside began a long series of papers in the English trade journal *The Electrician*[4]. It was in this series of papers that he began his use of vector notation in public writing. In a paper appearing in the *Philosophical Magazine* in

[3]These dates are from (Nahin, 2002, pp. 194-6).

[4](Nahin, 2002) in a section of the same title on page 101 and following has an interesting description of the journal *The Electrician*. It seems unlikely that Poincaré would have seen anything written there by Heaviside, unless someone specifically brought it to his attention.

1885 Heaviside (quoted in (Nahin, 2002, p. 196)) wrote, “Owing to the extraordinary complexity of the investigation when written out in Cartesian form (which I began doing, but gave up aghast), some abbreviated method of expression becomes desirable... I therefore adopt with some simplification, the method of vectors, which seems indeed the only proper method.”

When Heaviside's book on electromagnetic theory was published in 1893, he devoted an entire chapter to the presentation of vector analysis (Heaviside, 1893, Chapter 3). Early in the chapter (page 138), Heaviside references “Prof. Gibbs's pamphlet” (Gibbs, 1884) and adds it is “an able and in some respects original little treatise on vector analysis, though too condensed and also too advanced for learners' use.” From this quote we can also see that even when Heaviside is offering praise, his words can have a sharp edge. He also ends the paragraph sharply: “As regards his notation, however, I do not like it.” This would seem to refer to the vector product, div and curl as discussed shortly.

Proceeding into this chapter, we find first his definition of the scalar product of two vectors (Heaviside, 1893, p. 149, eqn. 12) in terms of the magnitudes of the two vectors and the cosine of the angle between them. At this point, Heaviside does not introduce the components of the vectors going into his definition. He does however apply the product to unit vectors and prove the various properties. Then, seven pages later, he shows the expansion of a vector in terms of components and orthogonal unit vectors. This is clear and without unexpected turns, except perhaps for Heaviside's recommendation to leave the dot out of the product as one might do in scalar algebra.

Next, Heaviside defines the vector cross product, the vector product of two vectors, first in terms of the magnitude of the two vectors, the sine of the angle between them and the direction along the line perpendicular to the plane defined by the two vectors (Heaviside, 1893, p. 157, eqn. 34). He then gives the definition of the vector product in terms of the components of the two vectors and the unit vectors (Heaviside, 1893, p. 159, eqn. 41). Heaviside represented the scalar product of two vectors, the dot product, with a dot centered between the two vectors. In contrast, for the vector product, he does not place a symbol between the two vectors—like a cross, $\times$, for example—and places a V in front of the two vectors. In that way, Heaviside writes the vector product of vectors $\boldsymbol{A}$ and $\boldsymbol{B}$ as $\mathrm{V}\boldsymbol{AB}$.[5] This is of course not what we are familiar with for the notation for vector product. First, it isn't suggestive of a cross product and second one can easily imagine the potential for confusion between the symbol V for vector product and V used for some other magnitude. Föppl, as we'll discuss below, uses a typographically distinctive variant of V; that would offer one possibility for reducing confusion. Following the choice made by Gibbs and using the cross-product notation we're familiar with is a much better choice.

[5]This appears suggestive of the notation for the vector component of a quaternion product.

The next definition provided by Heaviside is the differential operator ∇ (Heaviside, 1893, p. 178, eqn 119):

$$\nabla = i\frac{\mathrm{d}}{\mathrm{d}x} + j\frac{\mathrm{d}}{\mathrm{d}y} + k\frac{\mathrm{d}}{\mathrm{d}z}.$$

which he notes, "is a fictitious vector, inasmuch as … its components are not magnitudes but differentiators."

With that notational machinery defined, Heaviside is then in the position to define first the divergence (Heaviside, 1893, p. 188, eqn. 143) and then the curl (Heaviside, 1893, p. 191, eqn. 149). Each definition is provided first in terms of the scalar and vector products (respectively) of ∇ and a vector. In that way, his definition of curl starts with $\mathrm{V}\ \nabla\ \boldsymbol{E}$, continues with the components and unit vectors, and ends with $\mathrm{curl}\boldsymbol{E}$.

Heaviside explains his choice to use div and curl on page 194, writing, "The scalar product of ∇ and $\boldsymbol{D}$ conveys no such distinct idea as does divergence; nor does the vector product of ∇ and $\boldsymbol{E}$ speak so plainly as the curl or rotation of E." We can take this as Heaviside's explanation for what he did not like about Gibbs's notation, which used $\nabla \cdot \boldsymbol{D}$ and $\nabla \times \boldsymbol{E}$. The div and curl notation used in Morse and Feschbach, described above, would therefore meet with Heaviside's approval.

A. Föppl

Following our look at Heaviside's notation, we turn our attention to Föppl's book (Föppl, 1894). This order is appropriate because Föppl's work benefits from Heaviside's work. In his Foreword (Föppl, 1894, S. VII), Föppl wrote, "In presenting calculation with vectors, and in many other respects, I followed most closely the pattern provided by O. Heaviside in his treatises—which have recently been made available in bookstores as a collection. My presentation is altogether more influenced by the work of this master than by that of any other physicist—excepting Maxwell himself, of course. I consider Heaviside to be the outstanding successor to Maxwell in speculative-critical respects, just as undoubtedly Hertz—whom we unfortunately lost so young—was his successor in experimentally and confirming respects."[6] I think we should assume that Föppl is referring to the published collection of papers that Heaviside had written for *The Electrician*, and not the 1893 book that we were just discussing. Either way, in light of this testimony and the notation used in the book we can conclude that this is not a wholly independent effort to understand and present Maxwell's work and is all the same a worthy effort to reach a larger audience. That audience appears to have included

[6]Translation from German provided by Ilse Andrews, personal communication, January 8, 2018.

Albert Einstein as this book is widely cited as the textbook from which he learned electricity and magnetism.

To our point in this chapter, Augustus Föppl is clear in emphasizing the significant advantage of using vector notation (Heaviside's in his case) instead of writing out Cartesian coordinates.

H. Hertz

Of secondary importance, there are three other physicists whose work on electricity and magnetism Poincaré certainly read. Let us look first at the notation of Heinrich Hertz. In various places Poincaré discusses Hertz's theory of electromagnetic radiation. Most likely, Poincaré would have encountered this in (Hertz, Die Kräft electrischer Schwingungen behandelt nach der Maxwell'schen Theorie, 1889). This article is also available in translation as part of a collection in (Hertz, Electric Waves, 1893). In that article, and in others in the collection, Hertz uses notation based on Cartesian coordinates similar to Poincaré and does not use vector analysis and vector differential operators.

J. Larmor

Next, look at (Larmor, 1893). Poincaré wrote a series of commentaries on this paper, however I haven't discussed them elsewhere in this book. There are very few mathematical equations in Larmor's work and none involve vectors. This would not have led Poincaré to read or consider a different notation

P. Langevin

Finally look at (Langevin, 1905). Poincaré for example summarizes this paper in (Poincaré, Sur la dynamique de l'électron, 1906 §5). Langevin makes limited use of vectors, although he does write out the vector potential and position, velocity and acceleration of a body in Cartesian coordinates.

H. Poincaré: *Électricité et optique*

In the 1880s, Poincaré took up the study of Maxwell's work and in 1890 he published a book with his lectures from the second semester of 1888–89 on Maxwell's theory (Poincaré, Électricité et optique, I Les théories de Maxwell,

1890); he did not use vectors and vector analysis, instead sticking with Cartesian coordinates. Gibbs's book, *Elements of Vector Analysis*, was printed in 1881. Heaviside's book, *Electromagnetic Theory*, with a chapter on his independent vector analysis was published in 1893. And, Föppl's book in which he had two chapters on vector analysis based on Heaviside's work was published in 1894. The four—publishing books within a few years of each other, faced with understanding Maxwell's theory and working with the mathematical notation for electrodynamics—abandoned the use of W. Hamilton's quaternions that Maxwell had used in 1873 along with Cartesian components. In contrast to the other three Poincaré did not develop vector analysis independently or adopt it then or later.

First Edition

In these published lectures, Poincaré uses the same notation as (Poincaré, Sur la dynamique de l'électron, 1906, p. 132) 15 years later. In fact by using this notation (repeated here on page 229), it is easy to recognize that equation (3) (Poincaré, 1890, p. 15): $\frac{\mathrm{d}f}{\mathrm{d}x} + \frac{\mathrm{d}g}{\mathrm{d}y} + \frac{\mathrm{d}h}{\mathrm{d}z} = \rho$ states that divergence of the electric field is due to the charge density. Likewise (Poincaré, 1890, p. 146):

$$\frac{\mathrm{d}\alpha}{\mathrm{d}x} + \frac{\mathrm{d}\beta}{\mathrm{d}y} + \frac{\mathrm{d}\gamma}{\mathrm{d}z} = 0$$

states that the divergence of the magnetic field is zero.

Similarly, one can read in equation (3) on page 144:

$$\alpha = \frac{\mathrm{d}H}{\mathrm{d}y} - \frac{\mathrm{d}G}{\mathrm{d}z}$$
$$\beta = \frac{\mathrm{d}F}{\mathrm{d}z} - \frac{\mathrm{d}H}{\mathrm{d}x}$$
$$\gamma = \frac{\mathrm{d}G}{\mathrm{d}x} - \frac{\mathrm{d}F}{\mathrm{d}y}$$

meaning that the magnetic field is given by the curl of the vector potential. (Note that in the above equations all derivatives are partial derivatives.)

In these examples and in general skimming (Poincaré, 1890), while the notation appears bulky, it is not an impediment after some practice. In fact, written this way, one can by inspection see that the divergence of the curl of the vector potential is identically zero. Written as $\nabla \cdot \nabla \times \boldsymbol{A}$, such a verification by inspection is not possible. It is understandable that Poincaré could have felt comfortable with this notation using Cartesian coordinates and explicit derivatives in 1890 and therefore not been motivated to independently develop vector analysis as Gibbs and Heaviside did.

Second Edition

In 1901 a second edition of *Électricité et optique* was published (Poincaré, Électricité et optique, 1901), now including lectures from 1899. In a *Notice before the Introduction*, Poincaré states that the lecture notes from his courses in 1888 in 1890 are reprinted "with some reworking and modification" and some material deleted because it was superseded by the lectures from 1899. The new material is presented in the second part.

A comparison of the tables of contents from the two editions shows no changes in the subsection titles, the title of Chapter II is changed, a new subsection is added to Chapter III and the original Chapter XIII has been deleted. The deleted chapter is clearly the material Poincaré referred to in the Notice. During casual comparison of the other chapters, I found a couple of added sentences that clarified what was already there and did not add new content. The reworking is not an overhaul of the previous content and may only be responses to particular questions or requests for clarification from readers or students. The new subsection in Chapter III is a Remark providing a proof of an assumption made in the preceding calculations.

The equations reproduced and discussed above from pages 15, 146 and 144 (Poincaré, Électricité et optique, I Les théories de Maxwell, 1890) now appear on pages 15, 118 and 117 (Poincaré, Électricité et optique, 1901). These equations have not been changed from the first to second edition of *Électricité et optique*. Since there is no other indication that Poincaré made more than spot changes, the absence of changes in these equations only suggests that he was satisfied with the notation involving explicit Cartesian coordinates that he used 10 years later.

The second part containing the new material in the second edition (Poincaré, Électricité et optique, 1901) starts on page 229. Scanning a few pages suffices to show that he is continuing to use the same terminology. For example, at the top of page 241 there is the formula:

$$\frac{\mathrm{d}F}{\mathrm{d}x} + \frac{\mathrm{d}G}{\mathrm{d}y} + \frac{\mathrm{d}H}{\mathrm{d}z} = -\int \left(\frac{\mathrm{d}f}{\mathrm{d}x'}\mathrm{d}x' + \frac{\mathrm{d}f}{\mathrm{d}y'}\mathrm{d}y' + \frac{\mathrm{d}f}{\mathrm{d}z'}\mathrm{d}z' \right)$$

(where F, G and H are components of the vector potential and f is a function of the potential energy and separation of two current loops). It is more compact to write this as:

$$\nabla \cdot \boldsymbol{A} = -\int \nabla f \cdot \mathrm{d}\boldsymbol{x}.$$

In the first instance of its use with this meaning, on page 292, a summation sign ($\sum$) is used as shorthand for vector products. As an explanation, Poincaré writes with reference to equation (16 *bis*), "the sign indicates a cyclic permutation to be done on the letters α, β, γ; x, y, z, and F, G, H." This is the same convention for dot products as used in equation 1 (Poincaré, Sur la dynamique de l'électron, 1906) and repeated above on page 228; in particular it is discussed in the second observation on notational compactness.

The first line of equation (16 *bis*) as published is:

$$T = \frac{1}{8\pi} \int \sum \left(\frac{\mathrm{d}\gamma}{\mathrm{d}y} - \frac{\mathrm{d}\beta}{\mathrm{d}z} \right) F \mathrm{d}\tau$$

and following Poincaré's explanation, it should be expanded to:

$$T = \frac{1}{8\pi} \int \left[\left(\frac{\mathrm{d}\gamma}{\mathrm{d}y} - \frac{\mathrm{d}\beta}{\mathrm{d}z} \right) F + \left(\frac{\mathrm{d}\alpha}{\mathrm{d}z} - \frac{\mathrm{d}\gamma}{\mathrm{d}x} \right) G + \left(\frac{\mathrm{d}\beta}{\mathrm{d}x} - \frac{\mathrm{d}\alpha}{\mathrm{d}y} \right) H \right] \mathrm{d}\tau$$

where $\mathrm{d}\tau$ is an infinitesimal volume element.

In our familiar notation this is:

$$T = \frac{1}{8\pi} \int \nabla \times \boldsymbol{B} \cdot \boldsymbol{A} \mathrm{d}\tau$$

Poincaré uses integration by parts to show that this is equal to:

$$T = \frac{1}{8\pi} \int \boldsymbol{B} \cdot \nabla \times \boldsymbol{A} \mathrm{d}\tau = \frac{1}{8\pi} \int \boldsymbol{B} \cdot \boldsymbol{B} \mathrm{d}\tau$$

Explicitly writing the Cartesian coordinates, as Poincaré does, certainly makes it easier to see and verify the integration by parts. Perhaps Poincaré did see a practical advantage to staying with the familiar notation.

Poincaré's Notation, Again

If Poincaré had not seen vector analysis with differential operators earlier, he did see it when he studied (Lorentz, Electromagnetic phenomena in a system moving with any velocity smaller than that of, 1904), since Lorentz had used vectors and differential operators. Poincaré therefore did understand the notation. Even after this exposure, Poincaré continued to use in his writing largely the same notation for the partial differential equations of electrodynamics that he had used in 1890.

This consistency in his choice of notation makes it difficult to identify other occasions between 1890 and 1904 when Poincaré might have been exposed to vector analysis and differential operators. In his writing, Poincaré is sparing in his use of references and when they are provided, they appear in-line in brief form. Notation is clearly an area where Poincaré is conservative in his choices.

It is not only the notation that makes (Poincaré, 1890) seem somewhat unsatisfactory. Earlier I indicated that the effort to understand Maxwell required effort and three main components (above, page 233). Notation, just discussed, was the third of these components.

The first of the three components was focus. Relating to the focus in both editions of *Électricité et optique*, the most obvious issue is the use of the scalar and vector potentials (whose components are F, G, H in the above equations). Heaviside explicitly abandons the vector potential (Heaviside, 1893, p. 46), "[Maxwell] however, makes use of an auxiliary function, the vector potential of the electric current, and this rather complicates the matter, especially as regards the physical meaning of the process. It is always desirable when possible to keep as near as one can to first principles."

Another issue with focus is the use of auxiliary variables for "displacement velocity" and other quantities which conceal the time derivatives of electric and magnetic field components and also results in confusion with current density.

The strong point of Poincaré's work is the exposition; he has done a good job in these lectures of organizing and presenting Maxwell's theory for his students.

References

Crowe, M. J. (1985). *A History of Vector Analysis.* Mineola, New York: Dover Publications, Inc.

Föppl, A. (1894). *Einführung in die Maxwell'sche Theorie der Elektricität.* Leipzig: B. G. Teubner.

Gibbs, J. W. (1884). *Elements of Vector Analysis.* New Haven, Connecticut: Tuttle, Morehouse & Taylor.

Heaviside, O. (1893). *Electromagnetic Theory* (Vol. I). London: "The Electrician" Printing and Publishing Company Limited.

Hertz, H. (1889). Die Kräft electrischer Schwingungen behandelt nach der Maxwell'schen Theorie. *Annalen der Physik, ser. 3, 36*(1), 1–22.

Hertz, H. (1893). *Electric Waves.* (D. E. Jones, Trans.) London: Macmillan and Company.

Jackson, J. D. (1999). *Classical Electrodynamics* (Third ed.). New York: John Wiley & Sons.

Langevin, P. (1905). Sur l'origine des radiations et l'inertie électromagnétique. *J. Phys. Theor. Appl., 4*, 165–183. doi:https://doi.org/10.1051/jphystap:019050040016500

Larmor, J. (1893). A Dynamical Theory of the Electric and Luminiferous Medium. *Proceedings of the Royal Society of London, 54*, 438–461.

Lorentz, H. A. (1904). Electromagnetic phenomena in a system moving with any velocity smaller than that of. *Proceedings of the KNAW [Royal Netherlands Academy of Arts and Sciences], 6 (1903–4)*, 809–831.

Maxwell, J. C. (1873). *A Treatise on Electricity and Magnetism* (Vol. 1). London: MacMillan and Co.

Morse, P. M., & Feshbach, H. (1953). *Methods of Theoretical Physics, Part I.* New York: McGraw-Hill Book Company.

Nahin, P. J. (2002). *Oliver Heaviside.* Baltimore, Maryland: The Johns Hopkins University Press.

Poincaré, H. (1890). *Électricité et optique, I Les théories de Maxwell.* (J. Blondin, Éd.) Paris: Georges Carré.

Poincaré, H. (1901). *Électricité et optique* (éd. Deuxiène Édition, revue et complétée). (J. Blondin, & E. Néculéa, Éds.) Paris: Gauthier-Villars.

Poincaré, H. (1905). Sur la dynamique de l'électron. *Comptes rendus de l'Académie des Sciences, 140*, 1504–1508.

Poincaré, H. (1906). Sur la dynamique de l'électron. *Rendiconti del circolo matematico di Palermo, 21*, 129–176.

Wilson, E. B. (1901). *Vector Analysis: a Book for the Use of Students of Mathematics and Physics.* New York: Charles Scribner's Sons.

Chapter 13
Translation, Language and Culture

My Practice of Translation

My approach to translation in Part I of this book, in my previous translation of Poincaré (Poincaré, On the Three-Body Problem and the Equations of Dynamics, 2017), and in my commercial work, where I translate scientific and engineering documents including notably patents, is pragmatic. There may be some general value to describing how I work, so readers of my books may be interested to know how I have prepared these translations.

In my commercial work, my translations are commissioned—I receive a purchase order and a document in French to be translated, and I may also receive instructions and reference material. My job is to return a document in English respecting certain constraints. These constraints obviously include delivery time and compensation; they may include various instructions such as how to report errors found in the document to be translated, and whether to assure consistency with reference materials. In general, there is an expectation of quality and accuracy.

The engineering documents and procedures sent to me for translation may describe how something was done or what needs to be done. Patents describe an invention, set clear limits on what is included in the invention, describe an embodiment of the invention and disclose how to practice the invention. In all cases accuracy is important. If the document provided for translation is a production procedure that emphasizes limiting the quantity of acetaminophen in a production area, then a reader of my translation must understand this limitation.

In my commercial work in general, and especially in translating patents, I view my responsibility as providing a functional substitute, in English, for the documents sent to me, in French. National laws in France and the United States, and the European Patent Convention govern patent content and organization, filing and examination, and enforcement. Various treaties and agreements set multilateral recognition of other countries' patents. For this recognition to work in practice, a patent practitioner (patent attorney or patent agent), patent examiner or patent

B. D. Popp, *Henri Poincaré: Electrons to Special Relativity*,
https://doi.org/10.1007/978-3-030-48039-4_13

litigator in the US, for example, may need to know the content of the patent filed in France with its national patent office. The patent professional in the US then has a need to read and understand the dates, disclosure and claims written in French. To get that understanding, they request a translation into English since they can fully read and understand English, but not French. At that point the request is sent to me, or another patent translator, for translation of the patent from French into English. Upon receiving the translation, the professional can then draft an application for filing in the US, assess whether a recently filed application claims the same subject matter as the French patent, or prepare a brief in patent litigation involving a US patent claiming priority from the French patent. In all these cases the translation is used for the same purposes that are recognized for the original document; the one is a substitute for the other.

The translation of the French patent into English must be complete. If the translation omits something then the owner of the French patent could lose some part of their invention described in the French patent and not in the translation. Similarly, the translation into English cannot add something to the French patent since this would be an extension of the subject matter and such extension is not permissible under patent law. I also try to assure that nothing is omitted or added at a fairly low level of the text. Patent applications may be read and argued in detail and the meaning of specific words and phrases carefully considered. I work to maintain an equivalence even at that level.

It is important to understand that a translation of a French patent does not change it into a US patent. A French patent has been drafted, filed, examined and granted, as applicable, under French law. None of that changes when the patent is translated into English. In particular, a patent application drafted for filing in France follows requirements and conventions that are accepted in France but those same requirements and conventions would necessarily lead to an objection by a US Patent and Trademark Office (USPTO) examiner.

Consequently, preparing a French application for filing with the USPTO for patent protection in the US is therefore normally a two-step procedure. The first step is a translation of the application from French to English by a translator, and the second step is the preparation of a preliminary amendment (a revision of the application that addresses the objections that might be raised without adding or removing subject matter) by a patent practitioner.

Similar considerations apply to engineering documents and operating procedures, which often contain references to French or European standards and regulations. My role as a translator is to provide a functional substitute for the French document; it does not involve comparing AFNOR and ASTM standards, or EU and US regulation. A full, accurate translation from French into English does leave some reminders of its origin.

Unlike standards, regulations and statutes, the laws of physics are universal. Therefore, when translating scientific publications and research, some of the constraints are different but my underlying goal is still to present the research work or results accurately, without addition or removal.

My work with Poincaré is both similar and different.

I have translated and studied Poincaré's work, here and in my previous book, on my own time and for my own benefit. I first started because I wanted to carefully study what Poincaré had contributed to dynamical systems theory, and I continued here because I next wanted to understand what he wrote about electrons, electrodynamics and relativity. Translating his works was therefore my way of closely reading what he had written. This was done by going through Poincaré's writing sentence by sentence, word by word. As topics I found interesting came up, I would do additional reading and research to understand the content I was working on. And of course, I was writing the translation and laying out the equations as I progressed through the text. With both books I found examples in works written in English where Poincaré had not been correctly or fully understood. For example, there were questions about whether Poincaré had considered a system with fully chaotic behavior or held onto the concept of a pervasive luminiferous ether. I therefore became persuaded that it was worth undertaking the additional work to prepare my translation for publication. With this book it also became clear that during the additional reading and research I had found material that was worth writing up as chapters to accompany the translation.

My first goal here is therefore personal study and understanding of the work of Poincaré and the second is offering the fruit of my effort to other people to clarify the content and the scope of Poincaré's writing. As with my commercial translation work, I feel myself constrained (here ethically, there contractually or legally) to prepare the translation without addition or omission. For example, readers of my translation of (Poincaré, La théorie de Lorentz et le principe de réaction, 1900) in Part I can therefore assess whether Poincaré wrote about transfer of momentum by electromagnetic radiation or transfer of both momentum and mass and be confident that what they are reading does not misrepresent, by overstating or understating, what Poincaré wrote. (This example is specifically relevant to my discussion on page 219 of E. T. Whitaker's book (Whittaker, 1953).)

Reflecting a more literary consideration, I want my writing to bring Poincaré's voice through to another language. I am not trying to retell an important story in my own words. Instead, I am trying to serve Poincaré by helping his expression reach readers who know English and not French. This is served by matching Poincaré's consistency in word choice and syntactic structures with the corresponding consistency in my translation. Poincaré was also known for writing well and clearly and explaining patiently. In turn I strive to provide a translation that is also well written and clear. I have worked to understand the scientific context and content of the works I translate in order for that understanding to inform the many choices I make in translating each sentence with the hope that the result is seen as having clarity and fidelity.

Procedurally, I worked through each article I translated by reading and translating sentence by sentence and paragraph by paragraph, stopping to research terms and ideas as I felt necessary for either the translation or my own personal interest. Sometimes I went back to review what I had already done sometimes I reconsidered and globally changed choices for words or terms. Patiently, eventually, I arrived at a complete, rough translation. Then, after a break (possibly long), I read through the translation on the computer screen side-by-side with Poincaré's original writing, edited the translation and made any changes I thought were necessary to improve the writing, accuracy, or fidelity. I referred to the result of that round of translation followed by on-screen editing as a first draft. I then had this first draft printed with Poincaré's work in French on the left and my translation on the facing page. After a further break, I read through and edited the translation on paper marking it up in pen. Once the markup was transferred from paper to the electronic version, it became my second draft. I keep the rough translation to myself or trusted colleagues. I share the first draft with friendly readers. My aim with this procedure is to arrive at a second draft that is ready for submission to the publisher.

Availability of Sources

In my preparation of Part II, I have relied heavily on the work of various authors that was published in journals between roughly 1890 and 1910, and most especially the middle part of this range. In a book from even 15 or 20 years ago with this reliance on primary source material I would expect to see in the acknowledgments a thank you to librarians or libraries that made access to the source material possible. I would also expect that the author had spent untold hours in library rare book rooms or annexes consulting this material. That has not been my case. It is perhaps worthwhile discussing why and the significant difference that this is made for my work.

In brief, everything I wanted to consult was scanned and available on the internet. This was for me a tremendous resource and effectively made this book—or at least Part II—possible. In many cases, the content of the work has been digitized by optical character recognition and incorporated in the file as a text layer behind the page image. This makes it possible to search the document for particular keywords; this does need to be done with some caution since the recognition process is subject to error, especially with old yellowed paper. Just the same, the availability of scanned and digitized copies of 19th century scientific journals is a tremendous asset.

The same journal may be available on several websites and this is a point where significant differences appear. Many of the journals were scanned by Google and Google may have even indexed the digitized content. My experience using Google to search for references was fairly frustrating. When searching with Google for an article using a reference by journal name, year, volume and page, I was more likely to find other references to the article than to find the actual article I wanted. If I found a journal series, finding the right volume was still difficult. I abandoned the use of Google for this purpose.

I found archive.org to be somewhat easier to use but it was still the source of some frustration. In my experience the best was the Bibliothèque national de France (BnF).[1] I found on their website substantially all the articles from the Comptes rendus that I consulted. Their website was easy to use. The tools for picking a volume and number were straightforward and the tools for paging through a volume and downloading an article (or page range) were too. Further, the resolution of the scans prepared by BnF was also quite good; this matters for such things as reading small subscripts or superscripts in equations on old paper. The Biodiversity Heritage Library website also had a good interface and content. Websites for the Staats- und Universitätsbibliothek Göttingen and Deutsche National Bibliothek were also helpful and deserving of mention here. Certainly, every organization mentioned here is deserving of profound thanks and my sincere acknowledgment.

Language Biases

In preparing this book, I have frequently consulted the work of other authors on the history of science in this field and time. I have also sought out and read closely specific journal articles. In my earlier translation of Henri Poincaré's monograph on the three-body problem (Poincaré, On the Three-Body Problem and the Equations of Dynamics, 2017), I also looked at what various authors wrote—or might very appropriately have written, but didn't—about what Poincaré covered and concluded about dynamic systems and chaos. In reading those authors, I have seen what I perceived to be three language-related biases.

First, authors focus on sources published in languages that they can read. As an interesting example, relevant to this book, I suspect that Ernest Rutherford had a good command of written French. In the article (Rutherford, Uranium Radiation and the Electrical Conduction Produced by It, 1899) discussed in Part II, Chapter 8, he has 18 footnotes providing references to the Philosophical Magazine and Journal of Science or other United Kingdom journals, 10 with references to Comptes Rendus de l'Académie des sciences, and two with references to Annalen der Physik. (I did the tally this way, because the very first footnote lists eight articles by Henri Becquerel in the Comptes Rendus, which substantially belong together as an ongoing presentation of a smaller number of lines of research, and because I didn't eliminate multiple references to the same article.) The cited authors writing in French include Henri Becquerel, Pierre and Marie Curie, and Jean Perrin. After writing that article, Rutherford took a position at McGill University in Montreal. Even though McGill is an English-speaking university and Montreal had a substantially larger anglophone population before the Quebec separatist movement in the late 1960s

[1] On February 4, 2020, the website address was https://gallica.bnf.fr/accueil/en/content/accueil-en?mode=desktop

caused an exodus, Rutherford's willingness to accept a position in a predominantly French-speaking city also points to his comfort with the language.

This bias towards relying on sources that I can read easily is likely present in this book. While my ability to read and understand French is nearly as good as English, my ability to read German is only slightly better than no ability at all. I have attempted to compensate for this by asking translation colleagues for summaries of potentially interesting documents or for translation of specific paragraphs and documents. This has certainly improved my coverage of work by Max Abraham, Friedrich Giesel and Walter Kaufmann; it is also certain that several chapters would be substantially better if my ability to read German were more comparable to my ability to read English and French.

Second, instead of referring to primary sources in a language that they don't read, authors may use secondary sources in a language they do read as an indication of what was written in the primary source.

This is certainly the case in asking whether Poincaré understood fully chaotic behavior in a dynamical system. In his book (Lorenz, 1993, p. 118), Edward Lorenz asks, "Did [Poincaré] recognize the phenomenon of full chaos, where most solutions—not just special ones—are sensitively dependent and lack periodicity? He does not appear to have described his non-periodic solutions as being sensitively dependent, but he was quite aware of the general phenomenon of sensitive dependence." There is not an explicit citation of the work by Poincaré in that chapter. Looking at bibliography at the end of the book (op. cit. page 218) suggests that he relied on a chapter in a book by Poincaré that was written for a mass audience and translated into English soon after publication (Poincaré, The Foundations of Science, 1913). If, as I speculate, Lorenz relied on this translated chapter in a secondary work, he missed out on the large volume of writing on dynamical systems in French by Poincaré. The answer to the question quoted above is available in Poincaré's writing but to find it Lorenz would have had to have access to the content of the work in French by Poincaré just mentioned. Poincaré in *Sur le problème des trois corps et les équations de la dynamique* discusses the Duffing equation (without giving this name) (Poincaré, On the Three-Body Problem and the Equations of Dynamics, 2017, p. 157 et seq.). This is now used as an example of a non-linear oscillator and chaotic system. John Guckenheimer and Philip Holmes discusses the Duffing equation in Section 2.2 of (Guckenheimer and Holmes, 1983). Poincaré's discussion shows a clear understanding of a fully chaotic system.

Third, authors misunderstand or misrepresent what was written in primary material in a language that they don't know or don't fully understand.

Here it's harder to separate what might be a problem from misunderstanding another language from misunderstanding subject matter that is complicated and therefore intrinsically difficult to understand. Since separating is hard, I'm unwilling to suggest examples here that I may have seen.

French Historical Present Tense

Both (Poincaré, Sur la dynamique de l'électron, 1905) and (Poincaré, Sur la dynamique de l'électron, 1906) being with a discussion of evidence that absolute motion of the Earth relative to an omnipresent background (the ether) cannot be detected. This starting point is important because it shows that Poincaré accepts that it is well established that absolute motion of the Earth cannot be measured and that there is consequently a need for an explanation and a basis in theory.

This introductory discussion is written in the French *présent historique* (historical present) tense. Recognizing the use of this tense and correctly translating the tense into English are important for accurately conveying Poincaré's acceptance of this experimental fact and his motivation in preparing the papers.

Professional French-into-English translators commonly encounter the French *présent historique* tense in meeting minutes and reports of clinical cases written by doctors. The secretary prepares the meeting minutes after the meeting is over, and the doctor is not writing there at the bedside as the illness progresses—even though the use of the present tense might seem to suggest that. In these documents, the writers use the *présent historique* tense to describe events that occurred earlier. This convention allows the writer to avoid the use of the *passé composé* tense (a compound past tense); the convention is therefore similar to the journalistic recommendation in the US to avoid compound past tenses. The French does have a simple, meaning not compound, past tense called *passé simple* but it is largely limited to literary use; its use elsewhere would add considerable fussiness to the text. The use of the historical present in these documents allows a writer to avoid the compound past, on the one hand, and the simple past tense on the other.

Conventionally the French historical present tense is translated into a past tense in English. This reflects how the corresponding documents are written in the US by native English speakers. A native speaker might write sentences like, "The meeting was called to order and the minutes from the previous meeting approved." and "The patient, a 58-year-old female, was seen in the emergency department."

The misuse in English of the present tense for meeting minutes and medical reports translated from French can give a sense of immediacy that was not intended in the French writing.

However, in other situations the use in English of the present tense to describe past events can be a deliberate technique for adding vividness and immediacy. For example, this technique can even be used to create an atmospheric effect of a historical reenactment. And so, one can imagine on Patriots' Day in Massachusetts a present tense narration of the historical events: "We are standing next to the village green in Lexington. On one side Capt. Parker is steadying his company of colonial militia and on the other the vanguard of the 10th Regiment of Foot is marching into sight."

Poincaré in contrast is not narrating a reenactment of these experiments by his use of the *présent historique*, and the translation does not call for the immediacy of the use of the present tense in English.

In preparing the translation of the second paper (Poincaré, Sur la dynamique de l'électron, 1905), I found three published prior translations (the references are provided in the subsection Other Translations in the Preface).

When I first looked at these three translations, I quickly realized that they had a translation mistake in the first sentence.[2] The first sentence in Poincaré's paper is, "Il semble au premier abord que la lumière et les phénomènes optiques et électriques qui s'y rattachent vont nous fournir un moyen de déterminer le mouvement absolu de la Terre, ou plutôt son mouvement, non par rapport aux autres astres, mais par rapport à l'éther."

In one translation, this sentence reads, "It seems at first that the aberration of light and related optical and electrical phenomena will provide us with a means of determining the absolute motion of the Earth, or rather its motion with respect to the ether, as opposed to its motion with respect to other celestial bodies."

In the next translation, the sentence reads, "It would seem at first sight that the aberration of light and the optical and electrical effects related thereto should afford a means of determining the absolute motion of the Earth, or rather its motion relative to the ether instead of relative to the other celestial bodies."

In the last translation, the sentence reads, "It seems at first sight that the aberration of light and the related optical and electrical phenomena will provide us with a means of determining the absolute motion of the Earth, or rather its motion not with respect to the other stars but with respect to the ether."

Poincaré's next two sentences, still written in the same tense, refer to two experiments respectively by Fresnel in the 1870s and Michelson in 1887 that tried to use "un moyen de déterminer" and produced conclusive, negative results. Therefore, historically Poincaré while writing in 1905 was describing an idea that might have been held in the late 1860s, but was now contradicted by experiment. The experiments are all well established and not immediate; therefore, in English the use of the present tense is not accurate.

Despite this historical clue, the people providing these three translations failed to realize that the historical present tense used by Poincaré should be rendered in English with the past tense. I translated the sentence as, "On first consideration it seemed that the aberration of light and the optical phenomena associated with it were going to provide us a means for determining the absolute motion of the Earth or more accurately its motion, not with respect to other stars, but with respect to the ether." When comparing these translations, the reader should keep in mind the considerable variation in word choice and style is expected between translations that in their own right are fully accurate and acceptable.

[2]In an effort to avoid interference or unintended copying, I have not looked at these translations beyond the first page.

Insisting on the correct choice of tense in English matters because Poincaré is introducing his paper on the dynamics of the electron with a summary of historical fact. Understanding the tense chosen by Poincaré and translating it correctly into English conveys Poincaré's intent. Poincaré is emphasizing that it had not been possible to detect a background physical medium (known as the ether) against which to measure absolute motion. Poincaré accepts the inability to detect the ether as meaning that there is no ether. Starting from this conclusion, Poincaré moves forward to rework electrodynamics without absolute motion and with electrons.

The *Académie Française*

Henri Poincaré was elected to the French *Académie des sciences* on January 31, 1887. He had not yet turned 33. He was elected to the *Académie Française* on March 5, 1908, shortly after turning 54.

People who chose to read this book because of his writings on physics and mathematics will not be surprised by the first of these recognitions. Poincaré's reputation in both fields has solidly stood the test of time and readers of this book will easily accept that he earned election to the French *Académie des sciences*. The same readers may in contrast find his election to the *Académie Française* surprising. It is however also well earned, and therefore information about the *Académie Française* and Poincaré's election to it may help those readers move past their surprise.

To accomplish that, it is perhaps best to start with a brief presentation of the *Académie Française* and move from there to why Poincaré had earned the recognition of membership.

The founding of the *Académie Française* goes back to 1635. (The *Académie des sciences* was founded some 30 years later.) The members were to watch over the purity of the French language and make it capable of the highest eloquence. The academy published the first edition of its dictionary in 1694. Today the academy remains a bastion of defense of the French language. The ninth edition of its dictionary is in the process of being published: the first volume appeared in 1992; it was followed by two more published volumes, and there are still more to come.

The successive editions of the dictionary of the *Académie Française* are a visible part of its effort to standardize the French language and keep it up-to-date. The third edition of its dictionary was published in 1740 and made reforms to the spelling and accents of about a third of the 18,000 words in the dictionary (Huchon, 2002). In addition to exemplifying the role of the *Académie Française* in standardizing the spelling and use of the French language, one class of reforms applied to French spelling in the 1740 edition of the dictionary is important to physicists who want to correctly pronounce the name of the French physicist Augustin-Jean Fresnel, who developed the lens commonly used in lighthouses that bears his name and whose work on aberration is cited by Poincaré in the papers translated in Part I. The spelling reform that I'm referring to introduced the circumflex to words like *hôpital* as a marker indicating that a silent consonant following the accented vowel was eliminated. This spelling change was not applied to some family names; thus, while the *s* in Fresnel had become silent in spoken French it was not dropped and replaced by a circumflex added in the physicist's family name. Continuing this digression just a little longer, many French-Canadian names contain silent *s*'s that were dropped from the root word in the 1740 edition of the *Académie Française* dictionary. (Québec was founded as a French colony, but ceded to England under the 1713 Treaty of Utrecht—27 years before this reform.) Examples include names like Duchesne and Lévesque with roots *chêne* (oak) and *évêque* (bishop); the *s* in these names as in Fresnel is silent.

This role of the *Académie Française* continues to this day. With the appearance of new discoveries, inventions, fashions and ideas there is a corresponding need for new words. For example, the *Académie Française* allowed the word *courriel* (with origins in Québec) into the French language in June 2003 meaning electronic mail.

In addition to the *Académie des sciences*, Poincaré was a member of numerous other societies and academies in many countries. During Poincaré's induction into the *Académie Française* on January 28, 1909, Frédéric Masson (Masson, 1909), responding to Poincaré's biographical speech on his predecessor in the seat in the *Académie Française*, welcomed Poincaré by noting that he was already a member of 35 academies. His prestige and renown were significant and widespread. No matter how exclusive, what group wouldn't want Poincaré as a member? An organization concerned with promoting the French language (and implicitly French honor) also wanted such a pillar of French achievement as a member.

Before Poincaré, other famous French scientists were members of the *Académie Française*. In the introduction mentioned above, Frédéric Masson indicated that the *Académie Française* has a tradition of accepting as members certain members of the *Académie des sciences* of exceptional merit. Masson went on to say that the motivation was to provide for "the active collaboration of scholars ready to clarify the meaning and use that words from the natural, physical and mathematical sciences contribute to the language." Poincaré's membership in the *Académie Française* fulfilled a practical need.

Beyond the merely practical need, some of Poincaré's predecessors as members of both the *Académie des sciences* and the *Académie Française* were very distinguished, and at least two still have substantial name recognition today. Louis Pasteur was elected first to the *Académie des sciences* in 1862 and then to the *Académie Française* in 1881. He is known for his work on microbiology that includes notably the method that now bears his name, pasteurization, and rabies vaccine, and for identifying chirality (handedness) in the molecular structure of tartrates. Pierre-Simon Laplace was elected first to the *Académie des sciences* in 1773 and then to the *Académie Française* in 1816. He is known for his work on celestial mechanics, tides and differential equations.

Prominent French public leaders are also elected to membership in the *Académie Française*. Henri Poincaré's cousin Raymond Poincaré was elected to the *Académie Française* in 1909 and went on to serve as president of France during the First World War. As a further example, former French President Valéry Giscard d'Estaing is a current member.

Prior to his election, Poincaré had published three best-selling books discussing philosophy of science and popularizing selected topics from then current science. Each book was soon translated into English. The books in both English and French are currently available in print. The first book, *Science and Hypothesis* (Poincaré, La science et l'hypothèse, 1902) was published in 1902 and the second edition was published in 1906. In his welcoming speech, Frédéric Masson (Masson, 1909) observed that it had sold 16,000 copies. Einstein is known to have read and studied the 1902 edition of *Science and Hypothesis* with friends. Beyond their popularity—or behind it—the books presented a variety of subjects in a clear and engaging way. Some subjects, like his discussion of probability are still relevant and provide a perspective that is still fresh and interesting.

Earlier Poincaré had written several books of significant importance in different areas of mathematics and physics. Notably these books included two editions of his class notes on electricity and optics, three volumes on celestial mechanics and a book on the three-body problem. There are more in other areas of mathematics and physics.

Writing three best-selling books over a span of a few years won't get anyone a place in the *Académie Française*. Writing several major, technical books in mathematics and physics won't do it either. It has to be noted that Poincaré wrote well. His writing was respected as clear, direct and precise. As Director of the *Académie Française*, Jules Claretie, wrote, in a eulogy for Henri Poincaré, "[the *Académie Française* members] were seduced by the singularly elegant speech, simple and limpid, of this master writer who knew everything, verified everything, illuminated with his definitions, lead with his observations and guided our research, the study of our language, with his advice." Writing French well, writing it in the way described by Claretie will earn a chair in the *Académie Française* and that is why Henri Poincaré was elected.

It is certainly that aspect of his writing that makes the content of his work—on its own very interesting—pleasant and attractive to translate into English, and demands a translation that accurately and effectively gives a voice in a new language to what Poincaré wrote in clear and polished French.

La Belle Époque

The articles translated in Part I of this book and the events discussed in the previous chapters of Part II cover roughly the time from December 1895 when Jean Perrin showed that cathode rays were charged particles and W. Röntgen took an x-ray of his wife's hand to the most recent article translated in Part I (Poincaré, La dynamique de l'électron, 1908). This falls within the time known in France as the Belle Époque. The name, given well after the traumatic and shattering conflict of 1914–18, reflects nostalgia for a time before the carnage and devastation.[3]

For a hint at the social context in which Henri Poincaré, Henri Becquerel, Pierre and Marie Curie worked it is worth looking at some of the markers distinguishing this period.

From the perspective of peace and conflict the Belle Époque could be said to last from the end of the Franco-Prussian war and suppression of the Paris commune in May 1871 to August 1914 when the battle of frontiers and the battle of Tannenberg started. These prominent markers suggest a clear demarcation of the beginning, peace, and end, conflict. Clear demarcations are deceptive. In the 1880s, rings of massive fortifications were built around Liege in Belgium and Verdun in France to protect against German attack. In the 1890s the French Army developed a new 75 mm field artillery piece with a very effective recoil damper. The damper allowed a much higher rate of accurate fire since the carriage for the gun did not jump during firing so that it did not need to be reaimed. This Modèle 75 (or Mle 75 which led to the nickname Mademoiselle Soixante-quinze) was to prove very effective against massed infantry. After this preparation for conflict, there was diplomatic conflict but no shots fired over French and German colonies in North Africa during the first Moroccan conflict in 1905 and 1906 and a shooting war between Russia and Japan a year earlier.[4] Things weren't as peaceful as they seemed in hindsight.

The time was also marked by the emergence of heavy industry and engineering. A particular example is iron; it contributes to many other things relevant to this period including heavy artillery and cars. As markers two events are worth noting. The first is the construction of the Eiffel Tower for the Exposition Universelle de Paris in 1889. The design dates to May 1884 and construction started in January 1887. It was opened to the public in May 1889 during the Exposition. It remains a triumph of iron and engineering. At the other end of the period is the Titanic. Its construction started in March 1909 and its maiden voyage started April 10, 1912. Five days later it struck

---

[3]For a detailed view of the sociology of this time see (Prost, 2019). Barbara Tuchman (The Proud Tower: A Portrait of the World Before the War, 1890–1914, 2011) famously wrote, "A phenomenon of such extended malignance as the Great War does not come out of a Golden Age." Her book is suggested for a wider view of the era. The movie *Midnight in Paris* (Allen, 2011) has its own view of golden age thinking ("nostalgia in denial"). Adriana, a French woman from the 1920s, tells Gil Pender (portrayed by Marion Cottillard and Owen Wilson respectively), "For me la belle époque Paris would have been perfect." and "It's the greatest most beautiful era Paris has ever known."

[4]For details and more conflicts see (MacMillan, 2013).

an iceberg, buckling the iron plates of the hull, and sank within 3 hours with a loss of over 1500 lives.

The aspect that seems to get the most attention in hindsight is the emergence of the petit bourgeois, a small but growing middle class with leisure time and disposable income. Below the established privileged class and above the classes laboring in fields and factories, the petit bourgeois saw enormous improvements in quality of life that in addition to leisure time included hygiene, health and life expectancy. Here it is entertainment paid for with disposable income that gained a prominent position in nostalgia. There are many examples such as Pierre-Auguste Renoir's *Le déjeuner des canotiers* ("Luncheon of the Boating Party") from 1881. Bars and cabarets like Le chat noir and the Moulin rouge, and the restaurant Maxim's followed. Henri de Toulouse-Lautrec painted scenes of the nightlife and denizens of the demi-monde. While rising labor unrest after 1910 and the rising influence of Jean Jaurès could be mentioned as a bracketing event, a better choice is the deeply symbolic Taxis of the Marne in September 1914. One of the developments of the Belle Époque was the emergence of the car and with it a fleet of taxis in Paris. As the German advance approached Paris in late August (and followed the withdrawing French army to the south on the east side of Paris), the French army moved two infantry regiments by train from another part of the front, where they had already been fighting, to Paris. To move the troops from the center of Paris to the front, General Galleni, the military commandant of the city, requisitioned the taxis September 6–7, 1914. The police spread the word to the taxi drivers who one-by-one dropped off their fares where they were and went to their garage to fill up and then to the rally point in Place des Invalides. By the morning of the 8th, some 3000 to 5000 troops were delivered to the front east of Paris at Nanteuil-le-Haudouin and Silly-le-Long. For the army the event is symbolic (and not material) because the number of soldiers moved was small compared to the size of the forces committed and because they were not actively involved in the fighting of the First Battle of the Marne over the next week. For the citizens of Paris, the taxis are symbolic because the citizens contributed to the defense of their city and gave up their comfort and privilege. (Hanc, 2014)

References

Allen, W. (Director). (2011). *Midnight in Paris* [Motion Picture].

Guckenheimer, J., & Holmes, P. (1983). *Nonlinear Oscillations, Dynamical Systems, and Bifurcations of Vector Fields.* New York: Springer Science+Business Media.

Hanc, J. (2014, July 24). *A Fleet of Taxis Did Not Really Save Paris From the Germans During World War I.* Retrieved February 4, 2020, from Smithsonian Magazine: https://www.smithsonianmag.com/history/fleet-taxis-did-not-really-save-paris-germans-during-world-war-i-180952140/

Huchon, M. (2002). *Histoire de la langue française.* Paris: Librairie générale française.

Lorenz, E. N. (1993). *The Essence of Chaos.* Seattle: University of Washington Press.

MacMillan, M. (2013). *The War that Ended Peace; The Road to 1914.* New York: Random House.

Masson, F. (1909, January 28). *Réponse au discours de réception de Henri Poincaré.* Retrieved July 21, 2018, from Académie française: http://www.academie-francaise.fr/reponse-au-discours-de-reception-de-henri-poincare

Poincaré, H. (1900). La théorie de Lorentz et le principe de réaction. *Archives néerlandaises des sciences exactes et naturelles, 5*, 252–78.

Poincaré, H. (1902). *La science et l'hypothèse.* Paris: Ernest Flammarion.

Poincaré, H. (1905). Sur la dynamique de l'électron. *Comptes rendus de l'Académie des Sciences, 140*, 1504–1508.

Poincaré, H. (1906). Sur la dynamique de l'électron. *Rendiconti del circolo matematico di Palermo, 21*, 129–176.

Poincaré, H. (1908). La dynamique de l'électron. *Revue générale des sciences pures et appliquées, 19*, 386–402.

Poincaré, H. (1913). *The Foundations of Science.* (G. B. Halsted, Trans.) Lancaster, PA: The Science Press.

Poincaré, H. (2017). *On the Three-Body Problem and the Equations of Dynamics.* (B. D. Popp, Trans.) Springer.

Prost, A. (2019). *Les français de la belle époque.* Paris: Éditions Gallimard.

Rutherford, E. (1899). Uranium Radiation and the Electrical Conduction Produced by It. *Philosophical Magazine and Journal of Science, Fifth Series, 47*, 109–163.

Tuchman, B. W. (2011). *The Proud Tower: A Portrait of the World Before the War, 1890–1914.* Random House Trade Paperbacks.

Whittaker, E. (1953). *A History of the Theories of the Aether and Electricity, The Modern Theories.* London: Thomas Nelson and Sons Ltd.

Part III
Supplement, H. A. Lorentz

Chapter 14
Electromagnetic Phenomena in a System Moving with Any Velocity Smaller than That of Light

Physics. — "Electromagnetic phenomena in a system moving with any velocity smaller than that of light." By Prof. H. A. LORENTZ.

§ 1. The problem of determining the influence exerted on electric and optical phenomena by a translation, such as all systems have in virtue of the Earth's annual motion, admits of a comparatively simple solution, so long as only those terms need be taken into account, which are proportional to the first power of the ratio between the velocity of translation w and the velocity of light c. Cases in which quantities of the second order, i.e. of the order w^2/c^2, may be perceptible, present more difficulties. The first example of this kind is Michelson's well-known interference experiment, the negative result of which has led Fitz Gerald and myself to the conclusion that the dimensions of solid bodies are slightly altered by their motion through the ether.

Some new experiments in which a second order effect was sought for have recently been published. Rayleigh[1] and Brace[2] have examined the question whether the Earth's motion may cause a body to become doubly refracting; at first sight this might be expected, if the just mentioned change of dimensions is admitted. Both physicists have however come to a negative result.

Author: This article by Hendrik Antoon Lorentz originally appeared in Dutch as "Electromagntishe verschijnselen in een stelsel dat zich met willekeurige snelheid, Kleiner dan die van het licht, beweegt" in Verslagen van de gewone vergaderingen der Wis- en Natuurkundige Afdeeling of the Koninklijke Akademie van Wetenschappen (Netherlands), vol. 12, 1904 p. 986–1009 and in English as "Electromagnetic phenomena in a system moving with any velocity smaller than that of light." in KNAW, Proceedings, 6, 1903–1904, Amsterdam, 1904, p. 809–831.
It is provided here, reformatted for the convenience of the reader. It is discussed in Part II, Chapter 9; errors in equations 5, 7 and 9 are discussed there.

[1]Rayleigh, Phil. Mag. (6) 4 (1902), p. 678

[2]Brace, Phil. Mag. (6) 7 (1904), p. 317.

B. D. Popp, *Henri Poincaré: Electrons to Special Relativity*,
https://doi.org/10.1007/978-3-030-48039-4_14

In the second place Trouton and Noble[3] have endeavored to detect a turning torque acting on a charged condenser, whose plates make a certain angle with the direction of translation. The theory of electrons, unless it be modified by some new hypothesis, would undoubtedly require the existence of such a torque. In order to see this, it will suffice to consider a condenser with ether as dielectricum. It may he shown that in every electrostatic system, moving with a velocity $\boldsymbol{w}$[4], there is a certain amount of "electromagnetic momentum". If we represent this, in direction and magnitude, by a vector $\boldsymbol{G}$, the torque in question will be determined by the vector product[5]

$$\boldsymbol{G} \times \boldsymbol{w}. \tag{1}$$

Now, if the axis of z is chosen perpendicular to the condenser plates, the velocity $\boldsymbol{w}$ having any direction we like, and if U is the energy of the condenser, calculated in the ordinary way, the components of $\boldsymbol{G}$ are given[6] by the following formulae, which are exact up to the first order

$$G_x = \frac{2U}{c^2} w_x, \quad G_y = \frac{2U}{c^2} w_y, \quad G_z = 0.$$

Substituting these values in (1), we get for the components of the torque, up to terms of the second order,

$$\frac{2U}{c^2} w_y w_z, \quad -\frac{2U}{c^2} w_x w_z, \quad 0.$$

These expressions show that the axis of the torque lies in the plane of the plates, perpendicular to the translation. If α is the angle between the velocity and the normal to the plates, the moment of the torque will be $\frac{U}{c^2} w^2 \sin 2\alpha$; it tends to turn the condenser into such a position that the plates are parallel to the Earth's motion.

In the apparatus of Trouton and Noble the condenser was fixed to the beam of a torsion-balance, sufficiently delicate to be deflected by a torque of the above order of magnitude. No effect could however be observed.

[3]Trouton and Noble, London Roy. Soc. Trans. A 202 (1903), p. 165.

[4]A vector will be denoted by a bold letter, its magnitude by the corresponding non-bold letter.

[5]See my article: Weiterbiklung der Maxwell'schen Theorie. Electronentheorie in the Mathem. Encyclopadie V 14, § 21, a. (This article will be referenced as M. E.)

[6]M. E. § 56, c.

§ 2. The experiments of which I have spoken are not the only reason for which a new examination of the problems connected with the motion of the Earth is desirable. Poincaré[7] has objected to the existing theory of electric and optical phenomena in moving bodies that, in order to explain Michelson's negative result, the introduction of a new hypothesis has been required, and that the same necessity may occur each time new facts will he brought to light. Surely, this course of inventing special hypotheses for each new experimental result is somewhat artificial. It would be more satisfactory, if it were possible to show, by means of certain fundamental assumptions, and without neglecting terms of one order of magnitude or another, that many electromagnetic actions are entirely independent of the motion of the system. Some years ago, I have already sought to frame a theory of this kind[8]. I believe now to be able to treat the subject with a better result. The only restriction as regards the velocity will be that it be smaller than that of light.

§ 3. I shall start from the fundamental equations of the theory of electrons[9]. Let $\boldsymbol{D}$ be the dielectric displacement in the ether, $\boldsymbol{H}$ the magnetic force, ρ the volume-density of the charge of an electron, $\boldsymbol{v}$ the velocity of a point of such a particle, and $\boldsymbol{F}$ the electric force, i.e. the force, reckoned per unit charge, which is exerted by the ether on a volume-element of an electron. Then, if we use a fixed system of coordinates,

$$\begin{aligned} \nabla \cdot \boldsymbol{D} &= \rho, \quad \nabla \cdot \boldsymbol{H} = 0, \\ \nabla \times \boldsymbol{H} &= \frac{1}{c}\left(\frac{\partial \boldsymbol{D}}{\partial t} + \rho \boldsymbol{v}\right), \\ \nabla \times \boldsymbol{D} &= \frac{1}{c}\frac{\partial \boldsymbol{H}}{\partial t}, \\ \boldsymbol{F} &= \boldsymbol{D} + \frac{1}{c}\boldsymbol{v} \times \boldsymbol{H}. \end{aligned} \tag{2}$$

I shall now suppose that the system as a whole moves in the direction of x with a constant velocity w, and I shall denote by $\boldsymbol{u}$ any velocity a point of an electron may have in addition to this, so that

$$\boldsymbol{v}_x = \mathrm{w} + \boldsymbol{u}_x, \quad \boldsymbol{v}_y = \boldsymbol{u}_y, \quad \boldsymbol{v}_z = \boldsymbol{u}_z.$$

[7]Poincaré, Rapports du Congrès de physique de 1900, Paris, 1, p. 22, 23.
[8]Lorentz, Ziltingsverslag Akad, v. Wet., 7 (1899), p. 507, Amsterdam Proc., 1898–99, p. 427.
[9]M.E., §2.

If the equations (2) are at the same time referred to axes moving with the system, they become

$$\begin{aligned}
&\operatorname{div}\boldsymbol{D} = \rho, \quad \operatorname{div}\boldsymbol{H} = 0,\\
&\frac{\partial H_z}{\partial y} - \frac{\partial H_y}{\partial z} = \frac{1}{c}\left(\frac{\partial}{\partial t} - w\frac{\partial}{\partial x}\right)D_x + \frac{1}{c}\rho(w + u_x),\\
&\frac{\partial H_x}{\partial z} - \frac{\partial H_z}{\partial x} = \frac{1}{c}\left(\frac{\partial}{\partial t} - w\frac{\partial}{\partial x}\right)D_y + \frac{1}{c}\rho u_y,\\
&\frac{\partial H_y}{\partial x} - \frac{\partial H_x}{\partial y} = \frac{1}{c}\left(\frac{\partial}{\partial t} - w\frac{\partial}{\partial x}\right)D_z + \frac{1}{c}\rho u_y,\\
&\frac{\partial D_z}{\partial y} - \frac{\partial D_y}{\partial z} = -\frac{1}{c}\left(\frac{\partial}{\partial t} - w\frac{\partial}{\partial x}\right)H_x,\\
&\frac{\partial D_x}{\partial z} - \frac{\partial D_z}{\partial x} = -\frac{1}{c}\left(\frac{\partial}{\partial t} - w\frac{\partial}{\partial x}\right)H_y,\\
&\frac{\partial D_y}{\partial x} - \frac{\partial D_x}{\partial y} = -\frac{1}{c}\left(\frac{\partial}{\partial t} - w\frac{\partial}{\partial x}\right)H_z,\\
&f_x = D_x + \frac{1}{c}\left(u_y H_z - u_z H_y\right)\\
&f_y = D_y - wH_z + \frac{1}{c}\left(u_z H_x - u_x H_z\right)\\
&f_z = D_z - wH_y + \frac{1}{c}\left(u_z H_y - u_y H_x\right)
\end{aligned}$$

§ 4. We shall further transform these formulae by a change of variables. Putting

$$\frac{c^2}{c^2 - w^2} = k^2, \tag{3}$$

and understanding by l another numerical quantity, to be determined further on, I take as new independent variables

$$x' = klx, \quad y' = ly, \quad z' = lz \tag{4}$$

$$t' = \frac{l}{k}t - kl\frac{w}{c^2}x, \tag{5}$$

and I define two new vectors $\boldsymbol{D}'$ and $\boldsymbol{H}'$ by the formulae

$$\begin{aligned}
D'_x &= \frac{1}{l^2}D_x, \quad D'_y = \frac{k}{l^2}\left(D_y - \frac{w}{c}H_z\right), \quad D'_z = \frac{k}{l^2}\left(D_z + \frac{w}{c}H_y\right),\\
H'_x &= \frac{1}{l^2}H_x, \quad H'_y = \frac{k}{l^2}\left(H_y + \frac{w}{c}D_z\right), \quad H'_z = \frac{k}{l^2}\left(H_z - \frac{w}{c}D_y\right),
\end{aligned}$$

for which, on account of (3), we may also write

$$
\begin{aligned}
D_x &= l^2 D'_x, \quad D_y = kl^2\left(D'_y + \frac{w}{c}H'_z\right), \quad D_z = kl^2\left(D'_z - \frac{w}{c}H'_y\right),\\
H_x &= l^2 H'_x, \quad H_y = kl^2\left(H'_y + \frac{w}{c}D'_z\right), \quad H_z = kl^2\left(H'_z + \frac{w}{c}D'_y\right),
\end{aligned} \tag{6}
$$

As to the coefficient l, it is to be considered as a function of w, whose value is 1 for $w = 0$, and which, for small values of w, differs from unity no more than by an amount of the second order.

The variable t' may be called the "local time"; indeed, for $k = 1$, $l = 1$ it becomes identical with what I have formerly understood by this name.

If, finally, we put

$$
\frac{1}{kl^3}\rho = \rho', \tag{7}
$$

$$
k^2 u_x = u'_x, \quad ku_y = u'_y, \quad ku_z = u'_z, \tag{8}
$$

these latter quantities being considered as the components of a new vector $\boldsymbol{u}'$, the equations take the following form:

$$
\begin{aligned}
\mathrm{div}'\boldsymbol{D}' &= \left(1 - \frac{wu'_x}{c^2}\right)\rho', \mathrm{div}'\boldsymbol{H}' = 0,\\
\mathrm{rot}'\boldsymbol{H}' &= \frac{1}{c}\left(\frac{\partial \boldsymbol{D}'}{\partial t'} + \rho'\boldsymbol{u}'\right),\\
\mathrm{rot}'\boldsymbol{D}' &= -\frac{1}{c}\frac{\partial \boldsymbol{H}'}{\partial t'}
\end{aligned} \tag{9}
$$

$$
\begin{aligned}
f_x &= l^2 D'_x + l^2 \cdot \frac{1}{c}\left(u'_y H'_z - u'_z H'_y\right) + l^2 \cdot \frac{w}{c^2}\left(u'_y D'_y + u'_z D'_z\right)\\
f_y &= \frac{l^2}{k}D'_y + \frac{l^2}{k}\cdot\frac{1}{c}\left(u'_z H'_x - u'_x H'_z\right) - \frac{l^2}{k}\cdot\frac{w}{c^2}u'_x D'_y\\
f_z &= \frac{l^2}{k}D'_z + \frac{l^2}{k}\cdot\frac{1}{c}\left(u'_x H'_y - u'_y H'_x\right) - \frac{l^2}{k}\cdot\frac{w}{c^2}u'_x D'_z
\end{aligned} \tag{10}
$$

The meaning of the symbols d*iv*$'$ and *rot*$'$ in (9) is similar to that of d*iv* and *rot* in (2); only, the differentiations with respect to x, y, z are to be replaced by the corresponding ones with respect to x', y', z'.§ 5. The equations (9) lead to the conclusion that the vectors $\boldsymbol{D}'$ and $\boldsymbol{H}'$ may be represented by means of a scalar potential φ' and a vector potential $\boldsymbol{A}'$. These potentials satisfy the equations[10]

[10]M. E., §§ 4 and 10.

$$\Delta'\varphi' - \frac{1}{c^2}\frac{\partial^2\varphi'}{\partial t'^2} = -\rho', \tag{11}$$

$$\Delta'\mathrm{A}' - \frac{1}{c^2}\frac{\partial^2\mathrm{A}'}{\partial t'^2} = -\frac{1}{c}\rho' u', \tag{12}$$

and in terms of them $\boldsymbol{D}'$ and $\boldsymbol{H}'$ are given by

$$\boldsymbol{D}' = -\frac{1}{c}\frac{\partial \boldsymbol{A}'}{\partial t'} - \mathrm{grad}'\varphi' + \frac{w}{c}\mathrm{grad}' A'_x, \tag{13}$$

$$\mathrm{H}' = \mathrm{rot}'\mathrm{A}'. \tag{14}$$

The symbol Δ' is an abbreviation for $\frac{\partial^2}{\partial x'^2} + \frac{\partial^2}{\partial y'^2} + \frac{\partial^2}{\partial z'^2}$, and $\mathrm{grad}'\varphi'$ denotes a vector whose components are $\frac{\partial\varphi'}{\partial x'}, \frac{\partial\varphi'}{\partial y'}, \frac{\partial\varphi'}{\partial z'}$. The expression $\mathrm{grad}' A'_x$ has a similar meaning.

In order to obtain the solution of (11) and (12) in a simple form, we may take x', y', z' as the coordinates of a point P' in a space S', and ascribe to this point, for each value of t', the values of ρ', u', φ', $\boldsymbol{A}'$, belonging to the corresponding point P $(x, y, z,)$ of the electromagnetic system. For a definite value t' of the fourth independent variable, the potentials φ' and $\boldsymbol{A}'$ in the point P of the system or in the corresponding point P' of the space S', are given by[11]

$$\varphi' = \frac{1}{4\pi}\int \frac{[\rho']}{r'}\mathrm{d}S' \tag{15}$$

$$\boldsymbol{A}' = \frac{1}{4\pi c}\int \frac{[\rho'\boldsymbol{u}']}{r'}\mathrm{d}S'. \tag{16}$$

Here $\mathrm{d}S'$ is an element of the space S', r' its distance from P' and the brackets serve to denote the quantity ρ' and the vector $\rho'\boldsymbol{u}'$, such as they are in the element $\mathrm{d}S'$, for the value $t' - {r'}/{c}$ of the fourth independent variable.

Instead of (15) and (16) we may also write, taking into account (4) and (7),

$$\varphi' = \frac{1}{4\pi}\int \frac{[\rho]}{r'}\mathrm{d}S' \tag{17}$$

$$\boldsymbol{A}' = \frac{1}{4\pi c}\int \frac{[\rho\boldsymbol{u}]}{r}\mathrm{d}S, \tag{18}$$

the integrations now extending over the electromagnetic system itself. It should be kept in mind that in these formulae r' does not denote the distance between the element $\mathrm{d}S$ and the point (x, y, z) for which the calculation is to be performed. If the element lies at the point (x_1, y_1, z_1), we must take

[11] M. E., §§ 5 and 10.

$$r' = l\sqrt{k^2(x - x_1)^2 + (y - y_1)^2 + (z - z_1)^2}.$$

It is also to be remembered that, if we wish to determine φ' and $\boldsymbol{A}'$ for the instant, at which the local time in P is t', we must take ρ and $\rho\boldsymbol{u}'$, such as they are in the element dS at the instant at which the local time of that element is $t' - r'/_c$.§ 6. It will suffice for our purpose to consider two special cases. The first is that of an electrostatic system, i.e. a system having no other motion but the translation with the velocity w. In this case $\boldsymbol{u}' = 0$, and therefore, by (12), $\boldsymbol{A}' = 0$. Also, φ' is independent of t', so that the equations (11), (13) and (14) reduce to

$$\begin{aligned} &\Delta'\varphi' = -\rho', \\ &\boldsymbol{D}' = -\text{grad}'\varphi', \quad \boldsymbol{H}' = 0\,. \end{aligned} \tag{19}$$

After having determined the vector $\boldsymbol{D}'$ by means of these equations, we know also the electric force acting on electrons that belong to the system. For these the formulae (10) become, since $\boldsymbol{u}' = 0$,

$$\boldsymbol{f}_x = l^2\boldsymbol{D}'_x, \quad \boldsymbol{f}_y = \frac{l^2}{k}\boldsymbol{D}'_y, \quad \boldsymbol{f}_z = \frac{l^2}{k}\boldsymbol{D}'_z, \tag{20}$$

The result may be put in a simple form if we compare the moving system Σ with which we are concerned, to another electrostatic system Σ' which remains at rest and into which Σ is changed, if the dimensions parallel to the axis of x are multiplied by kl, and the dimensions which have the direction of y or that of z, by l a deformation for which (kl, l, l) is an appropriate symbol. In this new system, which we may suppose to be placed in the above-mentioned space S', we shall give to the density the value ρ', determined by (7), so that the charges of corresponding elements of volume and of corresponding electrons are the same in Σ and Σ'. Then we shall obtain the forces acting on the electrons of the moving system Σ, if we first determine the corresponding forces in Σ', and next multiply their components in the direction of the axis of x by l^2, and their components perpendicular to that axis by $l^2/_k$. This is conveniently expressed by the formula

$$\boldsymbol{F}(\Sigma) = \left(l^2, \frac{l^2}{k}, \frac{l^2}{k}\right)\boldsymbol{F}(\Sigma'). \tag{21}$$

It is further to be remarked that, after having found $\boldsymbol{D}'$ by (19), we can easily calculate the electromagnetic momentum in the moving system, or rather its component in the direction of the motion. Indeed, the formula

$$\boldsymbol{G} = \frac{1}{1}\int \boldsymbol{D} \times \boldsymbol{H} \mathrm{d}S$$

shows that

$$G_x = \frac{1}{c}\int \left(D_y H_z - D_z H_y\right) \mathrm{d}S$$

Therefore, by (6), since $\boldsymbol{H}' = 0$

$$G_x = \frac{k^2 l^4 w}{c^2}\int \left(D_y'^2 + D_z'^2\right) \mathrm{d}S = \frac{klw}{c^2}\int \left(D_y'^2 + D_z'^2\right) \mathrm{d}S'. \tag{22}$$

§ 7. Our second special case is that of a particle having an electric moment, i.e. a small space S, with a total charge $\int \rho \mathrm{d}S = 0$, but with such a distribution of density, that the integrals $\int \rho x \mathrm{d}S$, $\int \rho y \mathrm{d}S$ and $\int \rho z \mathrm{d}S$ have values differing from 0.

Let x, y, z be the coordinates, taken relatively to a fixed point A of the particle, which may be called its center, and let the electric moment be defined as a vector $\boldsymbol{p}$ whose components are

$$\int \rho x \mathrm{d}S, \quad \int \rho y \mathrm{d}S, \quad \int \rho z \mathrm{d}S. \tag{23}$$

Then

$$\frac{\mathrm{d}p_x}{\mathrm{d}t} = \int \rho u_x \mathrm{d}S, \quad \frac{\mathrm{d}p_y}{\mathrm{d}t} = \int \rho u_y \mathrm{d}S, \quad \frac{\mathrm{d}p_z}{\mathrm{d}t} = \int \rho u_z \mathrm{d}S. \tag{24}$$

Of course, if x, y, z are treated as infinitely small, u_x, u_y, u_z must be so likewise. We shall neglect squares and products of these six quantities.

We shall now apply the equation (17) to the determination of the scalar potential φ' for an exterior point $P(x, y, z)$, at finite distance from the polarized particle, and for the instant at which the local time of this point has some definite value t'. In doing so, we shall give the symbol $[\rho]$, which, in (17), relates to the instant at which the local time in dS is $t' - r'/c$, a slightly different meaning. Distinguishing by r_0' the value of r' for the center A, we shall understand by $[\rho]$ the value of the density existing in the element dS at the point (x, y, z), at the instant t_0 at which the local time of A is $t' - r_0'/c$.

It may be seen from (5) that this instant precedes that for which we have to take the numerator in (17) by

$$k^2 \frac{w}{c^2}\widehat{x} + \frac{k}{l}\frac{r_0' - r'}{c} = k^2 \frac{w}{c^2}\widehat{x} + \frac{k}{l}\frac{1}{c}\left(\widehat{x}\frac{\partial r'}{\partial x} + \widehat{y}\frac{\partial r'}{\partial y} + \widehat{z}\frac{\partial r'}{\partial z}\right)$$

units of time. In this last expression we may put for the differential coefficients their values at the point A.

In (17) we have now to replace $[\rho]$ by

$$[\rho] + k^2 \frac{w}{c^2} \widehat{x} \left[\frac{\partial \rho}{\partial t}\right] + \frac{k}{l} \frac{1}{c} \left(\widehat{x} \frac{\partial r'}{\partial x} + \widehat{y} \frac{\partial r'}{\partial y} + \widehat{z} \frac{\partial r'}{\partial z}\right) \left[\frac{\partial \rho}{\partial t}\right], \tag{25}$$

where $\left[\frac{\partial \rho}{\partial t}\right]$ relates again to the time t_0. Now, the value of t' for which the calculations are to be performed having been chosen, this time t_0 will be a function of the coordinates x, y, z of the exterior point P. The value of $[\rho]$ will therefore depend on these coordinates in such a way that

$$\frac{\partial [\rho]}{\partial x} = -\frac{k}{l} \frac{1}{c} \frac{\partial r'}{\partial x} \left[\frac{\partial \rho}{\partial t}\right], \text{etc.}$$

by which (25) becomes

$$[\rho] + k^2 \frac{w}{c^2} \widehat{x} \left[\frac{\partial \rho}{\partial t}\right] - \left(\widehat{x} \frac{\partial [\rho]}{\partial x} + \widehat{y} \frac{\partial [\rho]}{\partial y} + \widehat{z} \frac{\partial [\rho]}{\partial z}\right).$$

Again, if henceforth we understand by r' what has above been called r'_0 the factor $1/r'$ must be replaced by

$$\frac{1}{r'} - \widehat{x} \frac{\partial}{\partial x}\left(\frac{1}{r'}\right) - \widehat{y} \frac{\partial}{\partial y}\left(\frac{1}{r'}\right) - \widehat{z} \frac{\partial}{\partial z}\left(\frac{1}{r'}\right),$$

so that after all, in the integral (17), the element $\mathrm{d}S$ is multiplied by

$$\frac{[\rho]}{r'} + k^2 \frac{w}{c^2} \frac{\widehat{x}}{r'} \left[\frac{\partial \rho}{\partial t}\right] - \frac{\partial}{\partial x} \frac{\widehat{x}[\rho]}{r'} - \frac{\partial}{\partial y} \frac{\widehat{y}[\rho]}{r'} - \frac{\partial}{\partial z} \frac{\widehat{z}[\rho]}{r'}.$$

This is simpler than the primitive form, because neither r', nor the time for which the quantities enclosed in brackets are to be taken, depend on x, y, z. Using (23) and remembering that $\int \rho \mathrm{d}S = 0$, we get

$$\varphi' = k^2 \frac{w}{4\pi c^2 r'} \left[\frac{\partial p_x}{\partial t}\right] - \frac{1}{4\pi} \left\{\frac{\partial}{\partial x} \frac{[p_x]}{r'} + \frac{\partial}{\partial y} \frac{[p_y]}{r'} + \frac{\partial}{\partial z} \frac{[p_z]}{r'}\right\},$$

a formula in which all the enclosed quantities are to be taken for the instant at which the local time of the center of the particle is $t' - r'/c$.

We shall conclude these calculations by introducing a new vector $\boldsymbol{p}'$ whose components are

$$p'_x = klp_x, \quad p'_y = lp_y, \quad p'_z = lp_z, \tag{26}$$

passing at the same time to x', y', z', t' as independent variables. The final result is

$$\varphi' = \frac{w}{4\pi c^2 r'} \frac{\partial [p'_x]}{\partial t'} - \frac{1}{4\pi} \left\{ \frac{\partial}{\partial x'} \frac{[p'_x]}{r'} + \frac{\partial}{\partial y'} \frac{[p'_y]}{r'} + \frac{\partial}{\partial z'} \frac{[p'_z]}{r'} \right\},$$

As to the formula (18) for the vector potential, its transformation is less complicated, because it contains the infinitely small vector $\boldsymbol{u}'$. Having regard to (8), (24), (26) and (5), I find

$$\boldsymbol{A}' = \frac{w}{4\pi c^2 r'} \frac{\partial [\boldsymbol{p}']}{\partial t'}$$

The field produced by the polarized particle is now wholly determined. The formula (13) leads to

$$\boldsymbol{D}' = -\frac{1}{4\pi c^2} \frac{\partial^2 [\boldsymbol{p}']}{\partial t'^2} + \frac{1}{4\pi} \mathbf{grad}' \left\{ \frac{\partial}{\partial x'} \frac{[p'_x]}{r'} + \frac{\partial}{\partial y'} \frac{[p'_y]}{r'} + \frac{\partial}{\partial z'} \frac{[p'_z]}{r'} \right\} \tag{27}$$

and the vector $\boldsymbol{H}'$ is given by (14). We may further use the equations (20), instead of the original formulae (10), if we wish to consider the forces exerted by the polarized particle on a similar one placed at some distance. Indeed, in the second particle, as well as in the first, the velocities $\boldsymbol{u}$ may he held to be infinitely small.

It is to be remarked that the formulae for a system without translation are implied in what precedes. For such a system the quantities with accents become identical to the corresponding ones without accents; also $k = 1$ and $l = 1$. The components of (27) are at the same time those of the electric force which is exerted by one polarized particle on another.

§ 8. Thus far we have only used the fundamental equations without any new assumptions. I shall now suppose *that the electrons, which I take to be spheres of radius R in the state of rest, have their dimensions changed by the effect of a translation, the dimensions in the direction of motion becoming kl times and those in perpendicular directions l times smaller.*

In this deformation, which may be represented by $\left(\frac{1}{kl}, \frac{1}{l}, \frac{1}{l}\right)$ each element of volume is understood to preserve its charge.

Our assumption amounts to saying that in an electrostatic system Σ, moving with a velocity w, all electrons are flattened ellipsoids with their smaller axes in the direction of motion. If now, in order to apply the theorem of § 6, we subject the system to the deformation (kl, l, l), we shall have again spherical electrons of radius R.

Hence, if we alter the relative position of the centers of the electrons in Σ by applying the deformation (kl, l, l), and if, in the points thus obtained, we place the centers of electrons that remain at rest, we shall get a system, identical to the imaginary system Σ', of which we have spoken in § 6. The forces in this system and those in Σ' will bear to each other the relation expressed by (21).

In the second place I shall suppose *that the forces between uncharged particles, as well as those between such particles and electrons, are influenced by a translation in quite the same way as the electric forces in an electrostatic system*. In other terms, whatever be the nature of the particles composing a ponderable body, so long as they do not move relatively to each other, we shall have between the forces acting in a system (Σ') without, and the same system (Σ) with a translation, the relation specified in (21), if, as regards the relative position of the particles, Σ' is got from Σ by the deformation (kl, l, l), or Σ from Σ' by the deformation $\left(\frac{1}{kl}, \frac{1}{l}, \frac{1}{l}\right)$.

We see by this that, as soon as the resulting force is 0 for a particle in Σ', the same must be true for the corresponding particle in Σ. Consequently, if, neglecting the effects of molecular motion, we suppose each particle of a solid body to be in equilibrium under the action of the attractions and repulsions exerted by its neighbors, and if we take for granted that there is but one configuration of equilibrium, we may draw the conclusion that the system Σ', if the velocity $\mathbf{w}$ is imparted to it, will *of itself* change into the system Σ. In other terms, the translation will produce the deformation $\left(\frac{1}{kl}, \frac{1}{l}, \frac{1}{l}\right)$

The case of molecular motion will be considered in § 12.

It will easily be seen that the hypothesis that has formerly been made in connection with Michelson's experiment, is implied in what has now been said. However, the present hypothesis is more general because the only limitation imposed on the motion is that its velocity be smaller than that of light.

§ 9. We are now in a position to calculate the electromagnetic momentum of a single electron. For simplicity's sake I shall suppose the charge e to be uniformly distributed over the surface, so long as the electron remains at rest. Then, a distribution of the same kind will exist in the system Σ' with which we are concerned in the last integral of (22). Hence

$$\int \left(D_y'^2 + D_z'^2\right) \mathrm{d}S' = \frac{2}{3} \int D'^2 \mathrm{d}S' = \frac{e^2}{6\pi} \int_R^\infty \frac{\mathrm{d}r}{r^2} = \frac{e^2}{6\pi R^3}$$

and

$$G_x = \frac{e^2}{6\pi c^2 R} klw.$$

It must be observed that the product kl is a function of w and that, for reasons of symmetry, the vector $\boldsymbol{G}$ has the direction of the translation. In general, representing by $\boldsymbol{w}$ the velocity of this motion, we have the vector equation

$$\boldsymbol{G} = \frac{e^2}{6\pi c^2 R} kl\boldsymbol{w}. \tag{28}$$

Now, every change in the motion of a system will entail a corresponding change in the electromagnetic momentum and will therefore require a certain force, which is given in direction and magnitude by

$$\boldsymbol{F} = \frac{\mathrm{d}\boldsymbol{G}}{\mathrm{d}t}. \tag{29}$$

Strictly speaking, the formula (28) may only be applied in the case of a uniform rectilinear translation. On account of this circumstance—though (29) is always true—the theory of rapidly varying motions of an electron becomes very complicated, the more so, because the hypothesis of § 8 would imply that the direction and amount of the deformation are continually changing. It is even hardly probable that the form of the electron will be determined solely by the velocity existing at the moment considered.

Nevertheless, provided the changes in the state of motion be sufficiently slow, we shall get a satisfactory approximation by using (28) at every instant. The application of (29) to such a *quasi-stationary* translation, as it has been called by Abraham[12], is a very simple matter. Let, at a certain instant, j_1 be the acceleration in the direction of the path, and j_2 the acceleration perpendicular to it. Then the force $\boldsymbol{F}$ will consist of two components, having the directions of these accelerations and which are given by

$$F_1 = m_1 j_1 \textit{ and } F_2 = m_2 j_2,$$

if

$$m_1 = \frac{e^2}{6\pi c^2 R}\frac{\mathrm{d}(klw)}{\mathrm{d}w} \text{ and } m_2 = \frac{e^2}{6\pi c^2 R} kl. \tag{30}$$

Hence, in phenomena in which there is an acceleration in the direction of motion, the electron behaves as if it had a mass m_1, in those in which the acceleration is normal to the path, as if the mass were m_2. These quantities m_1 and m_2, may therefore

[12] Abraham, Wied. Ann, 10 (1903), p. 105.

properly be called the "longitudinal" and "transverse" electromagnetic masses of the electron. I shall suppose that there is no other, *no "true" or "material" mass.*

Since k and l differ from unity by quantities of the order w^2/c^2, we find for very small velocities

$$m_1 = m_2 = \frac{e^2}{6\pi c^2 R}.$$

This is the mass with which we are concerned, if there are small vibratory motions of the electrons in a system without translation. If, on the contrary, motions of this kind are going on in a body moving with the velocity w in the direction of the axis of x, we shall have to reckon with the mass m_1, as given by (30), if we consider the vibrations parallel to that axis, and with the mass m_2, if we treat of those that are parallel to OY or OZ. Therefore, in short terms, referring by the index Σ to a moving system and by Σ' to one that remains at rest,

$$m(\Sigma) = \left(\frac{\mathrm{d}(klw)}{\mathrm{d}w}, kl, kl\right) m(\Sigma'). \tag{31}$$

§ 10. We can now proceed to examine the influence of the Earth's motion on optical phenomena in a system of transparent bodies. In discussing this problem, we shall fix our attention on the variable electric moments in the particles or "atoms" of the system. To these moments we may apply what has been said in § 7 For the sake of simplicity we shall suppose that, in each particle, the charge is concentrated in a certain number of separate electrons, and that the "elastic" forces that act on one of these and, conjointly with the electric forces, determine its motion, have their origin within the bounds of the *same* atom.

I shall show that, if we start from any given state of motion in a system without translation, we may deduce from it a corresponding state that can exist in the same system after a translation has been imparted to it, the kind of correspondence being as specified in what follows.

a. Let A_1', A_2', A_3' etc. be the centers of the particles in the system without translation (Σ'); neglecting molecular motions we shall take these points to remain at rest. The system of points A_1, A_2, A_3 etc., formed by the centers of the particles in the moving system Σ, is obtained from A_1', A_2', A_3' etc. by means of a deformation $\left(\frac{1}{kl}, \frac{1}{l}, \frac{1}{l}\right)$. According to what has been said in § 8, the centers will of themselves take these positions A_1', A_2', A_3' etc. if originally, before there was a translation, they occupied the positions A_1, A_2, A_3 etc.

We may conceive any point P' in the space of the system Σ' to be displaced by the above deformation, so that a definite point P of Σ corresponds to it. For two corresponding points P' and P we shall define corresponding instants, the one belonging to P', the other to P, by stating that the true time at the first instant is equal to the local time, as determined by (5) for the point P, at the second instant. By corresponding times for two corresponding *particles* we shall understand times that may be said to correspond, if we fix our attention on the *centers* A' and A of these particles.

b. As regards the interior state of the atoms, we shall assume that the configuration of a particle A in Σ at a certain time may be derived by means of the deformation $\left(\frac{1}{kl}, \frac{1}{l}, \frac{1}{l}\right)$ from the configuration of the corresponding particle in Σ', such as it is at the corresponding instant. In so far as this assumption relates to the form of the electrons themselves, it is implied in the first hypothesis of § 8.

Obviously, if we start from a state really existing in the system Σ', we have now completely defined a state of the moving system Σ. The question remains however, whether this state will likewise be a possible one.

In order to judge this, we may remark in the first place that the electric moments which we have supposed to exist in the moving system and which we shall denote by $\boldsymbol{p}$, will be certain definite functions of the coordinates x, y, z of the centers Λ of the particles, or, as we shall say, of the coordinates of the particles themselves, and of the time t. The equations which express the relations between $\boldsymbol{p}$ on one hand and x, y, z, t on the other, may be replaced by other equations, containing the vectors $\boldsymbol{p}'$ defined by (26) and the quantities x', y', z', t' defined by (4) and (5). Now, by the above assumptions a and b, if in a particle A of the moving system, whose coordinates are x, y, z, we find an electric moment $\boldsymbol{p}$ at the time t, or at the local time t', the vector $\boldsymbol{p}'$ given by (26) will be the moment which exists in the other system at the true time t' in a particle whose coordinates are x', y', z'. It appears in this way that the equations between $\boldsymbol{p}'$, x', y', z', t' are the same for both systems, the difference being only this, that for the system Σ' without translation these symbols indicate the moment, the coordinates and the true time, whereas their meaning is different for the moving system, $\boldsymbol{p}'$, x', y', z', t' being here related to the moment $\boldsymbol{p}$, the coordinates x, y, z and the general time t in the manner expressed by (26), (4) and (5).

It has already been stated that the equation (27) applies to both systems. The vector $\boldsymbol{D}'$ will therefore be the same in Σ' and Σ, provided we always compare corresponding places and times. However, this vector has not the same meaning in the two cases. In Σ' it represents the electric force, in Σ it is related to this force in the way expressed by (20). We may therefore conclude that the electric forces acting, in Σ and in Σ', on corresponding particles at corresponding instants, bear to each other the relation determined by (21). In virtue of our assumption b, taken in connection with the second hypothesis of § 8, the same relation will exist between the "elastic" forces; consequently, the formula (21) may also be regarded as indicating the relation between the total forces, acting on corresponding electrons, at corresponding instants.

It is clear that the state we have supposed to exist in the moving system will really be possible if, in Σ and Σ', the products of the mass m and the acceleration of an electron are to each other in the same relation as the forces, i.e. if

$$mj(\Sigma) = \left(l^2, \frac{l^2}{k}, \frac{l^2}{k}\right) mj(\Sigma'). \tag{32}$$

Now, we have for the accelerations

$$j(\Sigma) = \left(\frac{l}{k^3}, \frac{l}{k^2}, \frac{l}{k^2}\right) j(\Sigma'), \tag{33}$$

as may be deduced from (4) and (5), and combining this with (32), we find for the masses

$$m(\Sigma) = \left(k^3 l, kl, kl\right) m(\Sigma')$$

If this is compared to (31), it appears that, whatever be the value of l the condition is always satisfied, as regards the masses with which we have to reckon when we consider vibrations perpendicular to the translation. The only condition we have to impose on l is therefore

$$\frac{\mathrm{d}(klw)}{\mathrm{d}w} = k^3 l.$$

But, on account of (3),

$$\frac{\mathrm{d}(kw)}{\mathrm{d}w} = k^3,$$

so that we must put

$$\frac{\mathrm{d}l}{\mathrm{d}w} = 0, \quad l = const.$$

The value of the constant must be unity, because we know already that, for $w = 0$, $l = 1$.

We are therefore led to suppose *that the influence of a translation on the dimensions (of the separate electrons and of a ponderable body as a whole) is confined to those that have the direction of the motion, these becoming k times smaller than they are in the state of rest*. If this hypothesis is added to those we have already made, we may be sure that two states, the one in the moving system, the other in the same system while at rest, corresponding as stated above, may both be possible. Moreover, this correspondence is not limited to the electric momenta of the particles. In corresponding points that are situated either in the ether between the particles, or in that surrounding the ponderable bodies, we shall find at corresponding times the same vector $\boldsymbol{D}'$ and, as is easily shown, the same vector $\boldsymbol{H}'$. We may sum up by saying: If, in the system without translation, there is a state of motion in which, at a definite place, the components of $\boldsymbol{p}$, $\boldsymbol{D}$ and $\boldsymbol{H}$ are certain functions of the time, then the same system after it has been put in motion (and thereby deformed) can be the seat of a state of motion in which, at the corresponding place, the components of $\boldsymbol{p}'$, $\boldsymbol{D}'$ and $\boldsymbol{H}'$ are the same functions of the local time.

There is one point which requires further consideration. The values of the masses m_1 and m_2 having been deduced from the theory of quasi-stationary motion, the question arises, whether we are justified in reckoning with them in the case of the rapid vibrations of light. Now it is found on closer examination that the motion of an electron may be treated as quasi-stationary if it changes very little during the time a light-wave takes to travel over a distance equal to the diameter. This condition is fulfilled in optical phenomena, because the diameter of an electron is extremely small in comparison with the wavelength.

§ 11. It is easily seen that the proposed theory can account for a large number of facts.

Let us take in the first place the case of a system without translation, in some parts of which we have continually $\boldsymbol{p} = 0, \boldsymbol{D} = 0, \boldsymbol{H} = 0$. Then, in the corresponding state for the moving system, we shall have in corresponding parts (or, as we may say, in the same parts of the deformed system) $\boldsymbol{p}' = 0$, $\boldsymbol{D}' = 0$, $\boldsymbol{H}' = 0$. These equations implying $\boldsymbol{p} = 0, \boldsymbol{D} = 0, \boldsymbol{H} = 0$, as is seen by (26) and (6), it appears that those parts which are dark while the system is at rest, will remain so after it has been put in motion. It will therefore be impossible to detect an influence of the Earth's motion on any optical experiment, made with a terrestrial source of light, in which the geometrical distribution of light and darkness is observed. Many experiments on interference and diffraction belong to this class.

In the second place, if in two points of a system, rays of light of the same state of polarization are propagated in the same direction, the ratio between the amplitudes in these points may be shown not to be altered by a translation. The latter remark applies to those experiments in which the intensities in adjacent parts of the field of view are compared.

The above conclusions confirm the results I have formerly obtained by a similar train of reasoning, in which however the terms of the second order were neglected. They also contain an explanation of Michelson's negative result, more general and of somewhat different form than the one previously given, and they show why Rayleigh and Brace could find no signs of double refraction produced by the motion of the Earth.

As to the experiments of Trouton and Noble, their negative result becomes at once clear, if we admit the hypotheses of § 8. It may be inferred from these and from our last assumption (§ 10) that the only effect of the translation must have been a contraction of the whole system of electrons and other particles constituting the charged condenser and the beam and thread of the torsion-balance. Such a contraction does not give rise to a sensible change of direction.

It need hardly be said that the present theory is put forward with all due reserve. Though it seems to me that it can account for all well-established facts, it leads to some consequences that cannot as yet be put to the test of experiment. One of these is that the result of Michelson's experiment must remain negative, if the interfering rays of light are made to travel through some ponderable transparent body.

Our assumption about the contraction of the electrons cannot in itself be pronounced to be either plausible or inadmissible. What we know about the nature of electrons is very little and the only means of pushing our way farther will be to test such hypotheses as I have here made. Of course, there will be difficulties, e.g. as soon as we come to consider the rotation of electrons. Perhaps we shall have to suppose that in those phenomena in which, if there is no translation, spherical electrons rotate about a diameter, the points of the electrons in the moving system will describe elliptic paths, corresponding, in the manner specified in § 10, to the circular paths described in the other case.

§ 12 It remains to say some words about molecular motion. We may conceive that bodies in which this has a sensible influence or even predominates, undergo the same deformation as the systems of particles of constant relative position of which alone we have spoken till now. Indeed, in two systems of molecules Σ' and Σ, the first without and the second with a translation, we may imagine molecular motions corresponding to each other in such a way that, if a particle in Σ' has a certain position at a definite instant, a particle in Σ occupies at the corresponding instant the corresponding position. This being assumed, we may use the relation (33) between the accelerations in all those cases in which the velocity of molecular motion is very small as compared to w. In these cases the molecular forces may be taken to be determined by the relative positions, independently of the velocities of molecular motion. If, finally, we suppose these forces to be limited to such small distances that, for particles acting on each other, the difference of local times may be neglected, one of the particles, together with those which lie in its sphere of attraction or repulsion, will form a system which undergoes the often mentioned deformation. In virtue of the second hypothesis of § 8 we may therefore apply to the resulting molecular force acting on a particle, the equation (21). Consequently, the proper relation between the forces and the accelerations will exist in the two cases, if we suppose *that the masses of all particles are influenced by a translation to the same degree as the electromagnetic masses of the electrons.*

§ 13 The values (30) which I have found for the longitudinal and transverse masses of an electron, expressed in terms of its velocity, are not the same as those that have been formerly obtained by Abraham. The ground for this difference is solely to be sought in the circumstance that, in his theory, the electrons are treated as spheres of invariable dimensions. Now, as regards the transverse mass, the results of Abraham have been confirmed in a most remarkable way by Kaufmann's measurements of the deflection of radium-rays in electric and magnetic fields. Therefore, if there is not to be a most serious objection to the theory I have now proposed, it must be possible to show that those measurements agree with my values nearly as well as with those of Abraham.

I shall begin by discussing two of the series of measurements published by Kaufmann[13] in 1902. From each series he has deduced two quantities η and ζ, the "reduced" electric and magnetic deflections, which are related as follows to the ratio $\beta = w/c$:

$$\beta = k_1 \frac{\zeta}{\eta}, \quad \psi(\beta) = \frac{\eta}{k_2 \zeta^2}. \tag{34}$$

Here $\psi(\beta)$ is such a function, that the transverse mass is given by

$$m_2 = \frac{3}{4} \cdot \frac{e^2}{6\pi c^2 R} \psi(\beta), \tag{35}$$

whereas k_1 and k_2 are constant in each series.

It appears from the second of the formulae (30) that my theory leads likewise to an equation of the form (35); only Abraham's function $\psi(\beta)$ must be replaced by

$$\frac{4}{3} k = \frac{4}{3} \left(1 - \beta^2\right)^{-½}$$

Hence, my theory requires that, if we substitute this value for $\psi(\beta)$ in (34), these equations shall, still hold. Of course, in seeking to obtain a good agreement, we shall be justified in giving to k_1 and k_2 other values than those of Kaufmann, and in taking for every measurement a proper value of the velocity w, or of the ratio β. Writing sk_1, $\frac{3}{4} k_2'$ and β' for the new values, we may put (34) in the form

$$\beta' = sk_1 \frac{\zeta}{\eta} \tag{36}$$

and

$$\left(1 - \beta'^2\right)^{-½} = \frac{\eta}{k_2' \zeta^2}. \tag{37}$$

Kaufmann has tested his equations by choosing for k_1 such a value that, calculating β and k_2 by means of (34), he got values for this latter number that remained constant in each series as well as might be. This constancy was the proof of a sufficient agreement.

I have followed a similar method, using however some of the numbers calculated by Kaufmann. I have computed for each measurement the value of the expression

[13] Kaufmann, Physik. Zeitschr. 4 (1902), p. 55.

$$k_2' = \left(1 - \beta'^2\right)^{1/2} \psi(\beta) k_2, \tag{38}$$

that may be got from (37) combined with the second of the equations (34). The values of $\psi(\beta)$ and k_2 have been taken from Kaufmann's tables and for β' I have substituted the value he has found for β, multiplied by s, the latter coefficient being chosen with a view to obtaining a good constancy of (38). The results are contained in the following tables, corresponding to the tables III and IV in Kaufmann's paper.

III. $s = 0.933$

β	$\psi(\beta)$	k_2	β'	k_2'
0.851	2.147	1.721	0.794	2.246
0.766	1.86	1.736	0.715	2.258
0.727	1.78	1.725	0.678	2.256
0.6615	1.66	1.727	0.617	2.256
0.6075	1.595	1.655	0.567	2.175

IV. $s = 0.954$

β	$\psi(\beta)$	k_2	β'	k_2'
0.963	3.23	8.12	0.919	10.36
0.949	2.86	7.99	0.905	9.70
0.933	2.73	7.46	0.890	9.28
0.883	2.31	8.32	0.842	10.36
0.860	2.193	8.09	0.820	10.15
0.830	2.06	8.13	0.702	10.23
0.801	1.96	8.13	0.764	10.28
0.777	1.89	8.04	0.741	10.20
0.752	1.83	8.02	0.717	10.22
0.732	1.785	7.97	0.698	10.18

The constancy of k_2' is seen to come out no less satisfactory than that of k_2, the more so as in each case the value of s has been determined by means of only two measurements. The coefficient has been so chosen that for these two observations, which were in Table III the first and the last but one, and in Table IV the first and last, the values of k_2' should be proportional to those of k_2.

I shall next consider two series from a later publication by Kaufmann[14], which have been calculated by Runge[15] by means of the method of least squares, the coefficients k_1 and k_2 having been determined in such a way, that the values of η, calculated, for each observed ζ, from Kaufmann's equations (34), agree as closely as may be with the observed values of η.

[14]Kaufmann, Gött. Nachr. Math. phys. Kl., 1903, p. 90.

[15]Runge, ibidem, p. 326.

I have determined by the same condition, likewise using the method of least squares, the constants a and b in the formula

$$\eta^2 = a\zeta^2 + b\zeta^4,$$

which may be deduced from my equations (36) and (37). Knowing a and b, I find β for each measurement by means of the relation

$$\beta = \sqrt{a}\frac{\zeta}{\eta}$$

For two plates on which Kaufmann had measured the electric and magnetic deflections, the results are as follows, the deflections being given in centimeters.

I have not found time for calculating the other tables in Kaufmann's paper. As they begin, like the table for Plate 15, with a rather large negative difference between the values of η which have been deduced from the observations and calculated by Runge, we may expect a satisfactory agreement with my formulae.

Plate 15. a = 0.06489, b = 0.3039.

	η					β	
						Calculated by	
ζ	Observed.	Calculated by R.	Diff.	Calculated by L	Diff.	R.	L.
0.1495	0.0388	0.0404	−16	0.0400	−12	0.987	0.951
0.199	0.0548	0.0550	−2	0.0552	−4	0.964	0.018
0.2475	0.0716	0.0710	+6	0.0715	+1	0.030	0.881
0.296	0.0806	0.0887	+9	0.0895	+1	0.889	0.842
0.3435	0.1080	0.1081	−1	0.1090	−10	0.847	0.803
0.391	0.1290	0.1297	−7	0.1305	−15	0.804	0.763
0.437	0.1524	0.1527	−3	0.1532	−8	0.763	0.727
0.4825	0.1788	0.1777	+11	0.1777	+11	0.724	0.692
0.5265	0.2033	0.2039	−6	0.2033	0	0.688	0.660

Plate N°. 19. a = 0.05867, b = 0.2591.

	η					β	
						Calculated by	
ζ	Observed.	Calculated by R.	Diff.	Calculated by L.	Diff.	R.	L
0.1495	0.0404	0 0388	+16	0.0379	+25	0.990	0.954
0.199	0.0529	0 0527	+2	0 0522	+7	0.969	0.923
0.247	0.0678	0 0675	+3	0.0674	+4	0.939	0.888
0.296	0.0834	0.0842	−8	0.0844	−10	0.902	0.849
0.3435	0.1010	0.1022	−3	0.1026	−7	0.862	0.841

Plate N°. 19. a = 0.05867, b = 0.2591.

	η					β	
						Calculated by	
ζ	Observed.	Calculated by R.	Diff.	Calculated by L.	Diff.	R.	L
0.391	0.1219	0.1222	−3	0.1226	−7	0.822	0.773
0.437	0.1420	0.1434	−5	0.1437	−8	0.782	0.736
0.4825	0.1660	0.4665	−5	0.1664	−4	0.744	0.702
0.5265	0.1916	0.1906	+10	0.1902	+14	0.709	0.671

§ 14. I take this opportunity for mentioning an experiment that has been made by Trouton[16] at the suggestion of Fitz Gerald, and in which it was tried to observe the existence of a sudden impulse acting on a condenser at the moment of charging or discharging; for this purpose the condenser was suspended by a torsion-balance, with its plates parallel to the Earth's motion. For forming an estimate of the effect that may be expected, it will suffice to consider a condenser with ether as dielectricum. Now, if the apparatus is charged, there will be (§ 1) an electromagnetic momentum

$$\boldsymbol{G} = \frac{2U}{c^2}\boldsymbol{w}.$$

(Terms of the third and higher orders are here neglected). This momentum being produced at the moment of charging, and disappearing at that of discharging, the condenser must experience in the first case an impulse $-\boldsymbol{G}$ and in the second an impulse $+\boldsymbol{G}$.

However Trouton has not been able to observe these jerks.

I believe it may be shown (though his calculations have led him to a different conclusion) that the sensibility of the apparatus was far from sufficient for the object Trouton had in view.

Representing, as before, by U the energy of the charged condenser in the state of rest, and by $U + U'$ the energy in the stale of motion, we have by the formulae of this paper, up to the terms of the second order,

$$U' = \frac{2w^2}{c^2}U,$$

an expression, agreeing in order of magnitude with the value used by Trouton for estimating the effect.

The intensity of the sudden jerk or impulse will therefore be U'/w.

Now, supposing the apparatus to be initially at rest, we may compare the deflection α, produced by this impulse, to the deflection α' which may be given to

[16]Trouton, Dublin Roy. Soc. Trans. (2) 7 (1902), p. 379 (This paper may also be found in *The scientific writings of Fitz Gerald*, edited by Larmor, Dublin and London 1902, p. 557).

the torsion-balance by means of a constant torque K, acting during half the vibration time. We may also consider the case in which a swinging motion has already been set up; then the impulse, applied at the moment in which the apparatus passes through the position of equilibrium, will alter the amplitude by a certain amount β and a similar effect β' may be caused by letting the torque K act during the swing from one extreme position to the other. Let T be the period of swinging and l the distance from the condenser to the thread of the torsion-balance. Then it is easily found that

$$\frac{\alpha}{\alpha'} = \frac{\beta}{\beta'} = \frac{\pi U' l}{KTw}. \tag{39}$$

According to Trouton's statements U' amounted to one or two ergs, and the smallest torque by which a sensible deflection could be produced was estimated at 7.5 CGS-units. If we substitute this value for K and take into account that the velocity of the Earth's motion is 3×10^6 cm/sec., we immediately see that (39) must have been a very small fraction.

Editor's Notes

1. [Editor: Perhaps this should be Σ', not S'.]

Index

B. D Popp, *Henri Poincaré: Electrons to Special Relativity*,
https://doi.org/10.1007/978-3-030-48039-4

Printed in the United States
by Baker & Taylor Publisher Services